Plants: Their Response to Abiotic Stress

Plants: Their Response to Abiotic Stress

Edited by Zeke Adams

Syrawood Publishing House
New York

Published by Syrawood Publishing House,
750 Third Avenue, 9th Floor,
New York, NY 10017, USA
www.syrawoodpublishinghouse.com

Plants: Their Response to Abiotic Stress
Edited by Zeke Adams

International Standard Book Number: 978-1-64740-421-5 (Hardback)

Cataloging-in-publication Data

Plants : their response to abiotic stress / edited by Zeke Adams.
p. cm.
Includes bibliographical references and index.
ISBN 978-1-64740-421-5
1. Plants--Effect of stress on. 2. Plants--Adaptation. 3. Plant physiology.
4. Plant genomes. I. Adams, Zeke.
QK754 .P535 2023
581.7--dc23

TABLE OF CONTENTS

PREFACE

Over the recent decade, advancements and applications have progressed exponentially. This has led to the increased interest in this field and projects are being conducted to enhance knowledge. The main objective of this book is to present some of the critical challenges and provide insights into possible solutions. This book will answer the varied questions that arise in the field and also provide an increased scope for furthering studies.

Abiotic stress is defined as the negative impact of non-living factors on living organisms, such as plants, in a specific environment. It includes adverse effects of metal toxicity, high temperature, nutrient deficiency, drought, salinity and flooding. When plants are exposed to abiotic stresses for a long period, it results in metabolism alteration and damage to biomolecules. The plants usually develop certain defense mechanisms or responses in order to tolerate abiotic stresses by upregulation of osmolytes, osmoprotectants, and enzymatic and non-enzymatic antioxidants. Abiotic stresses greatly influence plant growth and crop production. The emergence of proteomics, deep sequencing technologies, epigenetics, and metabolomics have provided opportunities for understanding the biology of plants and developing stress tolerant varieties of crops. This book is a compilation of chapters that discuss the most vital concepts and emerging trends related to abiotic stress resistance in plants. It is appropriate for students seeking detailed information in this area of study as well as for experts.

I hope that this book, with its visionary approach, will be a valuable addition and will promote interest among readers. Each of the authors has provided their extraordinary competence in their specific fields by providing different perspectives as they come from diverse nations and regions. I thank them for their contributions.

Editor

1

Genome-Wide Identification of the ABA Receptors Genes and their Response to Abiotic Stress in Apple

Hongmin Hou [1,2,†], Lingling Lv [1,2,†], Heqiang Huo [3], Hongyi Dai [1] and Yugang Zhang [1,2,*]

1 College of Horticulture, Qingdao Agricultural University, Qingdao 266109, China; hmhou@qau.edu.cn (H.H.); 201301014@qau.edu.cn (L.L.); hydai@qau.edu.cn (H.D.)

2 Qingdao Key Laboratory of Genetic Development and Breeding in Horticultural Plants, Qingdao Agricultural University, Qingdao 266109, China

3 Mid-Florida Research and Education Center, University of Florida, Apopka, FL 32703, USA; hhuo@ufl.edu

* Correspondence: ygzhang@qau.edu.cn.

† These two authors contributed equally to this work.

Abstract: The pyrabactin resistance (PYR)/PYR1-like (PYL)/regulatory components of ABA receptor (RCAR) (known as PYLs for short) have been identified and characterized as the ABA receptors in some plants. However, little is known about the details regarding *PYL* family genes in the apple (*Malus domestica*). In this study, we identified 13 apple *PYLs*, termed *MdPYL1-13*, which could be classified into four groups according to structural features of the amino acid sequence. The gene structures and conserved motifs analysis found that the majority of *MdPYL*s had a similar number of exons and similar conserved motif profile in the same group. In addition, 11 gene pairs were identified to exhibit synteny by synteny analysis between the apple and *Arabidopsis*. Furthermore, we investigated *MdPYL*s transcript level in various organs of the red-fleshed apple (*Malus sieversii* f. Neidzwetzkyana (Dieck) Langenf) 'Xinjiang No.1'. The results suggested all *MdPYL*s within group I were expressed at relatively higher levels in all of the organs tested. However, the genes of group IV had little or no variation. Additionally, we found various hormone and stress-related cis-elements in the promoters of *MdPYLs* by analyzing cis-elements. Therefore, the expression levels of all *MdPYLs* were further detected under ABA, PEG, salt, and cold stresses in 'Xinjiang No.1' seedlings. We found that all *MdPYL*s except for *MdPYL11* were upregulated by ABA treatment, 10 genes were upregulated by PEG treatment, 12 genes were upregulated by NaCl treatment, and six genes were upregulated by cold treatment (4 °C) while seven genes were downregulated. Thus, these *MdPYLs* might be involved in the defense against abiotic stresses. In addition, the interaction between 13 MdPYLs and two 2C protein phosphatases in the apple (MdPP2C65 and MdPP2C72) was investigated in yeast two-hybrid assays. These results suggested that MdPYLs may bind to MdPP2C65 and MdPP2C72 in different manners and with different intensity. Our studies provide useful information for further investigating and researching the regulatory mechanisms of *PYL* family genes in response to abiotic stresses in the apple.

Keywords: *PYL* gene; abiotic stress; phylogenetic analysis; synteny analysis

1. Introduction

Abscisic acid (ABA) is an important phytohormone involved in plant growth, plant development, and protection of plants against abiotic stresses [1,2]. The core components of ABA signaling included the regulatory component of ABA receptors (PYLs), group A type 2C protein phosphatases (PP2Cs), and the sucrose nonfermenting 1-related protein kinases (SnRKs) [3–6]. The PYLs interacted with

PP2Cs to form PYL-PP2C heterodimer, which inhibited the phosphatase activity of key negative regulators PP2Cs, thereby leading to activation of pivotal positive regulators SnRK2s [7–10].

PYLs, as ABA receptors, belong to the StAR-related lipid-transfer (START) superfamily containing a ligand-binding pocket enclosed by four conserved loops, CL1-CL4 [4,6], which is important in ABA signal transduction pathway. To date, 14 PYLs containing the conserved structure have been recognized in the *Arabidopsis* genome [4,6,11], including *AtPYR1* (*Arabidopsis thaliana* pyrabactin resistance1) and 13 *AtPYLs* (*A. thaliana* PYR1-like1-13) [12]. Some *PYLs*, such as *AtPYR1*, *AtPYL1*, *AtPYL2*, and *AtPYL3* inhibited PP2Cs in an ABA-dependent manner, while others including *AtPYL4*, *AtPYL5*, *AtPYL6*, *AtPYL8*, *AtPYL9*, *AtPYL10*, and *AtPYL13* showed suppression of PP2Cs in the absence of ABA [10,12,13]. In addition, several of these genes were showed having diverse functions in the regulation of stomatal closure [14], drought resistance [5], leaf senescence [15], and root development [16–18].

Since the discovery of the *PYL* family in *Arabidopsis*, orthologous genes of *PYL* in other crops have been characterized at genome-wide levels, including six *PYLs* in sweet orange [19], eight *PYLs* in grape [20], 13 *PYLs* in rice [21], 14 *PYLs* in tomato [22], 14 *PYLs* in rubber tree [23], 24 *PYLs* in *Brassica rapa* [24], 27 *PYLs* in cotton [25], and 29 *PYLs* in tobacco [26]. Recently, the study regarding the roles of *PYLs* in other plant species besides *Arabidopsis* has begun to be reported, and the various function of the PYLs in plant development and stress resistance was emphasized. For instance, overexpression of rice *OsPYL3*, cotton *GhPYL10/12/26*, or maize *ZmPYL3/9/10/13* dramatically enhanced the sensitivity to ABA in transgenic *Arabidopsis* [27–29]. Similarly, overexpression of rice *OsPYL/RCAR5* or *OsPYL3/5/9/11* also increased the sensitivity to ABA and the tolerance to drought stress in transgenic rice [30–32]. Overexpressing Os*PYL9* significantly enhanced the tolerance to drought stress and leaf senescence induced by drought in transgenic *Arabidopsis* and rice plants [15]. Moreover, overexpression of *Arabidopsis AtPYL4* (A194T), *AtPYL5*, maize *ZmPYL8/9/12*, or cotton *GhPYL9-11A* greatly improved tolerance to drought in transgenic *Arabidopsis* plants, respectively [5,29,33,34]. In recent years, in addition to herbs, the function of the *PYLs* in woody and lianas plants has also been reported. Overexpression of *Populus trichocarpa PtPYRL1* or *PtPYRL5* in both *Arabidopsis* and poplar also enhanced ABA sensitivity and drought resistance [35]. In grape, *VvPYL1* could be combined with ABA and repress the ABI1 phosphatase activity [36]. The *VvPYL1* (*VvRCAR7*) exhibited high expression level in grape leaves treated by drought, salt, and cold stresses [37].

Unfortunately, in spite of an increase in reports in *Arabidopsis* and other plants such as maize, rice, and grape, the information concerning *PYLs* remains elusive in the apple. In this study, for the first time, we systematically identified the *MdPYLs* at the genome level in the *M. domestica* and performed their exon–intron structure and synteny analysis. Moreover, we investigated the spatiotemporal expression patterns of *MdPYLs* in various apple organs. Finally, we also examined expression patterns of *MdPYLs* under various abiotic stresses treatments. Our results will provide some clues for further investigating the function of *PYL* genes in apple tree growth and development, and resistance to abiotic stress.

2. Results

2.1. Identification and Expansion Patterns of MdPYLs in the M. domestica Genome

We determined that 13 putative *MdPYLs* were present in the *M. domestica* genome through BLASTP by using 14 AtPYL sequences as references. Then, we termed *MdPYL1* to *MdPYL13* based on their chromosomal order (Figure 1). Information regarding *MdPYLs* was provided in Table 1 consisting of the gene name, gene ID, length of coding DNA sequence (CDS), theoretical isoelectric point (pI), molecular weight (MW), and length of protein. The sequence analysis revealed that the *MdPYLs* lengths changed obviously in size, ranging from 558 bp (*MdPYL5*) to 3252 bp (*MdPYL1*), with one to four exons in each sequence. However, the lengths of CDS and the corresponding MdPYL proteins both had no significant difference, varying from 555 bp (*MdPYL1* and *MdPYL7*) to 777 bp (*MdPYL4*) and from 184 aa (MdPYL1 and MdPYL7) to 258 aa (MdPYL4), respectively. The predicted MW and pI values of these proteins ranged respectively from 20.13 kDa (MdPYL5) to 27.66 kDa (MdPYL4) and 5.02 (MdPYL10) to 8.30 (MdPYL11) (Table 1).

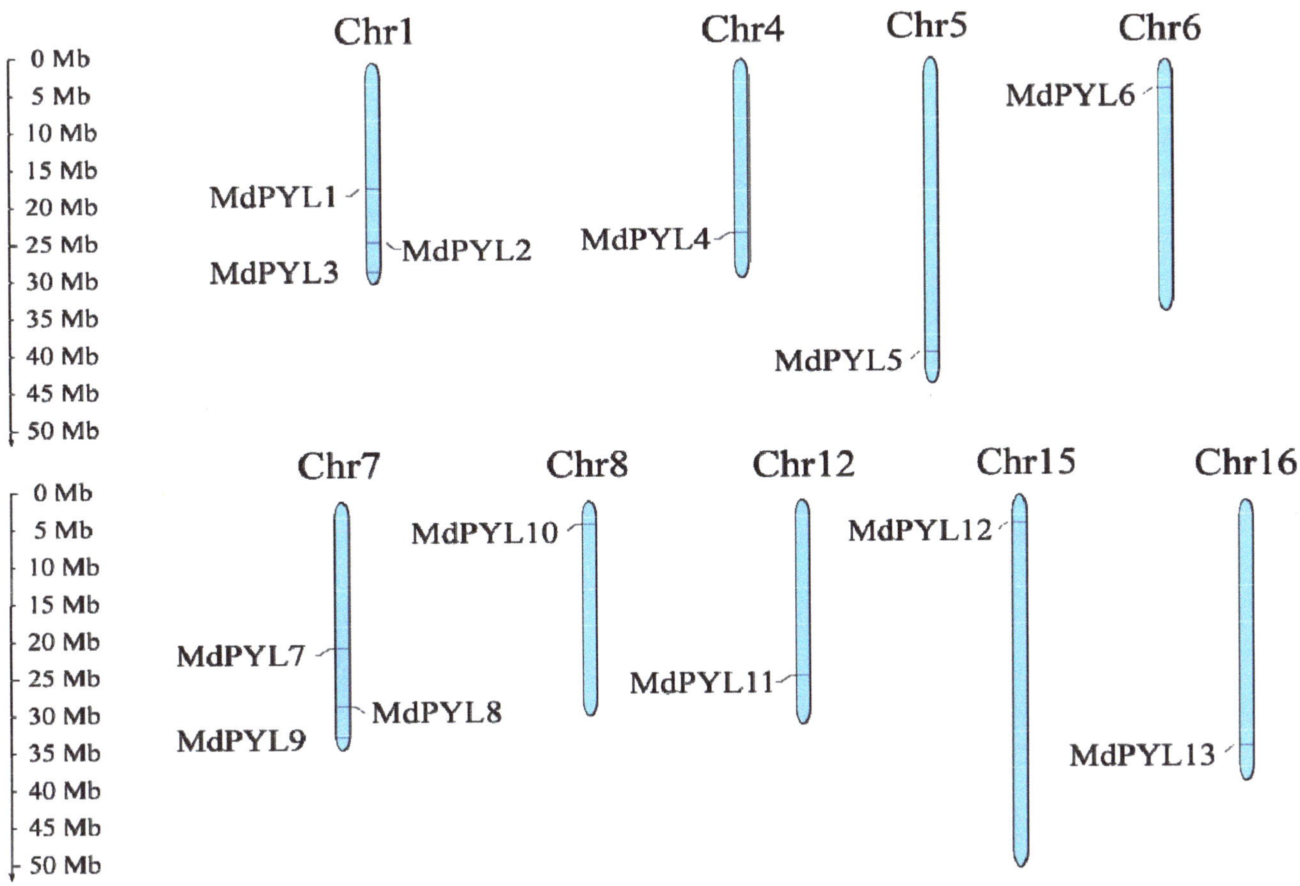

Figure 1. Chromosomal distribution and localization of *MdPYLs*. Only nine chromosomes contained *MdPYLs* are represented in this figure. The chromosome names are shown at the top of each chromosome. The chromosome scale is in millions of bases (Mb) on the left.

Table 1. Basic informations of 13 *MdPYLs* identified in apple.

Gene Name	Gene ID	Genomic (bp)	CDS (bp)	Exon	pI	MW (kDa)	Protein (aa)
MdPYL1	MD01G1078900	3252	555	3	6.45	20.80	184
MdPYL2	MD01G1158500	624	624	1	6.62	22.75	207
MdPYL3	MD01G1216100	3049	597	3	5.97	21.80	198
MdPYL4	MD04G1165000	1680	777	2	6.13	27.66	258
MdPYL5	MD05G1300200	558	558	1	5.82	20.13	185
MdPYL6	MD06G1034000	621	621	1	5.37	22.90	206
MdPYL7	MD07G1147700	2963	555	3	6.30	20.77	184
MdPYL8	MD07G1227100	624	624	1	6.44	22.74	207
MdPYL9	MD07G1286000	2576	615	3	6.06	22.33	204
MdPYL10	MD08G1043500	606	606	1	5.02	22.12	201
MdPYL11	MD12G1178800	2150	732	2	8.30	26.16	243
MdPYL12	MD15G1060800	609	609	1	5.20	22.04	202
MdPYL13	MD16G1274400	621	621	1	5.37	23.14	206

Moreover, according to available annotation information of the GDR database, 13 putative *MdPYLs* were distributed on nine chromosomes including chromosome 1, 4, 5, 6, 7, 8, 12, 15, and 16 (Figure 1). Three *MdPYLs* were present on chromosome 1 and 7, respectively, and one *MdPYL* on the remaining chromosomes (Figure 1).

2.2. Phylogenetic Analysis of the MdPYLs Family

We constructed a phylogenetic tree based on 109 protein sequences of *PYLs* from seven species, including 14 *AtPYLs* from *A. thaliana*, 13 *MdPYLs* from *M. domestica*, 7 *RcPYLs* from *R. communis*, 5 *VvPYLs* from *V. vinifera*, 9 *BdPYLs* from *B. distachyon*, 12 *OsPYLs* from *O. sativa*, 9 *TcPYLs* from *T. cacao*, and 40 *GhPYLs* from *G. hirsutum* (Table S1) via MEGA7 software using the neighbor-joining algorithm. In general, the 109 plant *PYLs* tested were classified into four subgroups and named as group I–IV (Figure 2, Tables S1 and S2). However, the plant PYL family evolutionary relationship was diverse in various species of plants. As expected, *MdPYLs* from apple generally showed closer genetic relationships to *PYLs* of dicotyledonous angiosperms (*V. vinifera*, *A. thaliana*, *G. hirsutum*, *R. communi*, and *T. cacao*) than monocotyledonous angiosperms (*O. sativa* and *B. distachyon*) (Figure 2). The results suggested plant *PYLs* classifying into the same groups might have similar functions.

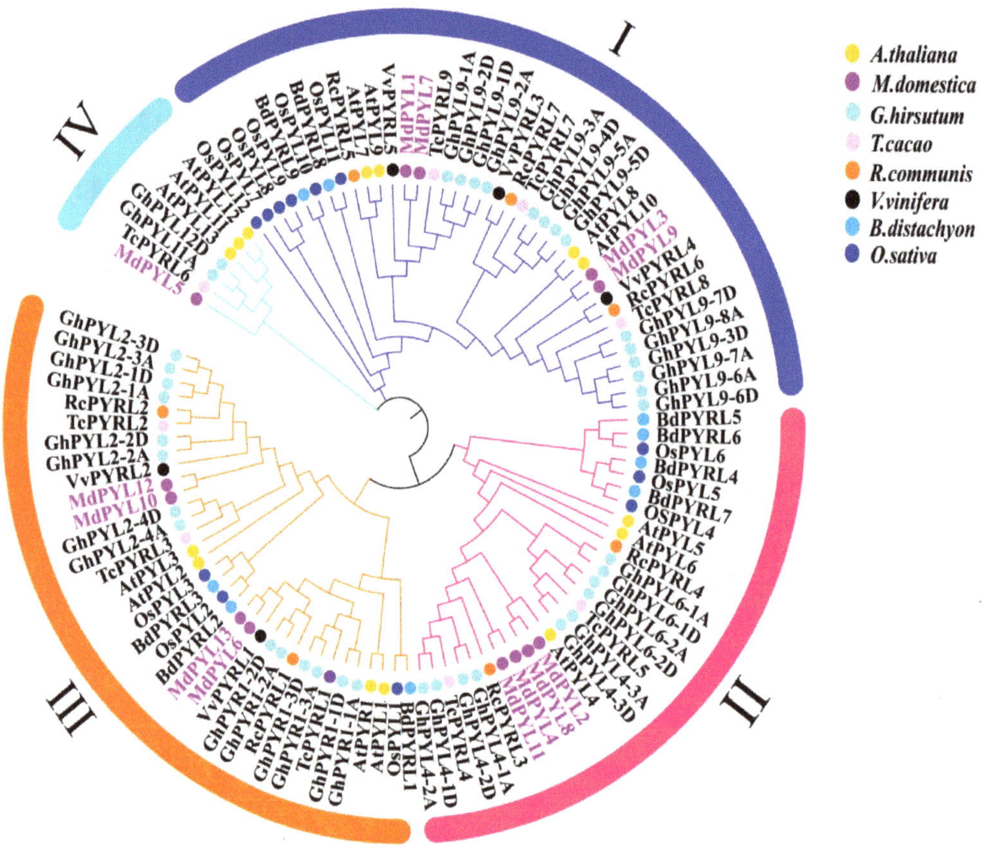

Figure 2. Phylogenetic analysis of *PYLs* from seven species. Different groups of *MdPYLs* are indicated in different colors bars. Circles of different colors denote *PYLs* from different species, including 14 *AtPYLs* from *A. thaliana* (yellow dot), 13 *MdPYLs* from *M. domestica* (purple dot), 7 *RcPYLs* from *R. communis* (orange dot), 5 *VvPYLs* from *V. vinifera* (black dot), 9 *BdPYLs* from *B. distachyon* (blue dot), 12 *OsPYLs* from *O. sativa* (navy blue dot), 9 *TcPYLs* from *T. cacao* (pink dot) and 40 *GhPYLs* from *G. hirsutum* (light blue dot). All amino acid sequences used in this analysis are listed in Supplementary Table S1.

2.3. Gene Structure and Duplication Analysis of MdPYLs

To study the structure and expansion mechanisms of *MdPYLs* gene family, we carried out the analysis of amino acid alignment, phylogenetic tree, exon–intron structures, the conserved motifs, and segmental duplications (Figure 3). In general, 13 MdPYLs protein all contained three α-helixes (α1–α4), seven-stranded β-sheet (β1–β7), and four highly conserved surface loops (CL1–CL4).(Figure 3a). The structure had been well characterized in the *AtPYLs* gene family, which involved in in ABA binding and inhibition of PP2Cs [38]. Moreover, the majority of *MdPYL*s had a similar profile of exons, and length of ORFs in the same group (Figure 3b,c). For example, all *MdPYL*s within group I included three exons, while *MdPYL*s in groups II and III had only one exon (no intron), except *MdPYL4* and *MdPYL11* both contained two exons (Figure 3b,c). Meanwhile, all of *MdPYL*s shared highly conserved motif 1, 2, and 3, and *MdPYL*s belonging to the same group seemed to have a similar conserved motif distribution. For example, in addition to the motif 1, 2, and 3, all of group I, II, and III respectively shared the highly conserved motifs 7 and 10, motif 1, 3d,e). To further understand the expansion mechanism of the *MdPYL*s, we also examined segmental and tandem duplications within the apple genome. No tandem duplications were identified and eight segmental duplication were found in 12 pairs of duplicated genomic regions, respectively, including (*MdPYL1*/*MdPYL3*, *MdPYL1/7*, *MdPYL2*/*MdPYL8*, *MdPYL3*/*MdPYL7*, *MdPYL3*/*MdPYL9*, *MdPYL4*/ *MdPYL11*, *MdPYL6*/*MdPYL13*, *MdPYL10*/*MdPYL12*) (Figure 4).

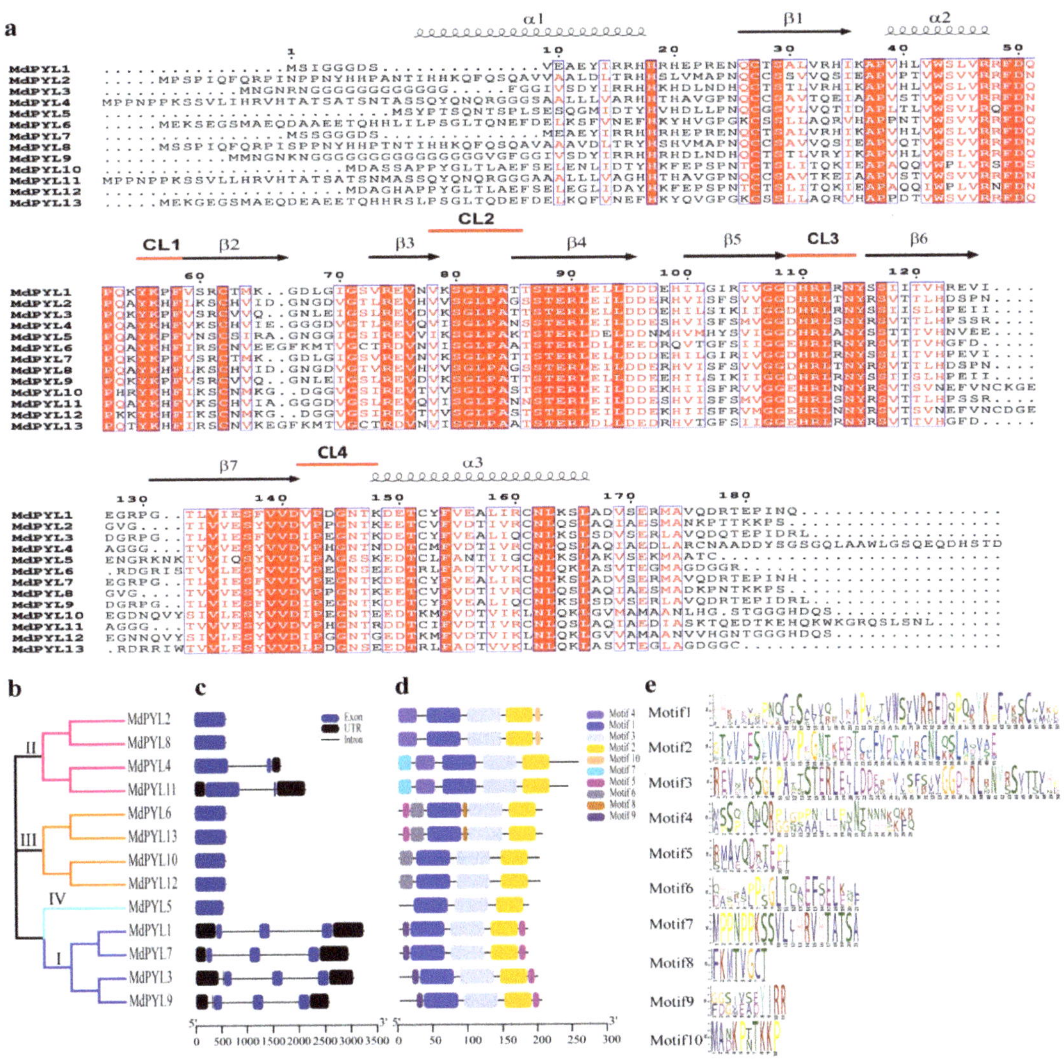

Figure 3. Amino acid sequence alignment, phylogenetic relationships, gene structure, and conserved motifs of *MdPYLs*. (**a**) Amino acid sequence alignment and secondary structural elements of the 13 MdPYLs. CL1–CL4 shown with red lines represent four conserved ABA receptor regions. (**b**) The phylogenetic tree of *MdPYLs*. Bars of different colors represent different groups of *PYLs*. (**c**) Exon–intron structure of apple *MdPYLs*. Black boxes denote untranslated regions; blue boxes denote exons; black lines denote introns. (**d**) Distributions of ten conserved motifs in *MdPYLs*. (**e**) The sequence logos of ten conserved motifs.

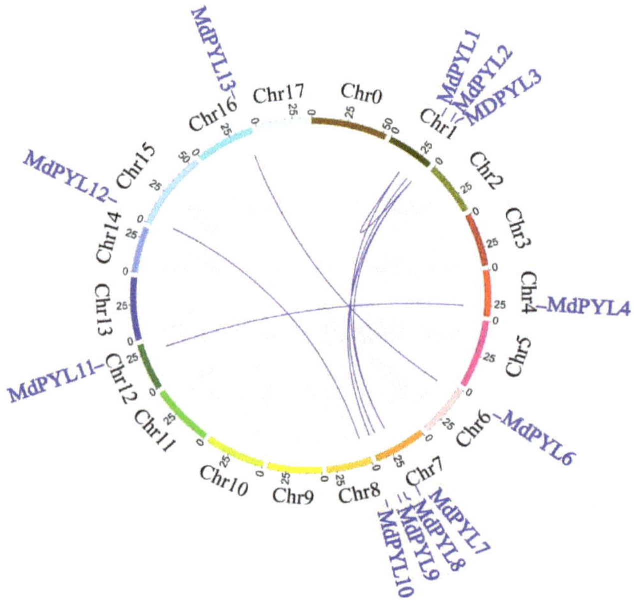

Figure 4. The synteny analysis of *MdPYLs* in the apple. Gray lines denote all synteny blocks in the apple genome, and the blue lines denote *MdPYL* duplicated gene pairs.

2.4. Evolutionary Relationship of PYLs Between Apple and Arabidopsis

Comparative genomic analysis between the apple and *Arabidopsis* genomes found the syntenies were clear and included the following ortholog pairs: *MdPYL1, 7-AtPYL9; MdPYL2, 8-AtPYL5, 6; MdPYL5-AtPYL13; MdPYL6, 13-AtPYL1; MdPYL11-AtPYL4* and *MdPYL12-AtPYL2* (Figure 5). Furthermore, *AtPYLs* and *MdPYLs* from the same group had similar exon–intron structures (Figure 6a,b). Meanwhile, all of the *AtPYLs* and *MdPYLs* shared highly conserved motifs 1, 2, and 3, except *AtPYL11, 12,* and *13* (Figure 6c). Significantly, many ortholog pairs between the apple and *Arabidopsis* genomes included nearly the same structure and conserved motif. For example, in *MdPYL6, 13-AtPYL1* gene pairs belonging to group III all had only one exon and conserved motif 6. *MdPYL1,7-AtPYL9* gene pairs belonging to group I all had three exons and conserved motif 8 (Figure 5,6). The result will provide further insight into the functions of apple *MdPYLs*.

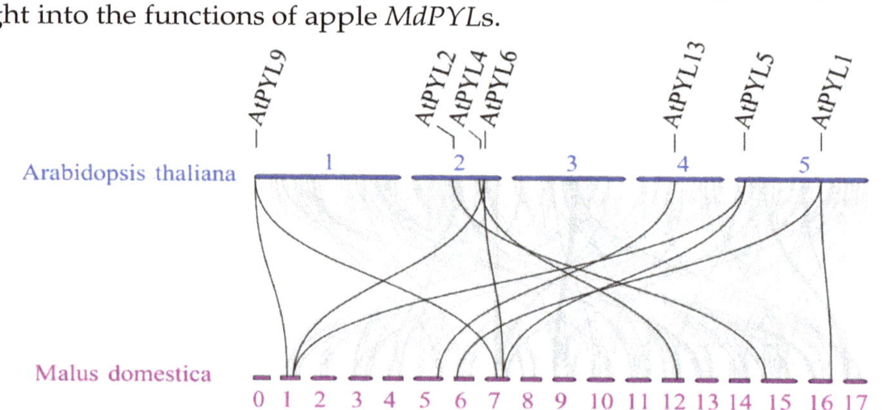

Figure 5. Synteny analysis of *PYLs* between apple and *A. thaliana*. Gray lines denote the collinear blocks between apple and *A. thaliana* genomes and the black lines denote the syntenic gene pairs of *PYLs*. Blue and purple lines represent respectively the apple chromosomes (0–17) and *Arabidopsis* chromosomes (1–5).

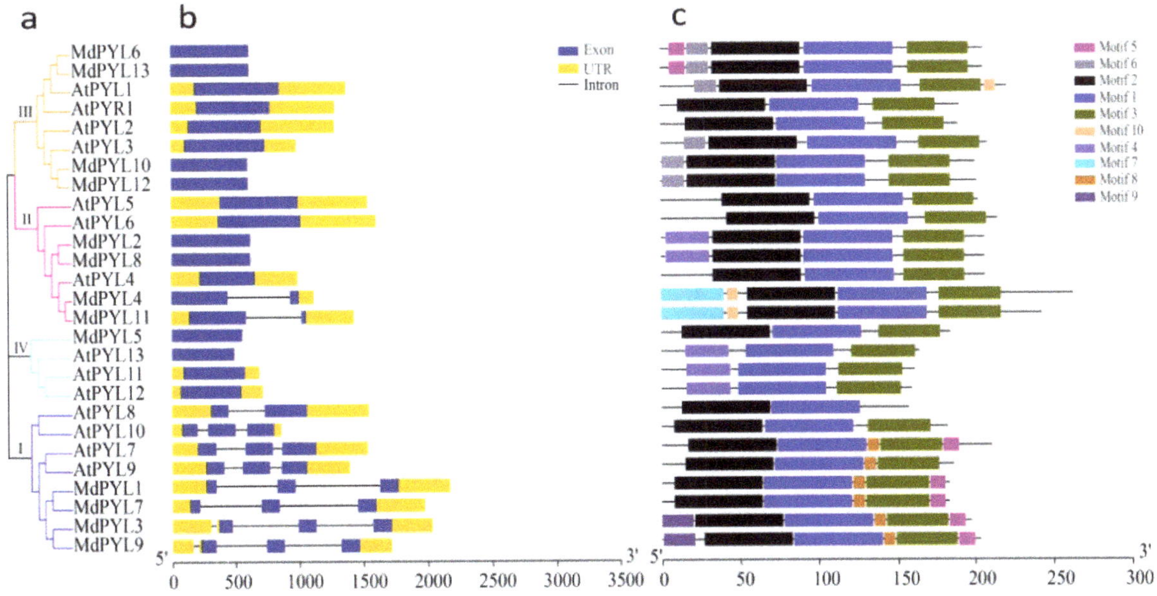

Figure 6. Phylogenetic relationships, gene structure, and conserved motifs of *MdPYLs* and *AtPYLs*. (**a**). The phylogenetic tree of *PYLs* from the apple and *Arabidopsis*. Bars of different colors represent different groups of *PYLs*. (**b**). Exon–intron structure of *MdPYL*s and *AtPYLs*. Yellow boxes denote untranslated regions; blue boxes denote exons; black lines denote introns. (**c**). Distributions of ten conserved motifs in *MdPYL*s and *AtPYL*s.

2.5. Cis-Element Analysis of the MdPYLs Promoter in Apple

Previous works have indicated that many *MdPYLs* involved in the response to different abiotic stresses. To further forecast the potential functions of *MdPYLs*, we analyzed the cis-acting elements involved in abiotic stresses responses in the *MdPYLs* promoter regions using PlantCARE software (Table S3,S4). In total, three abiotic stress-related and four hormone-related elements were discovered in the promoter of 13 *MdPYLs*, including drought-inducibility elements (MBS), defense and stress-responsive elements (TC-rich repeats), dehydration reaction elements (MYC), low-temperature-responsive elements (LTR), abscisic acid-responsive elements (ABRE), salicylic acid-responsive elements (TCA-element), ethylene-responsive elements (ERE), and MeJA-responsive elements (CGTCA-motif, TGACG-motif) (Figure 7). All *MdPYLs* promoter regions contained at least three cis-elements (Figure 7). For example, the *MdPYL7* promoter had seven stress-responsive elements including ABRE, MBS, MYC, LTR, TC-rich repeats, CGTCA, and TCA-element (Figure 7). The *MdPYL13* promoter region contained six stress-responsive elements including ABRE, MBS, MYC, LTR, ERE, and CGTCA-element. Moreover, the *MdPYL2, 5, 8* promoter included three, the *MdPYL1, 6, 7, 10, 11* promoter comprised four, and the *MdPYL3, 4, 12* promoter contained five stress-responsive elements. Furthermore, the *MdPYLs* promoter had two tissue-specifc cis-elements (root-specific regulatory element and seed-specific regulation element) (Figure 7). The results concluded that the expression of *MdPYLs* with different cis-elements may differ in response to various abiotic stresses.
results concluded that the expression of *MdPYLs* with different cis-elements may differ in response to various abiotic stresses.

To further understand the expression profiles of *MdPYL*s in apple development, we detected their expression levels in different tissues and organs from Chinese wild apple clone 'Xinjiang No.1' using qRT-PCR. In general, most *MdPYLs* were expressed differently in various organs, showing they might have different functions (Figure 8). Notably, *MdPYL5* (group IV) had little or almost no expression in the organs tested. In contrast, the expression levels of all *MdPYLs* within group I, including *MdPYL1*, *MdPYL3*, *MdPYL7*, and *MdPYL9*, were higher in all of the organs tested, and were more than 20 times of that in group IV. Furthermore, *MdPYLs* of group II, including *MdPYL2*, *MdPYL4*, *MdPYL8*, and *MdPYL11*, were preferentially expressed in the root and their expression level were more than fivefold of the other organs. At the same time, we found *MdPYLs* within group III expression exhibited the relatively higher levels than *MdPYLs* of group II in all tissue analyzed except for root. For example, the expression of *MdPYL10* and *MdPYL12* in peel, sarcocarp, and young fruit were over fivefold of all genes within group II (Figure 8). These results suggested *MdPYLs* might have a distinct effect on growth and development of the apple tree.

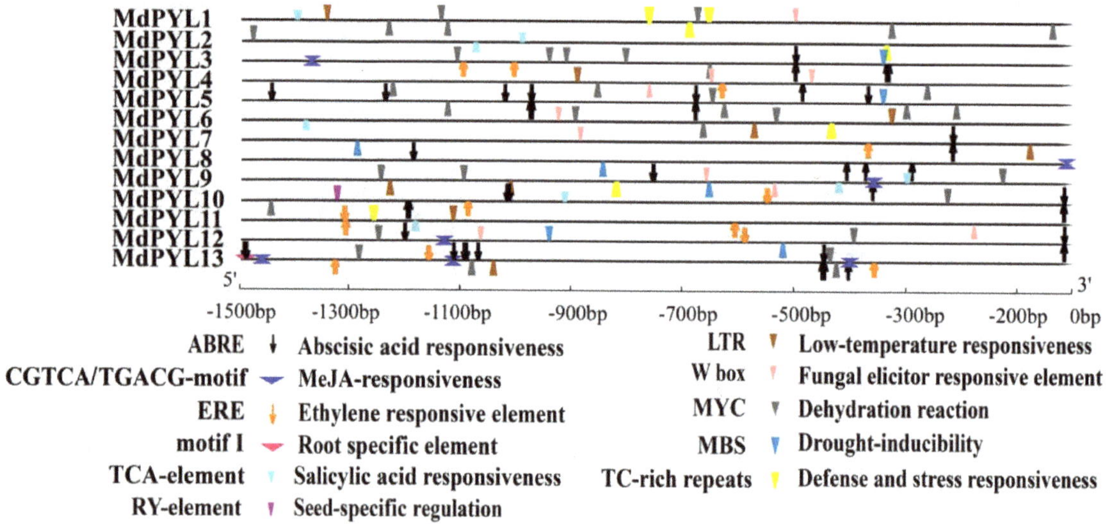

Figure 7. The cis-elements analysis in *MdPYL* promoters. The 1500 bp promoter sequences of 13 *MdPYLs* were analyzed by PlantCARE. The up and down directions respectively denote the cis-elements existing in the plus or minus strand of *MdPYL* promoters.

2.6. Expression Profiles of Apple MdPYLs in Diverse Organs

To further understand the expression profiles of *MdPYLs* in apple development, we detected their expression levels in different tissues and organs from Chinese wild apple clone 'Xinjiang No.1' using qRT-PCR. In general, most *MdPYLs* were expressed differently in various organs, showing they might have different functions (Figure 8). Notably, *MdPYL5* (group IV) had little or almost no expression in the organs tested. In contrast, the expression levels of all *MdPYLs* within group I, including *MdPYL1*, *MdPYL3*, *MdPYL7*, and *MdPYL9*, were higher in all of the organs tested, and were more than 20 times of that in group IV. Furthermore, *MdPYLs* of group II, including *MdPYL2*, *MdPYL4*, *MdPYL8*, and *MdPYL11*, were preferentially expressed in the root and their expression level were more than fivefold of the other organs. At the same time, we found *MdPYLs* within group III expression exhibited the relatively higher levels than *MdPYLs* of group II in all tissue analyzed except for root. For example, the expression of *MdPYL10* and *MdPYL12* in peel, sarcocarp, and young fruit were over fivefold of all genes within group II (Figure 8). These results suggested *MdPYLs* might have a distinct effect on growth and development of the apple tree.

The expression of *MdPYL2* in roots was set to 1 and *MdActin* was used as an internal control. The relative expression levels of genes were calculated based on the 2-ΔΔCt method, and values are the mean ± SD obtained from three biological replicates.

2.7. Expression Profiles of MdPYLs in Response to Various Abiotic Stresses

PYLs, as ABA receptors, might play a vital function in abiotic stress signaling pathways. Therefore, we detected the expression of all 13 apple *MdPYLs* in Chinese wild apple clone 'Xinjiang No.1' tissue-cultured seedlings upon various abiotic stress treatments (including ABA NaCl, PEG, cold) by carrying out qRT-PCR. Genes with expression levels that were increased or decreased more than two-fold, were used for the following analyses (Figure 9).

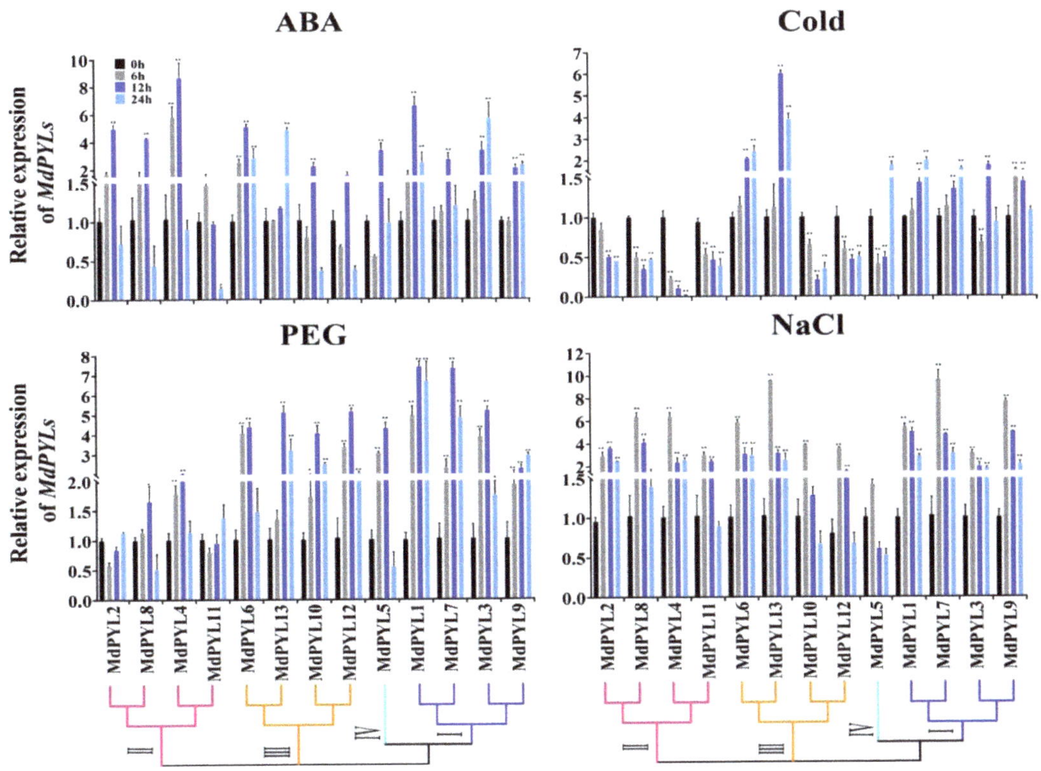

Figure 9. Expression of the apple *MdPYLs* under ABA, PEG, cold (4 °C), and salt treatment in the tissue-cultured apple seedlings (Xinjiang No.1). The gene relative expression was calculated using the 2-ΔΔCt method with *MdActin* as an internal control, and value represents mean ± SD of three biological replicates. Asterisks indicated values that are significantly different from CK (0 h) (* $p < 0.05$, ** $p < 0.01$, one-way ANOVA).

In our study, we found that the expression of almost all of *MdPYLs* were significantly changed by ABA, PEG, NaCl, and cold. For example, the expression of all *MdPYLs* except for *MdPYL11* exhibited significant increases by ABA treatment (Figure 9). Nine genes (*MdPYL1*, *MdPYL2*, *MdPYL4*, *MdPYL5*, *MdPYL6*, *MdPYL7*, *MdPYL8*, *MdPYL10*, and *MdPYL12*) reached the peak at 12 h after ABA treatment, while three genes (*MdPYL3*, *MdPYL9*, and *MdPYL13*) reached the peak at 24 h. Notably, *MdPYL1*, *MdPYL2*, *MdPYL3*, *MdPYL4*, *MdPYL6*, *MdPYL8*, and *MdPYL13* all were upregulated more than fourfold after 12 h or 24 treatment, respectively. Similarly, following with PEG treatment, 10 genes (*MdPYL1*, *MdPYL2*, *MdPYL3*, *MdPYL5*, *MdPYL6*, *MdPYL7*, *MdPYL9*, *MdPYL10*, *MdPYL12*, and *MdPYL13*) were sharply upregulated in expression after 6 h of treatment and reached the maximum at 12 h, while the expression of remaining three apple *MdPYLs* had no obvious changes (Figure 9). The expression of all *MdPYLs* within group I, group III, and group IV were over fourfold higher than CK (0 h) at 12 h after PEG treatment. After NaCl treatment, the expression of all *MdPYLs* except for *MdPYL5* showed significant increases and surged to the peak at 6 h (more than threefold of CK) (Figure 9). Following cold treatment (4 °C), all *MdPYLs* within group I, and *MdPYL6*, *13* within group III demonstrated increased expression. More remarkably, *MdPYL*13 was upregulated more than sixfold after 6 h treatment. In contrast, all *MdPYLs* within group II, and *MdPYL10*, *12* within group III were significantly downregulated (Figure 9).

2.8. Interaction between MdPYL and MdPP2C Proteins

PYLs can interact with PP2Cs to form PYL-ABA-PP2C triple complexes promoting the ABA signals transmission [1]. However, other studies suggested that PYLs can bind PP2Cs in an ABA-dependent or ABA-independent manner in yeast two-hybrid (Y2H) assays [4,12]. In this study, 13 MdPYLs and 2 MdPP2Cs were isolated (Figure S1) to detect their potential interactions in the yeast two-hybrid system. The full coding sequence of all MdPYLs were respectively fused to the DNA-binding domain (BD) of pGBKT7 vector, while two MdPP2Cs were respectively fused with activation domain (AD) of pGADT7 vector. After co-transformed into the Y2H Gold yeast strain, all of yeast cells were capable of growth on SD-Leu/-Trp medium in the absence or presence of ABA. However, the co-transformed yeast cells grew faster in medium with ABA (Figure 10). Interaction between 13 MdPYLs and two MdPP2Cs was determined by growth assay on SD-Leu/-Trp/-His/X-α-Gal/3-AT medium with and without 10 μM ABA. In this experiment, all MdPYLs interacted with MdPP2C65 in an ABA-dependent manner, and these interactions were considerably enhanced in presence of ABA. However, MdPYLs differentially interacted with MdPP2C72. MdPYL1, MdPYL3, MdPYL6, MdPYL10, MdPYL11, and MdPYL12 interacted with MdPP2C72 in an ABA-dependent manner. MdPYL4, MdPYL7, MdPYL8, and MdPYL9 showed weak interaction signals with MdPP2C72 in the absence of ABA. Furthermore, we found that MdPYL2,5,13 showed stronger interaction signals activating the reporter gene X-α-Gal with MdPP2C72 in an ABA-independent manner (Figure 10). In addition, all MdPYLs and MdPP2Cs had no obvious self-activation (Figure S2). These results suggested that MdPYLs may bind to MdPP2C65 and MdPP2C72 in different manners and with different affinities.

3. Discussion

3.1. Gene Functional Diversification of Apple MdPYLs

PYLs encoding an ABA receptor family contained a similar ligand-binding pocket embraced by four conserved surface loops (CL1–CL4). In the study, 13 *MdPYLs* shared the typical structure were identified in the apple, with an almost equal number of *AtPYLs* from *Arabidopsis* (Figures 2 and 3a). Comparative genomic analysis found that nine *MdPYLs* (*MdPYL1, 2, 5, 6, 7, 8, 11, 12, 13*) were positioned in syntenic regions of the apple and *Arabidopsis* genomes (Figure 5), and the other four *MdPYLs* (*MdPYL3, 4, 6, 9*) were formed by segment duplication events in the apple genome (Figure 4). Meanwhile, almost all of the ortholog pairs between apple and *Arabidopsis* genomes also contained similar exon–intron structures and conserved motifs. The results suggested that the function of *PYLs* gene might be relatively conserved between the apple and *Arabidopsis*. To date, the functions of many *AtPYLs* have been reported in *Arabidopsis* [15–18,22,39]. According to *MdPYLs* expression data (Figures 8 and 9) and the information known concerning their *Arabidopsis* counterparts, we will better predict the probable functions of apple *MdPYLs*.

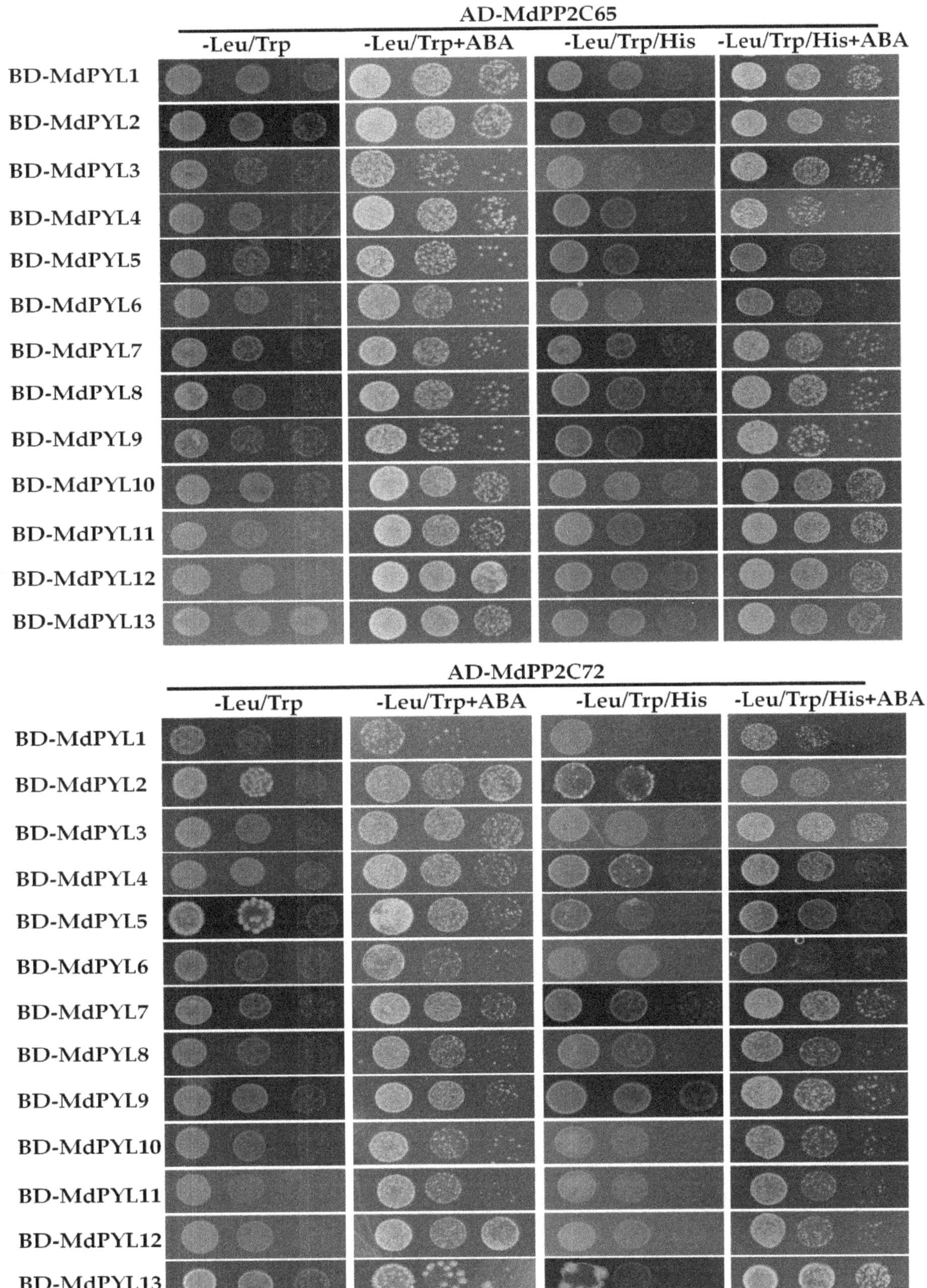

Figure 10. Interactions between MdPYLs and MdPP2Cs in yeast. Co-transformed positive yeast cells dilutions (10^{-1}, 10^{-2}, and 10^{-3}) were spotted onto the plates. Interaction was determined by growth assay on SD-Leu/-Trp/-His/X-α-Gal/3-AT medium in the presence and absence of 10 μM ABA.

3. Discussion

3.2. Expression Profiles of Apple PYLs and Their Potential Functions in Various Apple Organs

ABA is an important phytohormone for plant growth and development. PYL family members, as ABA receptors, have been identified and characterized in many plants including *Arabidopsis*, maize, rice, grape, rubber tree, and so on. To date, their expression profiles in various tissues have been revealed substantial differences among *PYL* genes for many plants. For example, some *PYLs* respectively displayed relatively higher levels in seeds of soybean, the callus of *B. rapa*, and the latex of rubber tree than other tissues [23,24,40,41]. Moreover, most *OsPYLs* were expressed in all organs of rice, *OsPYL3* and *OsPYL5* were predominantly expressed in leaves, and *OsPYL1* in roots [42]. Similarly, in maize, *ZmPYL11* was upregulated in leaves and *ZmPYL6* and *ZmPYL10* in roots [43]. In tomato, *PYR/PYL/RCAR* genes (10g085310 and 3g095780) had high levels of expression in root, and 1g095700 showed predominant expression in leaf, 12g055990 and 8g082180 in fruit tissues [22]. In grape, most of the *VyPYLs* genes, especially *VyPYL1*, *VyPYL4a*, and *VyPYL7-9* were expressed relatively strongly in roots, stems, and tendrils [44]. These reports indicated that *PYLs* play an important role in plant growth and development. The function of PYLs in *Arabidopsis* has been studied to support this idea. For example, overexpression of *RCAR11*, *RCAR12*, *RCAR13*, or *RCAR14* inhibited the germination and root growth of transgenic *Arabidopsis* in an ABA-treated condition [45]. *AtPYL9* promoted ABA-induced leaf senescence in transgenic *Arabidopsis* [15]. *AtPYL8* promoted lateral root growth by interacting with AtMYB77 and enhancing its transcriptional activity [18]. However, *MdPYLs* expression patterns and functions are still unclear in apple. Therefore, we speculated the roles of the apple *MdPYLs* based on qRT-PCR results in different tissues, and the information known of *Arabidopsis* counterparts (Figure 5) and homology genes of other species (Figure 2).

In general, the apple *MdPYLs* exhibited qualitatively and quantitatively distinct expression profiles. However, *MdPYLs* clustered to the same group, especially the eight pairs of genes in duplicated genomic regions, showed similar expression patterns, respectively (Figure 8), suggesting that they might have similar functions. For example, all *MdPYLs* within group I, including *MdPYL1*, *MdPYL3*, *MdPYL7*, and *MdPYL9*, exhibited relatively higher expression levels in all of the tissues tested compared to *PYLs* in other groups (Figure 8). Thus, we speculated that *MdPYLs* of the group I might function in the regulation of plant different growing and developmental stages. The function of homologous genes in other species has been studied to support this hypothesis. For instance, *AtPYL8* and *AtPYL9*, which are orthologs of *MdPYL1, 7*, are involved in controlling root sensitivity to ABA [16]. *OsPYL8* and *OsPYL9*, belonging to group I, were specifically expressed in the rice endosperms and positively regulated by ABA during seed germination [46]. Furthermore, *MdPYLs* of group II, including *MdPYL2*, *MdPYL4*, *MdPYL8*, and *MdPYL11* were preferentially expressed in the root (Figure 8), indicating that these genes might play a role in root development. Indeed, the *AtPYL5*/*AtPYL6* gene, as the ortholog of *MdPYL2*/*MdPYL8*, was expressed at the higher levels in root and might play a critical role for the regulation of root growth and root system architecture [5]. In short, these results indicated the diverse biological functions of different *PYLs* in plant growth and development.

3.3. Apple MdPYLs Were Responsive to Various Abiotic Stresses

To date, *PYLs*, as important ABA receptors, have been found to play crucial roles in responding to various abiotic stresses in plants. Overexpressing *pRD29A::PYL9* dramatically promoted drought resistance and leaf senescence induced by drought in *Arabidopsis* and rice plants [15]. Similarly, overexpression of *AtPYL4, 5, 7, 8,* and *13* also respectively enhanced drought tolerance and showed hypersensitivity to ABA during early seedling development in transgenic *Arabidopsis* [47]. Furthermore, overexpressing ABA receptors (*RCAR11-14)* also increased drought resistance and accelerated stress-responsive gene expression in transgenic *Arabidopsis* [45]. With the deepening study in *Arabidopsis*, the functions of orthologous PYL proteins' responses to stresses were reported in several plants, such as rice, cotton, tomato, and grape. In rice, ectopic expression of *OsPYL3* improved cold and drought tolerance in transgenic *Arabidopsis* [31]. Constitutive expression of *OsPYL/RCAR5* increased drought and salt stress tolerance in rice [30]. In *Gossypium*, the expression level of many *GhPYLs* were downregulated by ABA treatment and upregulated by osmotic stress [27,34]. Overexpression of *GhPYL9-11A*, *GhPYL10*, *GhPYL12*, and *GhPYL26* improved the tolerance to drought stress in transgenic *Arabidopsis* [27,34]. Heterologous expression of tomato *ABA receptors* (6g050500 or 3g007310, or 8g076960) and *Vitis yeshanensis VyPYL9* all enhanced transgenic *Arabidopsis* drought resistance [22,44]. However, there is only a handful information concerning these genes response to stress in apple. Thus, we investigated the response of all 13 *MdPYLs* to various abiotic stress conditions in the study (Figure 9), and found that 12 *MdPYLs* (except *MdPYL11*), 12 *MdPYLs* (except *MdPYL5*), and 10 *MdPYLs* (except *MdPYL4*, *MdPYL8*, and *MdPYL11*) had significant expression changes when treated with ABA, salinity, and PEG, respectively (Figure 9). The results were consistent with the function of PYLs in *Arabidopsis* and rice, which indicated they may involve in drought and salinity stresses response. In addition, *MdPYL6* and *MdPYL13*, which were located in duplicated genomic regions, exhibited significant increases in expression levels following cold treatment (4 °C) (Figure 9). Meanwhile, the promoter regions of *MdPYL6* and *MdPYL13* both contained low-temperature-responsive elements (LTR) (Figure 7, Tables S3,S4). This result insinuated that these two genes will be the key candidate ABA receptors responding to cold stresses in apple. Indeed, the rice *OsPYL3* and *OsPYL9* gene, which the ortholog of *MdPYL6/13* (Figure 2 and Table S2), was reported to be involved in enhancing cold tolerance in transgenic *Arabidopsis* and rice [28,42].

Furthermore, many evidences have shown that PYLs involve in plant response to abiotic stress by inhibiting PP2Cs in an ABA-dependent manner [3,5,6,48]. Different works have been reported that PYLs interacted with PP2Cs in an ABA-dependent or ABA-independent manner [5,8]. For instance, AtPYL8/RCAR3 interacted with *Fagus sylvatica* FsPP2C1 in an ABA-independent manner and positively regulates ABA signaling during abiotic stress responses [48]. AtPYL5 enhanced transgenic *Arabidopsis* resistance to drought by inhibited HAB1 phosphatase activity in an ABA-dependent manner [5]. In this work, we evaluated the interactions between 13 MdPYLs and two MdPP2Cs. Our results indicated that 13 MdPYLs selectively interacted with MdPP2Cs in ABA-dependent or ABA-independent manner (Figure 10), implying the function diversity among different *MdPYLs*. In conclusion, it seems that *PYLs* have various functions in the apple and play a key role in abiotic stress induced by ABA molecular signals.

4. Materials and Methods

4.1. Identification of MdPYL Gene Family and Chromosomal Location in Apple Genome

We used the protein sequence of 14 *AtPYLs* from the *Arabidopsis* genome to search for *MdPYLs* of the apple genome. The BLAST program was set to the default value (*e*-value < e^{-10}) [49]. Manual reanalysis was performed for suspicious genes with a PYL structure and low E-value. The amino acid number, isoelectric point (pI), and molecular weight (MW) were analyzed by using the ExPASy website. The localization of *MdPYLs* on the chromosome were mapped by using the Gene Structure Display Server.

4.2. Phylogenetic Analysis of PYL Gene Family

The phylogenetic tree was constructed using the PYLs amino acid sequences from *M. domestica*, *A. thaliana*, *B. distachyon*, *O. sativa*, *R. communis*, *T. cacao*, *V. vinifera*, and *G. hirsutum*. Multiple alignments of protein sequences were analyzed by the neighbor-join (NJ) algorithm from 1000 repeated by MEGA 5.2 software [50].

4.3. Co-linear Analysis of Apple MdPYLs

The homologous genes were searched by MCScanx software [51], and the collinearity of the *MdPYLs* was obtained using the Circos program [52].

4.4. Gene Structure and Motif Composition of Apple MdPYLs

The MdPYLs protein sequences were found from the apple genome, and their intron–exon structures were analyzed by GSDS [53]. The MdPYL protein motifs were analyzed by Multiple EM for Motif Elicitation version 4.11.4 (MEME) [54]. The maximum number of motifs was set to 10, and the motif length was set to 6–200 amino acids.

4.5. Analysis of Cis-Acting Elements in the Promoter of MdPYLs

The cis-acting elements were detected using the PLACE database in the 1500 base pair (bp) upstream of the gene initiation codon (ATG) of the *MdPYLs* (Tables S3 and S4).

4.6. Plant Materials and Treatments

The red-fleshed apple (*Malus sieversii* f. Neidzwetzkyana (Dieck) Langenf) 'Xinjiang No.1', a wild apple resource with resistance to abiotic stress, was from Xinjiang of China.

The apple 'Xinjiang No.1' tissue culture seedlings used for stress treatment was cultured in a specific medium with Murashigeand Skoog (MS) medium, 0.8% agar, 0.5 mmol L^{-1} indole-3-butytric acid (IBA) and 0.7 mmol L^{-1} 6-benzylaminopurine (6-BA). After 30 days of growth, the 'Xinjiang No.1' seedlings were transferred to the medium, which supplied with 10% polyethylene glycol (PEG) 6000 or 100 mmol L^{-1} abscisic acid (ABA) or 100 mmol L^{-1} NaCl to induce different abiotic stress, respectively. Furthermore, the temperature for apple tissue culture seedlings cultured in a specific medium was adjusted to 4 °C to induce cold stress. The stress-treated seedlings were harvested at 0, 6, 12, and 24 h, then immediately frozen with liquid nitrogen and stored in a −80 °C freezer for follow-up studies.

The different apple organs, including roots, stems, leaves, pedicels, calyxes, anthers, filaments, receptacles, petals, pistils, sarcocarps, peels, young fruits (20 days after flowering), and seeds were collected from seven-year old 'Xinjiang No.1', which was obtained from tissue culture seedlings transplanted to the Jiao Zhou Experimental Farm (longitude 120°39′ E, latitude 36°27′ N) of Qing Dao Agricultural University.

4.7. Quantitative Real-Time RT-PCR Analysis

RNA was extracted from the collected samples using the EASYspin Plant RNA Rapid Extraction Kit (YPHBIO, Beijing, China). The RNA concentration was measured with a NanoDrop 2000 (Gene Company Limited, Hong Kong, China) instrument. The cDNA was obtained according to a reverse transcription kit (Takara, Dalian, China). Specific primers were designed on NCBI-BLAST (Table S5). According to the manufacturer's instructions, quantitative real-time RT-PCR (qRT-PCR) was performed on a LightCycler® 480 real-time PCR instrument (Roche, Shanghai, China). The PCR program was as followed: 95 °C for 2 min, 40 cycles of 95 °C for 30 s, 56 °C for 30 s and 72 °C for 30 s with a final dissociation stage. Actin of apple was used as an internal control for normalized gene expression levels. Repeat at least three times for each sample. Gene relative expression levels were calculated using GraphPad.Prism.5.0 software.

4.8. Yeast Two-Hybrid Assay

Full-length sequences of 13 MdPYLs and two MdPP2Cs (MdPP2C72: XM_008359137.2 and MdPP2C65: XM_008371984.3) were respectively amplified by PCR from 'Xinjiang No.1' leaf using the specific primers with enzyme digestion sites (Table S5). The full coding sequence of all MdPYLs were respectively fused to the DNA-binding domain (BD) of pGBKT7 vector, while two MdPP2Cs were respectively fused with activation domain (AD) of pGADT7 vector via double digestion technique. Combinations of pGADT7-T with pGBKT7-53 and pGBKT7-Lam were used as positive and negative controls, respectively. Combinations of empty pGADT7 with pGBKT7-MdPYLs, and empty pGBKT7 with pGADT7-MdPP2Cs were used to investigate whether MdPYLs or MdPP2Cs protein had self-activation activity. Combinations of pGBKT7-MdPYLwith pGADT7-MdPP2Cs were used to determine the potential interactions between MdPYLs and MdPP2Cs.

After co-transformation into the Y2H Gold yeast strain, the yeasts were plated on a SD/-Leu/-Trp media. Then, the positive yeasts were grown in liquid SD-Leu/Trp medium for 12 h at 30 °C. Finally, 3 µL of the co-transformed yeast cells dilutions of 10^{-1}, 10^{-2}, and 10^{-3} were spread on plates containing SD/-Leu/-Trp, SD/-His/-Leu/-Trp/X-α-Gal medium with 10 mM 3-AT (3-amino-1, 2, 4-triazole) and SD/-His/-Leu/-Trp/X-α-Gal medium with 10 mM 3-AT (3-amino-1, 2, 4-triazole) and 10 µM ABA. The plates were photographed after four days of incubation at 30 °C. The experiments were repeated three times.

5. Conclusions

Thirteen *MdPYL*s were identified in *M. domestica* genome. Phylogenetic reconstruction and gene structure analysis demonstrated that *MdPYLs* could be divided into four groups, and that they had similar gene structures and high conserved motifs in the same group. Moreover, comparative genomic analysis showed that homologs of nine *MdPYLs* were located in corresponding syntenic regions of *Arabidopsis*. The MdPYLs expression analysis in various organs revealed distinct spatiotemporal patterns. Furthermore, gene expression analysis showed that *MdPYLs* are possibly involved in multiple abiotic stress responses (ABA, salt, PEG, and cold). The results presented here call for further research aiming at revealing the potential important functions of these genes in the apple.

Supplementary Materials
Table S1: Amino acid sequences of the plant *PYLs*. Table S2: The clustered groups of PYL proteins from eight species. Table S3: The 1500 bp promoter sequences of the 13 *MdPYL*s. Table S4: Cis-element analysisof the *MdPYL*s promoter regions in the apple. Table S5: Primers used in this study. Figure S1: The results of agarose gel electrophoresis test of 13 MdPYLs and two MdPP2Cs. Figure S2: Self activation analysis of 13 MdPYLs and two MdPP2Cs fusions.

Author Contributions: Y.Z. and H.H. (Hongmin Hou) designed the study. H.H. (Hongmin Hou), L.L., and Y.Z. performed data analysis. Y.Z. provided guidance on the whole study. H.H. (Hongmin Hou), L.L., H.D., H.H. (Heqiang Huo), and Y.Z. wrote and revised the manuscript. All authors have read and agreed to the published version of the manuscript.

References

1. Cutler, S.R.; Rodriguez, P.L.; Finkelstein, R.R.; Abrams, S.R. Abscisic acid: Emergence of a core signaling network. *Annu. Rev. Plant. Biol.* **2010**, *61*, 651–679. [CrossRef]
2. Lee, S.C.; Luan, S. ABA signal transduction at the crossroad of biotic and abiotic stress responses. *Plant. Cell Environ.* **2012**, *35*, 53–60. [CrossRef] [PubMed]
3. Fujii, H.; Chinnusamy, V.; Rodrigues, A.; Rubio, S.; Antoni, R.; Park, S.Y.; Cutler, S.R.; Sheen, J.; Rodriguez, P.L.; Zhu, J.K. In vitro reconstitution of an abscisic acid signalling pathway. *Nature* **2009**, *462*, 660–664. [CrossRef] [PubMed]
4. Park, S.Y.; Fung, P.; Nishimura, N.; Jensen, D.R.; Fujii, H.; Zhao, Y.; Lumba, S.; Santiago, J.; Rodrigues, A.; Chow, T.F.; et al. Abscisic acid inhibits type 2C protein phosphatases via the PYR/PYL family of START proteins. *Science* **2009**, *324*, 1068–1071. [CrossRef] [PubMed]
5. Santiago, J.; Rodrigues, A.; Saez, A.; Rubio, S.; Antoni, R.; Dupeux, F.; Park, S.Y.; Márquez, J.A.; Cutler, S.R.; Rodriguez, P.L. Modulation of drought resistance by the abscisic acid receptor PYL5 through inhibition of clade A PP2Cs. *Plant. J.* **2009**, *60*, 575–588. [CrossRef] [PubMed]
6. Ma, Y.; Szostkiewicz, I.; Korte, A.; Moes, D.; Yang, Y.; Christmann, A.; Grill, E. Regulators of PP2C phosphatase activity function as abscisic acid sensors. *Science* **2009**, *324*, 1064–1068. [CrossRef]
7. Miyazono, K.; Miyakawa, T.; Sawano, Y.; Kubota, K.; Kang, H.J.; Asano, A.; Miyauchi, Y.; Takahashi, M.; Zhi, Y.; Fujita, Y.; et al. Structural basis of abscisic acid signalling. *Nature* **2009**, *462*, 609–614. [CrossRef]

8. Nishimura, N.; Hitomi, K.; Arvai, A.S.; Rambo, R.P.; Hitomi, C.; Cutler, S.R.; Schroeder, J.I.; Getzoff, E.D. Structural mechanism of abscisic acid binding and signaling by dimeric PYR1. *Science* **2009**, *326*, 1373–1379. [CrossRef]
9. Peterson, F.C.; Burgie, E.S.; Park, S.Y.; Jensen, D.R.; Weiner, J.J.; Bingman, C.A.; Chang, C.E.; Cutler, S.R.; Phillips, G.N., Jr.; Volkman, B.F. Structural basis for selective activation of ABA receptors. *Nat. Struct. Mol. Biol.* **2010**, *17*, 1109–1113. [CrossRef]
10. Miyakawa, T.; Fujita, Y.; Yamaguchi-Shinozaki, K.; Tanokura, M. Structure and function of abscisic acid receptors. *Trends. Plant. Sci.* **2013**, *18*, 259–266. [CrossRef]
11. Santiago, J.; Dupeux, F.; Betz, K.; Antoni, R.; Gonzalez-Guzman, M.; Rodriguez, L.; Márquez, J.A.; Rodriguez, P.L. Structural insights into PYR/PYL/RCAR ABA receptors and PP2Cs. *Plant. Sci.* **2012**, *182*, 3–11. [CrossRef] [PubMed]
12. Hao, Q.; Yin, P.; Li, W.; Wang, L.; Yan, C.; Lin, Z.; Wu, J.Z.; Wang, J.; Yan, S.F.; Yan, N. The molecular basis of ABA-independent inhibition of PP2Cs by a subclass of PYL proteins. *Mol. Cell.* **2011**, *42*, 662–672. [CrossRef] [PubMed]
13. Li, W.; Wang, L.; Sheng, X.; Yan, C.; Zhou, R.; Hang, J.; Yin, P.; Yan, N. Molecular basis for the selective and ABA-independent inhibition of PP2CA by PYL13. *Cell Res.* **2013**, *23*, 1369–1379. [CrossRef] [PubMed]
14. Gonzalez-Guzman, M.; Pizzio, G.A.; Antoni, R.; Vera-Sirera, F.; Merilo, E.; Bassel, G.W.; Fernandez, M.A.; Holdsworth, M.J.; Perez-Amador, M.A.; Kollist, H.; et al. Arabidopsis PYR/PYL/RCAR receptors play a major role in quantitative regulation of stomatal aperture and transcriptional response to abscisic acid. *Plant. Cell* **2012**, *24*, 2483–2496. [CrossRef] [PubMed]
15. Zhao, Y.; Chan, Z.; Gao, J.; Xing, L.; Cao, M.; Yu, C.; Hu, Y.; You, J.; Shi, H.; Zhu, Y.; et al. ABA receptor PYL9 promotes drought resistance and leaf senescence. *Proc. Natl. Acad. Sci. USA* **2016**, *113*, 1949–1954. [CrossRef]
16. Antoni, R.; Gonzalez-Guzman, M.; Rodriguez, L.; Peirats-Llobet, M.; Pizzio, G.A.; Fernandez, M.A.; De Winne, N.; De Jaeger, G.; Dietrich, D.; Bennett, M.J.; et al. PYRABACTIN RESISTANCE1-LIKE8 plays an important role for the regulation of abscisic acid signaling in root. *Plant. Physiol.* **2013**, *161*, 931–941. [CrossRef]
17. Xing, L.; Zhao, Y.; Gao, J.; Xiang, C.; Zhu, J.K. The ABA receptor PYL9 together with PYL8 plays an important role in regulating lateral root growth. *Sci. Rep.* **2016**, *6*, 27177. [CrossRef]
18. Zhao, Y.; Xing, L.; Wang, X.; Hou, Y.J.; Gao, J.; Wang, P.; Duan, C.G.; Zhu, X.; Zhu, J.K. The ABA receptor PYL8 promotes lateral root growth by enhancing MYB77-dependent transcription of auxin-responsive genes. *Sci. Signal.* **2014**, *7*, ra53. [CrossRef]
19. Romero, P.; Lafuente, M.T.; Rodrigo, M.J. The Citrus ABA signalosome: Identification and transcriptional regulation during sweet orange fruit ripening and leaf dehydration. *J. Exp. Bot.* **2012**, *63*, 4931–4945. [CrossRef]
20. Boneh, U.; Biton, I.; Zheng, C.; Schwartz, A.; Ben-Ari, G. Characterization of potential ABA receptors in Vitis vinifera. *Plant. Cell Rep.* **2012**, *31*, 311–321. [CrossRef]
21. He, Y.; Hao, Q.; Li, W.; Yan, C.; Yan, N.; Yin, P. Identification and characterization of ABA receptors in Oryza sativa. *PLoS ONE* **2014**, *9*, e95246. [CrossRef] [PubMed]
22. González-Guzmán, M.; Rodríguez, L.; Lorenzo-Orts, L.; Pons, C.; Sarrión-Perdigones, A.; Fernández, M.A.; Peirats-Llobet, M.; Forment, J.; Moreno-Alvero, M.; Cutler, S.R.; et al. Tomato PYR/PYL/RCAR abscisic acid receptors show high expression in root, differential sensitivity to the abscisic acid agonist quinabactin, and the capability to enhance plant drought resistance. *J. Exp. Bot.* **2014**, *65*, 4451–4464. [CrossRef] [PubMed]
23. Guo, D.; Zhou, Y.; Li, H.L.; Zhu, J.H.; Wang, Y.; Chen, X.T.; Peng, S.Q. Identification and characterization of the abscisic acid (ABA) receptor gene family and its expression in response to hormones in the rubber tree. *Sci. Rep.* **2017**, *7*, 45157. [CrossRef] [PubMed]
24. Li, Y.; Wang, D.; Sun, C.; Hu, X.; Mu, X.; Hu, J.; Yang, Y.; Zhang, Y.; Xie, C.G.; Zhou, X. Molecular characterization of an AtPYL1-like protein, BrPYL1, as a putative ABA receptor in Brassica rapa. *Biochem. Biophys. Res. Commun.* **2017**, *487*, 684–689. [CrossRef]
25. Zhang, G.; Lu, T.; Miao, W.; Sun, L.; Tian, M.; Wang, J.; Hao, F. Genome-wide identification of ABA receptor PYL family and expression analysis of PYLs in response to ABA and osmotic stress in Gossypium. *Peer J.* **2017**, *5*, e4126. [CrossRef]

26. Bai, G.; Xie, H.; Yao, H.; Li, F.; Chen, X.; Zhang, Y.; Xiao, B.; Yang, J.; Li, Y.; Yang, D.H. Genome-wide identification and characterization of ABA receptor PYL/RCAR gene family reveals evolution and roles in drought stress in Nicotiana tabacum. *BMC Genom.* **2019**, *20*, 575. [CrossRef]
27. Chen, Y.; Feng, L.; Wei, N.; Liu, Z.H.; Hu, S.; Li, X.B. Overexpression of cotton PYL genes in Arabidopsis enhances the transgenic plant tolerance to drought stress. *Plant. Physiol. Biochem.* **2017**, *115*, 229–238. [CrossRef]
28. Lenka, S.K.; Muthusamy, S.K.; Chinnusamy, V.; Bansal, K.C. Ectopic Expression of Rice PYL3 Enhances Cold and Drought Tolerance in Arabidopsis thaliana. *Mol. Biotechnol.* **2018**, *60*, 350–361. [CrossRef]
29. He, Z.; Zhong, J.; Sun, X.; Wang, B.; Terzaghi, W.; Dai, M. The Maize ABA Receptors ZmPYL8, 9, and 12 Facilitate Plant Drought Resistance. *Front. Plant. Sci.* **2018**, *9*, 422. [CrossRef]
30. Kim, H.; Hwang, H.; Hong, J.W.; Lee, Y.N.; Ahn, I.P.; Yoon, I.S.; Yoo, S.D.; Lee, S.; Lee, S.C.; Kim, B.G. A rice orthologue of the ABA receptor, OsPYL/RCAR5, is a positive regulator of the ABA signal transduction pathway in seed germination and early seedling growth. *J. Exp. Bot.* **2012**, *63*, 1013–1024. [CrossRef]
31. Kim, H.; Lee, K.; Hwang, H.; Bhatnagar, N.; Kim, D.Y.; Yoon, I.S.; Byun, M.O.; Kim, S.T.; Jung, K.H.; Kim, B.G. Overexpression of PYL5 in rice enhances drought tolerance, inhibits growth, and modulates gene expression. *J. Exp. Bot.* **2014**, *65*, 453–464. [CrossRef] [PubMed]
32. Miao, C.; Xiao, L.; Hua, K.; Zou, C.; Zhao, Y.; Bressan, R.A.; Zhu, J.K. Mutations in a subfamily of abscisic acid receptor genes promote rice growth and productivity. *Proc. Natl. Acad. Sci. USA* **2018**, *115*, 6058–6063. [CrossRef] [PubMed]
33. Pizzio, G.A.; Rodriguez, L.; Antoni, R.; Gonzalez-Guzman, M.; Yunta, C.; Merilo, E.; Kollist, H.; Albert, A.; Rodriguez, P.L. The PYL4 A194T mutant uncovers a key role of PYR1-LIKE4/PROTEIN PHOSPHATASE 2CA interaction for abscisic acid signaling and plant drought resistance. *Plant. Physiol.* **2013**, *163*, 441–455. [CrossRef] [PubMed]
34. Liang, C.; Liu, Y.; Li, Y.; Meng, Z.; Yan, R.; Zhu, T.; Wang, Y.; Kang, S.; Ali Abid, M.; Malik, W.; et al. Activation of ABA Receptors Gene GhPYL9-11A Is Positively Correlated with Cotton Drought Tolerance in Transgenic Arabidopsis. *Front. Plant. Sci.* **2017**, *8*, 1453. [CrossRef]
35. Yu, J.; Yang, L.; Liu, X.; Tang, R.; Wang, Y.; Ge, H.; Wu, M.; Zhang, J.; Zhao, F.; Luan, S.; et al. Overexpression of Poplar Pyrabactin Resistance-Like Abscisic Acid Receptors Promotes Abscisic Acid Sensitivity and Drought Resistance in Transgenic Arabidopsis. *PLoS ONE* **2016**, *11*, e0168040. [CrossRef]
36. Li, G.; Xin, H.; Zheng, X.F.; Li, S.; Hu, Z. Identification of the abscisic acid receptor VvPYL1 in Vitis vinifera. *Plant. Biol. (Stuttg)* **2012**, *14*, 244–248. [CrossRef]
37. Boneh, U.; Biton, I.; Schwartz, A.; Ben-Ari, G. Characterization of the ABA signal transduction pathway in Vitis vinifera. *Plant. Sci.* **2012**, *187*, 89–96. [CrossRef]
38. Yin, P.; Fan, H.; Hao, Q.; Yuan, X.; Wu, D.; Pang, Y.; Yan, C.; Li, W.; Wang, J.; Yan, N. Structural insights into the mechanism of abscisic acid signaling by PYL proteins. *Nat. Struct. Mol. Biol.* **2009**, *16*, 1230–1236. [CrossRef]
39. Fuchs, S.; Tischer, S.V.; Wunschel, C.; Christmann, A.; Grill, E. Abscisic acid sensor RCAR7/PYL13, specific regulator of protein phosphatase coreceptors. *Proc. Natl. Acad. Sci. USA* **2014**, *111*, 5741–5746. [CrossRef]
40. Bai, G.; Yang, D.H.; Zhao, Y.; Ha, S.; Yang, F.; Ma, J.; Gao, X.S.; Wang, Z.M.; Zhu, J.K. Interactions between soybean ABA receptors and type 2C protein phosphatases. *Plant. Mol. Biol.* **2013**, *83*, 651–664. [CrossRef]
41. Di, F.; Jian, H.; Wang, T.; Chen, X.; Ding, Y.; Du, H.; Lu, K.; Li, J.; Liu, L. Genome-Wide Analysis of the PYL Gene Family and Identification of PYL Genes That Respond to Abiotic Stress in Brassica napus. *Genes* **2018**, *9*, 156. [CrossRef] [PubMed]
42. Tian, X.; Wang, Z.; Li, X.; Lv, T.; Liu, H.; Wang, L.; Niu, H.; Bu, Q. Characterization and Functional Analysis of Pyrabactin Resistance-Like Abscisic Acid Receptor Family in Rice. *Rice* **2015**, *8*, 28. [CrossRef] [PubMed]
43. Fan, W.; Zhao, M.; Li, S.; Bai, X.; Li, J.; Meng, H.; Mu, Z. Contrasting transcriptional responses of PYR1/PYL/RCAR ABA receptors to ABA or dehydration stress between maize seedling leaves and roots. *BMC Plant. Biol.* **2016**, *16*, 99. [CrossRef] [PubMed]
44. Liu, J.; Zhao, F.L.; Guo, Y.; Fan, X.C.; Wang, Y.j.; Wen, Y.Q. The ABA receptor-like gene VyPYL9 from drought-resistance wild grapevine confers drought tolerance and ABA hypersensitivity in Arabidopsis. *Plant. Cell Tissue Organ. Cult.* **2019**, *138*, 543–558. [CrossRef]

45. Li, X.; Li, G.; Li, Y.; Kong, X.; Zhang, L.; Wang, J.; Li, X.; Yang, Y. ABA Receptor Subfamily III Enhances Abscisic Acid Sensitivity and Improves the Drought Tolerance of Arabidopsis. *Int. J. Mol. Sci.* **2018**, *19*, 1938. [CrossRef]
46. Chen, Z.; Kong, L.; Zhou, Y.; Chen, Z.; Tian, D.; Lin, Y.; Wang, F.; Chen, S. Endosperm-specific OsPYL8 and OsPYL9 act as positive regulators of the ABA signaling pathway in rice seed germination. *Funct. Plant. Biol.* **2017**, *44*, 635–645. [CrossRef]
47. Han, Y.L.; Jang, G.; Um, T.; Kim, J.K.; Lee, J.S.; Yang, D.C. The soluble ABA receptor PYL8 regulates drought resistance by controlling ABA signaling in Arabidopsis. *Plant. Biotechnol. Rep.* **2015**, *9*, 319–330.
48. Saavedra, X.; Modrego, A.; Rodríguez, D.; González-García, M.P.; Sanz, L.; Nicolás, G.; Lorenzo, O. The nuclear interactor PYL8/RCAR3 of Fagus sylvatica FsPP2C1 is a positive regulator of abscisic acid signaling in seeds and stress. *Plant. Physiol.* **2010**, *152*, 133–150. [CrossRef]
49. Altschul, S.F.; Madden, T.L.; Schaffer, A.A.; Zhang, J.; Zhang, Z.; Miller, W.; Lipman, D.J. Gapped BLAST and PSI-BLAST: A new generation of protein database search programs. *Nucleic Acids Res.* **1997**, *25*, 3389–3402. [CrossRef]
50. Felsenstein, J. Confidence Limits on Phylogenies: An Approach Using the Bootstrap. *Evolution* **1985**, *39*, 783–791. [CrossRef]
51. Wang, Y.; Tang, H.; Debarry, J.D.; Tan, X.; Li, J.; Wang, X.; Lee, T.H.; Jin, H.; Marler, B.; Guo, H.; et al. MCScanX: A toolkit for detection and evolutionary analysis of gene synteny and collinearity. *Nucleic Acids Res.* **2012**, *40*, e49. [CrossRef]
52. Krzywinski, M.; Schein, J.; Birol, I.; Connors, J.; Gascoyne, R.; Horsman, D.; Jones, S.J.; Marra, M.A. Circos: An information aesthetic for comparative genomics. *Genome Res.* **2009**, *19*, 1639–1645. [CrossRef] [PubMed]
53. Hu, B.; Jin, J.; Guo, A.Y.; Zhang, H.; Luo, J.; Gao, G. GSDS 2.0: An upgraded gene feature visualization server. *Bioinformatics* **2015**, *31*, 1296–1297. [CrossRef] [PubMed]
54. Bailey, T.L.; Boden, M.; Buske, F.A.; Frith, M.; Grant, C.E.; Clementi, L.; Ren, J.; Li, W.W.; Noble, W.S. MEME SUITE: Tools for motif discovery and searching. *Nucleic Acids Res.* **2009**, *37*, W202–W208. [CrossRef] [PubMed]

2

Strawberry *FaNAC2* Enhances Tolerance to Abiotic Stress by Regulating Proline Metabolism

Jiahui Liang [1], Jing Zheng [1], Ze Wu [2] and Hongqing Wang [1,*]

[1] Department of Fruit Science, College of Horticulture, China Agricultural University, Beijing 100193, China; bs20173170791@cau.edu.cn (J.L.); sy20193172524@cau.edu.cn (J.Z.)

[2] Key Laboratory of Landscaping Agriculture, Ministry of Agriculture and Rural Affairs, College of Horticulture, Nanjing Agricultural University, Nanjing 210095, China; wuze@njau.edu.cn

[*] Correspondence: wanghq@cau.edu.cn.

Abstract: The quality and yields of strawberry plants are seriously affected by abiotic stress every year. NAC (NAM, ATAF, CUC) transcription factors are plant-specific, having various functions in plant development and response to stress. In our study, *FaNAC2* from strawberry (*Fragaria* × *ananassa*, cultivar "Benihoppe") was isolated and found to be a member of the ATAF sub-family, belonging to the NAC family of transcription factors. *FaNAC2* was strongly expressed in the shoot apical meristem and older leaves of strawberries, and was induced by cold, high salinity, and drought stress. To investigate how *FaNAC2* functions in plant responses to abiotic stress, transgenic *Nicotiana benthamiana* plants ectopically overexpressing *FaNAC2* were generated. The transgenic plants grew better under salt and cold stress, and, during simulated drought treatment, these transgenic lines not only grew better, but also showed higher seed germination rates than wild-type plants. Gene expression analysis revealed that key genes in proline biosynthesis pathways were up-regulated in *FaNAC2* overexpression lines, while its catabolic pathway genes were down-regulated and proline was accumulated more with the overexpression of *FaNAC2* after stress treatments. Furthermore, the gene expression of abscisic acid biosynthesis was also promoted. Our results demonstrate that *FaNAC2* plays an important positive role in response to different abiotic stresses and may be further utilized to improve the stress tolerance of strawberry plants.

Keywords: strawberry; ATAF; *FaNAC2*; abiotic stress

---

1. Introduction

Agricultural crops grow in a constantly changing environment and are often subjected to abiotic stresses such as drought, heat, cold, and high salinity. These stresses are associated with increased accumulation of certain deleterious chemicals like reactive oxygen species (ROS), which affect the stability of cell membranes and the structure of proteins, finally leading to reduced crop yield and even death [1,2]. To adapt to environmental stress and complete their life cycle, plants have evolved a complex mechanism that tightly regulates gene expression through precise signaling. Until now, many stress-induced proteins including transcription factors (TFs), osmotic stress-adaptive proteins and key enzymes in abscisic acid (ABA) biosynthesis and signaling pathways have been reported [3–5].

The NAC [No apical meristem (NAM), Arabidopsis transcription activation factor (ATAF), and Cup-shaped cotyledon (CUC)] superfamily is one of the largest groups of plant-specific TFs, which not only play an important role in various stages of plant growth and development, but also participate in responses to biotic and abiotic stress [6–10]. Many NAC genes have been identified through their function in response to drought, cold, and high salinity stress [5]. Drought-induced genes *ANAC019*, *ANAC055*, and *RD26/ANAC072*, when overexpressed in Arabidopsis, improved drought tolerance [11,12]. Overexpression of *SlNAM1* from tomato could improve chilling tolerance of transgenic tobacco [13]. Besides, it has also been shown that *OsNAC6* (from rice), *SNAC2* (from rice), *TaNAC4* (from wheat), *TaNAC8* (from wheat), and *CarNAC1* (from chickpea), function as transcriptional activators in response to various abiotic stresses [9,14–17].

The ATAF TFs comprise a sub-family of NAC proteins. The first report of a stress-inducible ATAF-like gene was *StNAC* from potato [18]. Elicited *ATAF1* and *ATAF2* are considered to function as repressors of responsive genes under biotic and abiotic stress, because *ataf1* and *ataf2* mutants showed high stress resistance [19–21]. However, overexpression of *ATAF1* in Arabidopsis also exhibited enhanced plant tolerance to drought [22]. Subsequently, more ATAF family members from different species were explored in response to abiotic stress. *OsNAC52* from rice belongs to the ATAF sub-family, and functions as an important transcriptional activator in ABA-inducible gene expression [23]. *GmNAC2* plays a negative regulatory role in abiotic stress in *Glycine max*, and participates in the ROS signaling pathway by regulating the expression of ROS removal genes [24]. Overexpression of *SlNAC2* (from tomato) in Arabidopsis resulted in enhanced tolerance to salinity stress [25]. In addition, *SlNAC11* from tomato plays a stress-inducible TF role, depicting a positive response to abiotic stress tolerance [26]. *DgNAC1* from chrysanthemum worked as a positive regulator in responses to salt stress, as its overexpression in transgenic chrysanthemum showed lower levels of MDA (malondialdehyde) and reactive oxygen species (H_2O_2 and O^{2-}), greater activities of SOD (superoxide dismutase), POD (peroxidase) and CAT (catalase), as well as more proline content than wild-type (WT) under salt stress [27]. In addition, it was reported that *CsATAF1* (from cucumber) was a crucial activator of the drought stress response via an ABA-dependent pathway, and inhibited ROS accumulation [28].

Strawberries are one of the most economically valuable crops in the world, which often suffers from water deficit, high salinity, flooding, and extreme temperature, leading to yield reduction [29–31]. Although many studies on responses of strawberry plants to abiotic stress have been reported, few data have showed NAC family members participating in abiotic stress resistance. Zhang et al. [31] showed that there are five *FvNAC* genes significantly contributing to various abiotic and biotic stress responses in woodland strawberry (*Fragaria vesca*), but their regulatory mechanisms are largely unknown.

In our study, *FaNAC2* was isolated from the cultivar strawberry (*Fragaria* × *ananassa*, cultivar "Benihoppe"); it was highly expressed in shoot apical meristem and old leaves of strawberry plants, and showed especially high expression in the guard cells of leaves. *FaNAC2* was induced under abiotic stress treatment, and many cis-acting elements that are responsive to abiotic stress were predicted in the *FaNAC2* promoter. We overexpressed *FaNAC2* in *Nicotiana benthamiana* (*N. benthamiana*), and found that the transgenic plants showed higher drought, cold and salt tolerance. These results indicated that *FaNAC2* might play a positive role in plant responses to abiotic stress.

2. Results

2.1. FaNAC2 Encodes an ATAF Protein that Belongs to the NAC Family

FaNAC2 encodes a protein of 289 amino acids, and is a member of a plant-specific NAC family of transcription factors. We used the cDNA sequence of *FaNAC2* as a query to perform a BLAST search against 122 NAC family members of Arabidopsis, using MEGA 7.0 software. It was found that *FaNAC2* has the closest relationship with *AT1G01720*, which encodes an ATAF sub-group protein, and is also named *AtNAC2* (Figure 1A). A multiple sequence alignment of ATAF homologues from *Arabidopsis*, rose, *Suaeda liaotungensis*, soybean, cucumber and strawberry was performed. As shown in Figure 1B, high sequence similarities between *FaNAC2* and other plant ATAF proteins were found in the N-terminus, which contained several distinguishable conserved domains, and five sub-domains. These results indicated that *FaNAC2* encoded an ATAF1 protein and was a typical member of the NAC transcription factors.

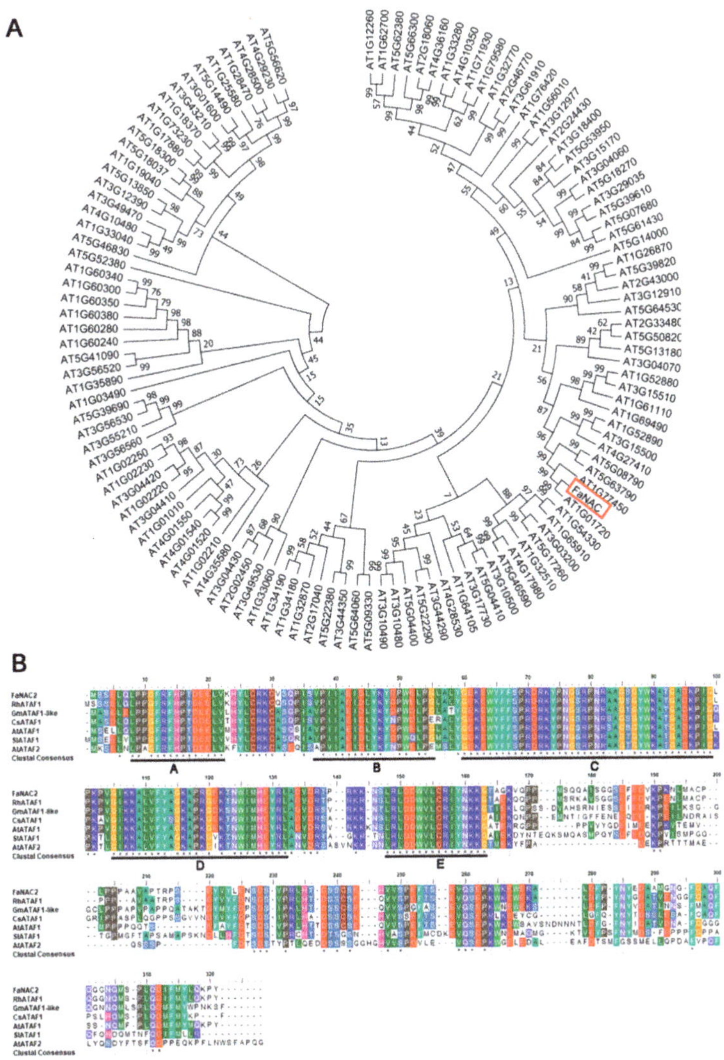

Figure 1. Phylogenetic analysis and amino acid sequence alignment of *FaNAC2*. (**A**) Phylogenetic relationship between *FaNAC2* (red boxed) from strawberry and NAC family members in Arabidopsis. MEGA 7.0 software was used to construct the Neighbor-Joining tree. The nearest *NAC* gene is *AT1G01720* (*AtATAF1* or *AtNAC2*). (**B**) Protein sequence alignment of ATAFs. The black underlines indicate the conserved N-terminal domain of NAC family. A–E represent five conserved sub-domains. Accession Numbers: Rosa hybrid cultivar, RhATAF1 (AXT99858.1); Arabidopsis thaliana, AtATAF1 (AT1G01720), AtATAF2 (AT5G08790); Suaeda liaotungensis K., SlNAC2 (JX860282.1); Glycine max, GmATAF1-like/GmNAC2 (AAX85979.1); Cucumis sativus L., CsATAF1 (Csa4M361820.1).

2.2. Expression Pattern of FaNAC2

To explore the function of *FaNAC2*, we first analyzed its spatial and temporal expression patterns in strawberry plants. The qRT-PCR (quantitative RT-PCR) results showed that *FaNAC2* was expressed at higher levels in shoot apical meristem, old leaves and flowers, compared to roots and fruits (Figure 2A). For different stages of leaves, the expression of *FaNAC2* in older leaves was more than that in younger leaves and mature leaves, indicating that the accumulation of *FaNAC2* might increase with the senescence of leaves. Amongst the different sizes of flower buds, *FaNAC2* had a high level of expression in the late petals and early stages of pistils (Figure 2A).

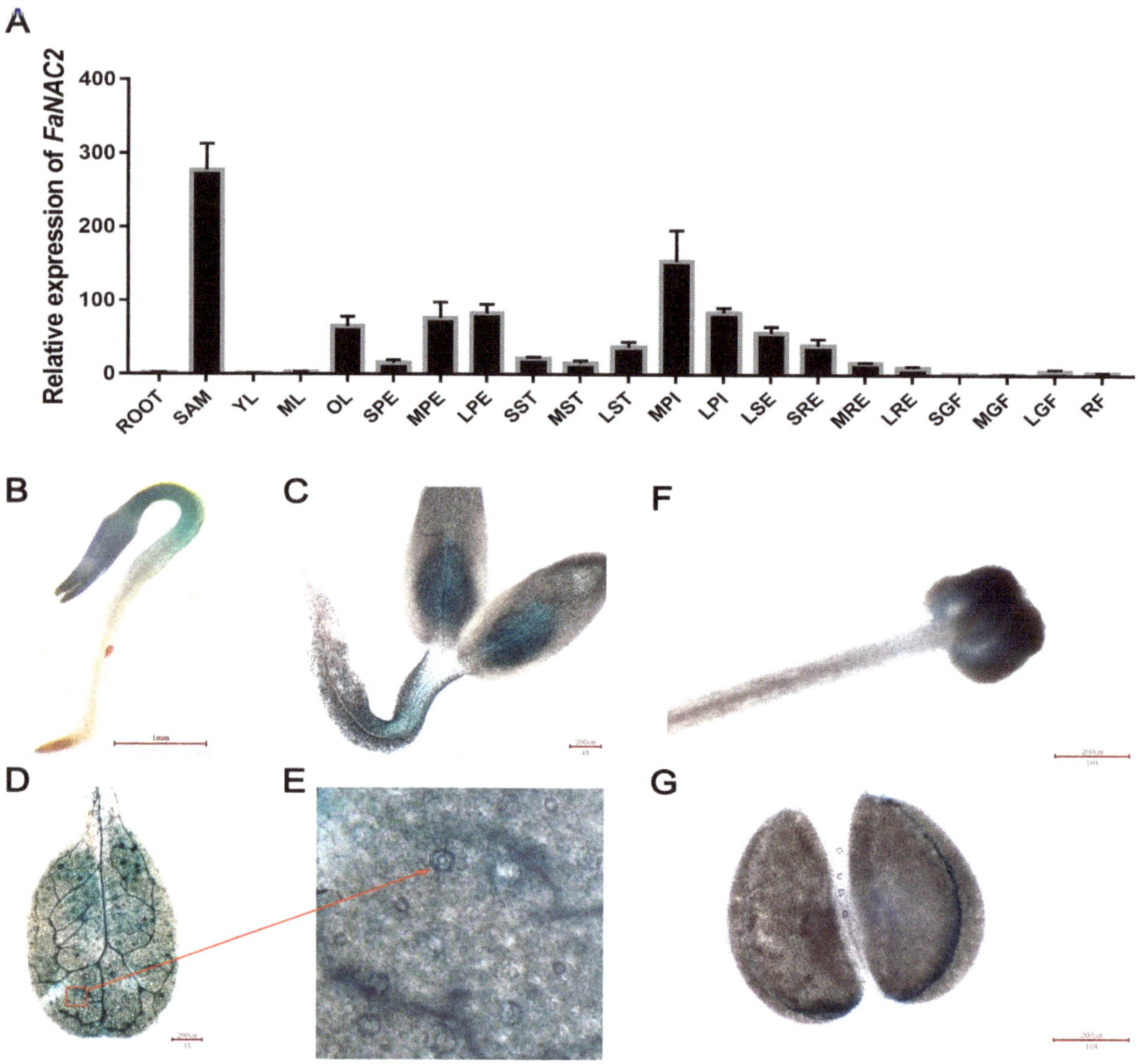

Figure 2. Expression pattern of *FaNAC2*. (**A**) qRT-PCR analysis of strawberry. ROOT: roots; SAM: shoot apical meristem; YL: young leaves (the second folded leaf); ML: mature leaves (the third or fourth fully expanded leaf); OL: old leaves (below the seventh leaf); SPE: petals from small flower buds (length <0.5 cm); MPE: petals from middle flower buds (length 0.5–0.8 cm); LPE: petals from large opened flower; SST: stamen from small flower buds; MST: stamen from middle flower buds; LST: stamen from large opened flower; MPI: pistils from middle flower buds; LPI: pistils from large flower buds; LSE: sepals from large opened flower; SRE: receptacles from small flower buds; MRE: receptacles from middle flower buds; LRE: receptacles from large opened flower; SGF: small green fruits; MGF: middle green fruits; LGF: large green fruits; RF: red fruits. The *FaACTIN* gene was used as an internal reference to normalize the expression data. Data are presented as

β-glucuronidase (GUS) analysis of 5 DAG (days after germination) Pro*FaNAC2*–GUS *N. benthamiana (Nicotiana benthamiana)* seedlings. (**D**,**E**) GUS analysis of true leaf of 13 DAG seedling and its enlarged *N. benthamiana (Nicotiana benthamiana)* seedlings. GUS analysis of stigma and anther of 35 DAG and its enlarged view. The red arrow points to guard cells in (E). (**F**,**G**) GUS analysis of stigma and anther of 35 DAG Pro*FaNAC2*–GUS lines after planting on soil.

To better understand the expression pattern of *FaNAC2*, the transgenic *N. benthamiana* contained a *β-glucuronidase* (*GUS*) reporter gene which, under the control of *FaNAC2* promoter, was produced and detected. The results revealed that the GUS signals were detected in cotyledons and true leaves of young plants (Figure 2B,C), and it showed strong expression in the guard cells of true leaves (Figure 2D,E), suggesting that *FaNAC2* might have a function in controlling stoma. In addition, *GUS* expression was also be seen in the stigmas and anthers of the flowers (Figure 2F,G).

2.3. *FaNAC2 Is Induced by Cold, Salinity, and Drought Treatment*

Since many *cis*-elements related to abiotic stress, such as ABRE (response to abscisic acid), LTR (response to cold) and MBS (response to drought), were found on the promoter sequence of *FaNAC2* (Table 1), we hypothesized that *FaNAC2* would also respond to abiotic stress. Thus, we performed stress treatments, including 200 mM NaCl, 20% polyethylene glycol (PEG) 6000, and 4 °C, to detect the expression changes of *FaNAC2* using 3 MAC (months after cutting node bud from runners) tissue cultured seedlings of strawberry for qRT-PCR assay. The results revealed that its expression was strongly up-regulated in both SAM (shoot apical meristem) and leaves in response to abiotic stress (Figure 3A,B). During cold and salt treatment, the expression of *FaNAC2* in the SAM was induced at 3 h and then gradually decreased compared to the initial expression levels, while the expression of *FaNAC2* from the leaves was a little slower, being induced only at 6 h following stress. Under the condition of simulated drought treatment, *FaNAC2* showed a great difference in expression pattern in different tissues; although *FaNAC2* was induced at 9 h, the expression of *FaNAC2* in the SAM was gradually decreased from 12 h after induction, while *FaNAC2* expression in the leaves increased at all the times tested, which indicated that *FaNAC2* might participate in drought stress response, mainly in the leaves.

Table 1. Predicted *cis*-elements in the promoter of *FaNAC2*.

***Cis*-elements**	**Sequence**	**Number**	**Character**
ABRE	ACGTG	4 (+)	Response to abscisic acid
ARE	AAACCA	1 (+)	Response to anaerobic process
Box-4	ATTAAT	1 (+)	Response to light
CGTCA-motif	CGTCA	3 (+)	Response to Jasmonic Acid
G-box	CACGTG	7 (+)	Response to light reaction
GCN4-motif	TGAGTCA	1 (+)	Involved in endosperm expression
LTR	CCGAAA	1 (+)	Response to cold
MBS	CAACTG	1 (+)	Response to drought
RY-element	CATGCATG	1 (+)	Specific seeds' regulation
TCA-element	TCAGAAGAGG	2 (+)	Response to salicylic acid
TGA-element	AACGAC	2 (+)	Response to auxin

To confirm the expression pattern, different stress treatments were performed at 7 DAG (days after germination) of Pro*FaNAC2*–GUS *N. benthamiana*, and the results showed that the *GUS* expression became stronger under cold, simulated drought and salt stress, indicating that *FaNAC2* could be induced by different abiotic stresses (Figure 3C). Taking all these results together, it could be inferred that *FaNAC2* might function in different tissues to cope with different stress conditions.

2.4. *Overexpression of FaNAC2 Improves Stress Tolerance in Transgenic N. benthamiana*

To further investigate how *FaNAC2* plays roles in abiotic stress resistance, *FaNAC2* was ectopically transformed into *N. benthamiana* under the control of a CaMV (Cauliflower Mosaic Virus)–35S promoter, and ten positive lines were obtained through screening using kanamycin and RT-PCR analysis. By observing the growth potential of germination of 13 DAG *35S::FaNAC2* and wild-type lines under salt treatments, we found that *35S::FaNAC2* lines exhibited better growth under salt stress; for example, the leaf areas of transgenic plants were larger than WT following salt treatments (Figure 4A,B), indicating that overexpression of *FaNAC2* promoted the salt tolerance of plants.

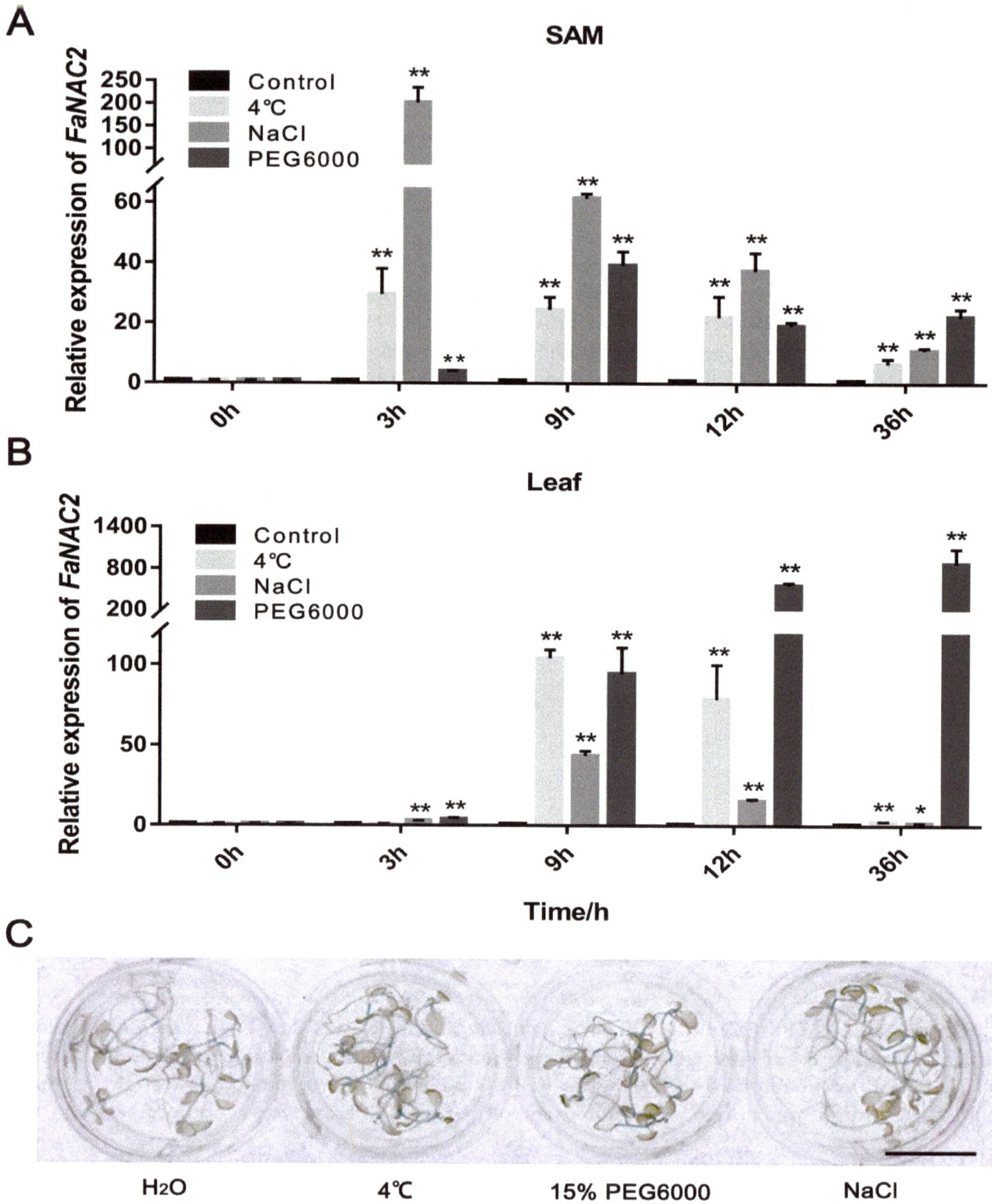

Figure 3. *FaNAC2* is induced by cold, drought, and salinity stress. (**A**) The expression level of *FaNAC2* in the SAM (shoot apical meristem) of strawberry under different abiotic stresses. (**B**) The expression level of *FaNAC2* in the leaves of strawberry under different abiotic stresses. (**C**) Different abiotic stress treatments to 7 DAG (days after germination) Pro*FaNAC2*–GUS seedlings. Salinity treatment used 200 mM NaCl. Three independent experiments were performed and error bars indicate standard deviation (Student's *t*–test; * $p < 0.05$; ** $p < 0.01$). The scale bar represents 1 cm.

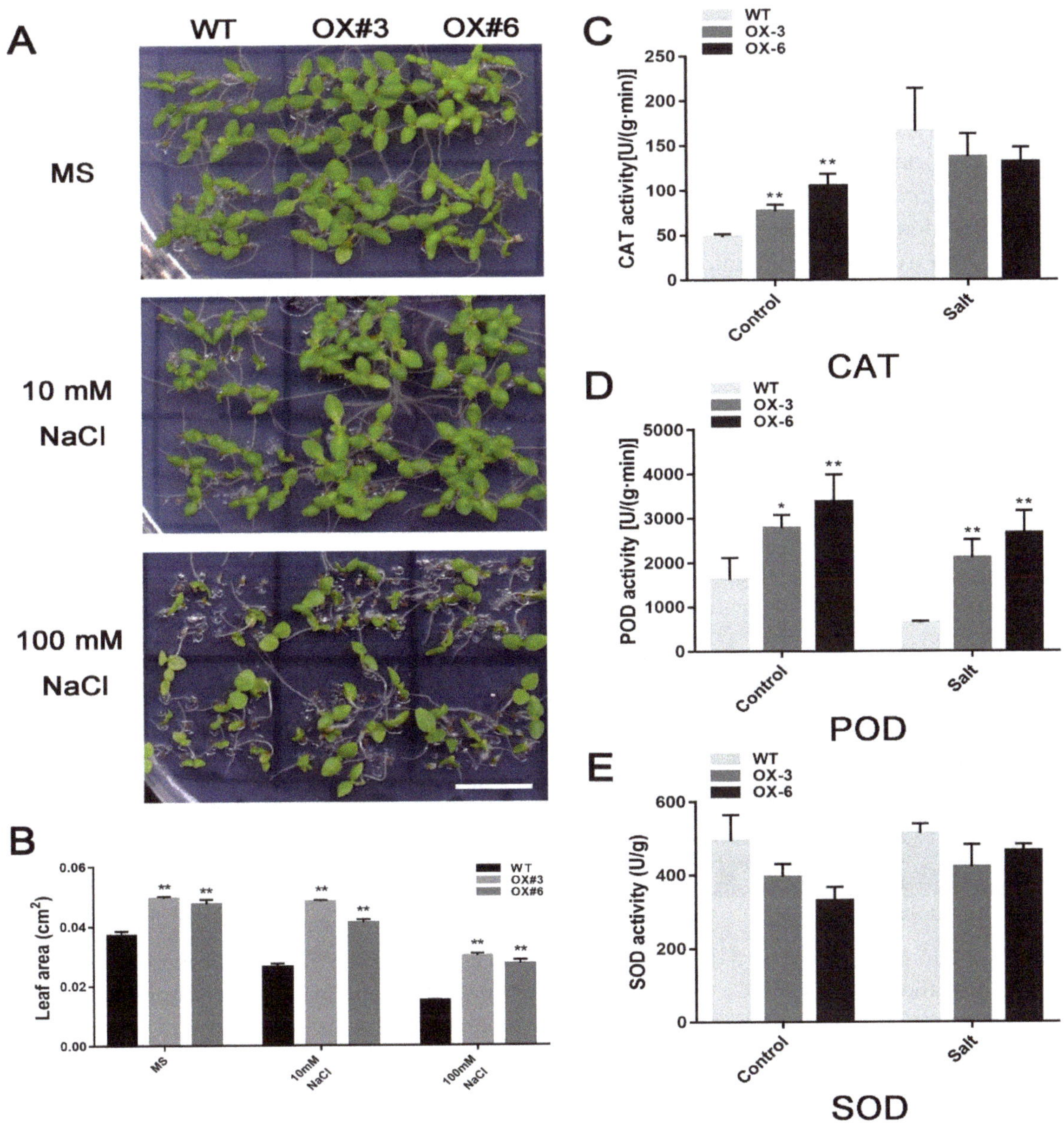

Figure 4. Comparison of resistance to salinity of *35S::FaNAC2* and wild-type (WT) lines. (**A**) Analysis of 13 DAG seedlings of *35S::FaNAC2* and WT seeds under salt and control treatments. OX: overexpression. (**B**) Leaf area of *35S::FaNAC2* and WT 13 DAG seedlings under salt and control treatments. (**C–E**) Enzyme activity of several enzymes related to plant resistance to salt. CAT: Catalase;

As the plants get stronger, the 40 DAG *35S::FaNAC2* lines and WT plants were irrigated with 300 mM NaCl for one week, with normal watering as a control. By comparing the enzyme activity of antioxidant enzymes including catalase (CAT), peroxidase (POD) and superoxide dismutase (SOD) from two lines under different conditions, it was found that CAT enzyme activity in *35S::FaNAC2* lines was significantly higher than that in WT lines under control conditions, but there was no significant difference after salt treatment (Figure 4C). POD activity of *35S::FaNAC2* plants was higher than that of WT lines under both control and salt treatment, while there was no significant change in SOD enzyme activity (Figure 4D,E). These results suggested that *FaNAC2* might promote plant salt tolerance by partially affecting the activity of some antioxidant enzymes.

Through different treatments for seed germination, we found that there was no difference in seed germination rates between *35S::FaNAC2* and WT lines under normal conditions. However, the seeds of *35S::FaNAC2* lines had a 10% higher germination rates than that of WT plants under the treatment of simulated drought (10% PEG 6000; Figure 5A,B). Subsequently, we conducted drought treatment on 40 DAG *35S::FaNAC2* and WT plants; *35S::FaNAC2* plants had a higher recovery rate after rehydration (Figure 5C). Comparing the WLR (water loss rate) from leaves of the two lines, it was found that the dryness of *35S::FaNAC2* leaves was lower, suggesting that they retained leaf water more easily (Figure 5D). Besides, a key gene for ABA biosynthesis, *NbNCED1* (*9-cis-epoxycarotenoid dioxygenase 1*), was also detected in both the normal and drought treatment conditions, and *NbNCED1* expression in *35S::FaNAC2* lines was significantly higher than that of WT (Figure 5E). Taking all these results together, our data suggested that *FaNAC2* might exist as a positive regulator of drought stress tolerance.

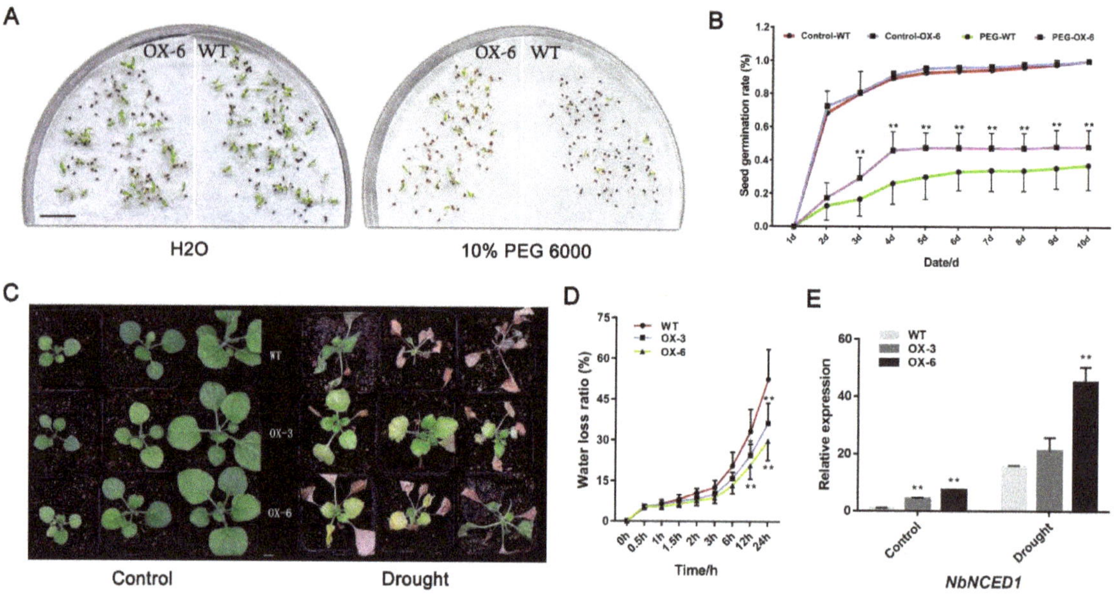

Figure 5. Comparison the rehydration rate of *35S::FaNAC2* and WT lines under drought stress. (**A**,**B**) Germination rates of *35S::FaNAC2* and WT seeds under water and simulated drought conditions. (**C**) Phenotypes of 40 DAG *35S::FaNAC2* and WT lines under control and rehydration conditions. (**D**) Water loss rate (WLR) of leaves from *35S::FaNAC2* and WT lines. (**E**) The qRT-PCR analysis of *NbNCED1* expression, related to ABA biosynthesis pathways. The scale bar represents 1 cm. Bars are means (± S.D.) of three independent experiments (Student's *t*-test; * $p < 0.05$; ** $p < 0.01$).

In order to understand the function of *FaNAC2* in cold stress, *35S::FaNAC2* lines and WT plants were subjected to cold treatment, which was performed as follows: 4 °C for 2 h, 0 °C for 1 h, −5 °C for 1 h and 4 °C for 1 h. The results showed that *35S::FaNAC2* lines were more cold-resistant than WT plants, showing less damage (Figure 6A). Meanwhile, a key gene, *NbNPK1* (*Nicotiana protein kinase 1*), which was involved in cold resistance signal transmission in plants, was found to show increased expression in *35S::FaNAC2* lines compared to WT plants under control treatment, while it was significantly higher than that in WT plants under cold treatment (Figure 6B). All of these results demonstrated that *FaNAC2* might play a positive role in response to abiotic stress.

2.5. FaNAC2 Promotes Plant Abiotic Stress Tolerance via Regulating Proline Metabolism

To further investigate how *FaNAC2* regulates plant stress tolerance, we detected the expression changes of key genes in proline biosynthesis, *NbP5CS1* (*Pyrroline-5 carboxylate synthetase 1*), and catabolism, *NbP5CDH* (*P5C dehydrogenase*) and *NbproDH2* (*Proline dehydrogenase 2*) from *35S::FaNAC2* and WT plants (Figure 7). Compared with WT, *NbP5CS1* expression from *35S::FaNAC2* lines was increased either in control or salt stress conditions (Figure 7A,B). *NbproDH2* expression was also higher in transgenic lines than WT in control conditions, while its expression significantly decreased after salt treatment. The expression of another key gene in the proline catabolism pathway, *NbP5CDH*, in *35S::FaNAC2* lines was also down-regulated after salt treatment (Figure 7B).

Figure 6. Comparison of *35S::FaNAC2* and WT plants under cold stress. (**A**) Phenotypes of *35S::FaNAC2* lines and WT under control and cold conditions. (**B**) Expression of *NbNPK1* related to plant resistance to cold pathways in *35S::FaNAC2* lines and WT under cold treatment. The scale bar represents 1 cm. Bars are means (±S.D.) of three independent experiments (Student's *t*-test; ** $p < 0.01$).

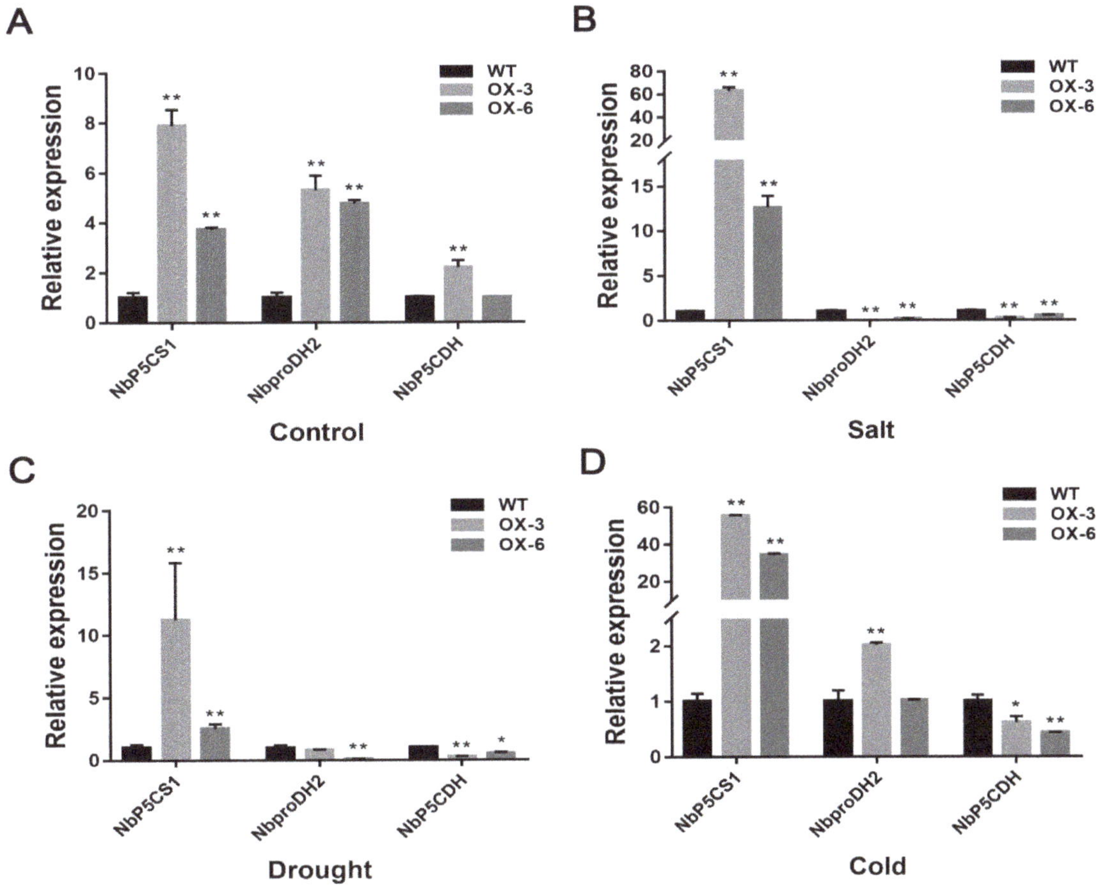

Figure 7. The qRT-PCR analysis of several genes related to proline matabolism in *35S::FaNAC2* lines and WT under different abiotic stress treatments. (**A**) Expression of genes in *35S::FaNAC2* lines and WT under control conditions. (**B**) Expression of genes in *35S::FaNAC2* lines and WT under salt stress. C) Expression of genes in *35S::FaNAC2* lines and WT under drought stress. (**D**) Expression of genes in *35S::FaNAC2* lines and WT under cold stress. Bars are means (±S.D.) of three biological replicantes experiments (Student's *t*-test; * $p < 0.05$; ** $p < 0.01$).

Following drought treatment, the expression of *NbP5CS1* in *35S::FaNAC2* was significantly higher than that of WT, while the expression of *NbP5CDH* and *NbproDH2* was decreased compared with that of WT (Figure 7C).

Similar to the two previous stress treatments, further analysis revealed that the expression of *NbP5CS1* was up-regulated in *35S::FaNAC2* lines under both control and cold treatment. The *NbproDH2* expression of *35S::FaNAC2* lines was more than five times as high as that in WT plants under the control treatment, whereas it was twice as high as that in WT plants under the cold stress, suggesting that *NbproDH2* of *35S::FaNAC2* was decreased during the cold resistance compared to control condition. Meanwhile, *NbP5CDH* expression was relatively high in the control treatment, but lower in the WT plants after cold treatment (Figure 7D).

In addition, proline levels were detected under control and abiotic stress. It was found that proline content in *35S::FaNAC2* lines was slightly higher than WT plants in the control environment. With different abiotic stress treatments on the plants, proline content in *35S::FaNAC2* strains was significantly higher than WT strains (Figure 8). Due to both the proline biosynthesis gene and the proline catabolism gene being up-regulated under normal conditions, it was speculated that *FaNAC2* might promote proline metabolism to sustain proline hemostasis for normal growth of transgenic plants, but *FaNAC2* could improve proline levels, in response to different abiotic stresses, for increased tolerance.

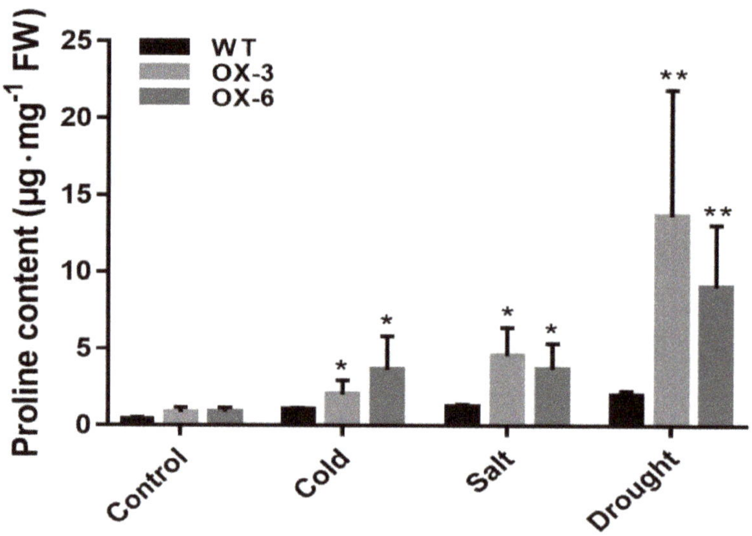

Figure 8. The proline content in *35S::FaNAC2* lines and WT leaves under different abiotic stress treatments. The fifth leaf was collected for proline content determination. Bars are means (±S.D.) of three independent experiments (Student's *t*-test; * $p < 0.05$; ** $p < 0.01$).

Taking all of these results into account, we concluded that, in general, *FaNAC2* might be involved in plant abiotic stress tolerance by regulating proline accumulation and catabolism.
involved in plant abiotic stress tolerance by regulating proline accumulation and catabolism.

3. Discussion

3.1. The Expression Pattern of FaNAC2

Many NAC family members have been reported to be involved in plant growth and development processes, such as SAM (shoot apical meristem) establishment, lateral root development, leaf senescence and cell wall formation [32–35]. Thus, we speculated that *FaNAC2* may also be involved in many plant developmental processes. Our data show that *FaNAC2* might be involved in leaf senescence, as it expressed highly in old leaves (Figure 2A). These results are similar to the function of other members of the NAC family. Overexpression of *OsNAC2* has been shown to as our data shown (Figure 5), thus, whether *FaNAC2* promotes leaf senescence via ABA synthesis needs further investigation.

GUS analysis showed high levels of *FaNAC2* promoter activity in guard cells of transgenic *N. benthamiana* leaves (Figure 2D,E). Considering these results, it is possible that there is a function for *FaNAC2* in plant development. Stomata can affect transpiration and photosynthesis by regulating their closure with the changing environment, via sensing ABA signals under adverse conditions [37–40]. Thus, *FaNAC2* may regulate transpiration and leaf water retention by adjusting guard cells. Besides, although the theory that root-sourced ABA can act as a signal to regulate stomatal aperture gained widespread acceptance, ABA biosynthetic mutants showed that stomatal aperture is predominantly regulated by leaf-sourced ABA [41–44]. In our results (Figure 5E), *FaNAC2* could promote the key gene in ABA synthesis pathways in the leaf, thus we speculated that *FaNAC2* may regulate stomatal closure by participating in ABA biosynthesis in leaves, further improving the drought resistance of plants.

3.2. FaNAC2 Functions as a Positive Regulator in Response to Abiotic Stress

Abiotic stress is an important factor that threatens the yield and quality of strawberry. To alleviate the damage of abiotic stress, plants usually initiate complex adaptation via genetic mechanisms including regulation of gene expression and increased concentration of osmolytes [5,27]. There are several reports revealing that the ATAF sub-group of TFs belonging to the NAC family play important roles in response to abiotic stress; however, their function is still under debate. In Arabidopsis, ATAF1 was reported to negatively regulate stress-responsive gene expression during drought stress, because *ataf1* mutants displayed higher recovery rates than WT under drought [20]. Overexpression of *GmNAC2* reduces abiotic stress tolerance in *Glycine max*, which also functions as a negative regulator by participating in ROS signaling pathways [24]. It has been reported that *AtATAF1* overexpression in transgenic lines enhanced drought tolerance [22]. Subsequently, *OsNAC52* from rice, *SlNAC2* from tomato, *DgNAC1* from chrysanthemum and *CsATAF1* from cucumber were reported to function as positive regulators in response to abiotic stress [23,25,27,28].

In our study, GUS analysis showed that the promoter activity of strawberry *FaNAC2* was induced by drought, salt and cold stress (Figure 3C), and *FaNAC2* expression levels exhibited the same trend that was verified by qRT-PCR (Figure 3A,B) in strawberry. Besides, ectopic overexpression of *FaNAC2* in *N. benthamiana* plants showed higher tolerance to salinity, drought and cold stress (Figures 4–6). Members of the ATAF sub-family in dicotyledons have conserved domains and can be identified by some conserved regions that respond positively to stress. Overexpressed transgenic lines of *ANAC019*, *ANAC055* and *ANAC072* improved the drought resistance of plants, and the conserved *cis*-elements CATGT and CACG, for their binding, were identified [12]. However, the central function of *FaNAC2* in response to stress in strawberry is still unknown, and further studies on the abiotic stress pathway involving *FaNAC2* are needed.

Proline accumulation has been reported to occur after biotic and abiotic stress [45–47]. It varies across different species under stress and can be more than 100 times higher than that under control conditions [48]. In our study, both the synthesis and catabolism pathways of proline were induced in *35S::FaNAC2* lines (Figure 7A), suggesting that the overexpression of *FaNAC2* can promote the metabolism of proline. However, in a stress environment, the expression levels of *NbP5CS1* in *35S::FaNAC2* were still higher than that in WT, and both *NbproDH2* and *NbP5CDH* were decreased compared to control condition (Figure 7B–D). Further, the proline content of 35S::*FaNAC2* was higher than WT under the abiotic stress condition (Figure 8). We speculate that *FaNAC2* might promote the accumulation of proline under adverse conditions by activating proline synthesis and inhibiting proline degradation, so as to promote the stress tolerance of plants.

Taken together, *FaNAC2* from strawberry might play a positive role in response to abiotic stress by regulating proline metabolism. Although the role of *FaNAC2* in stress tolerance needs to be further validated in strawberries, we demonstrate that it can serve as a candidate gene to enhance stress tolerance, as long as the spatial–temporal expression is controlled.

4. Materials and Methods

4.1. Plant Materials, Growth Conditions

The strawberry cultivar "Benihoppe" (*Fragaria* × *ananassa* Duch.) was used in this study and maintained in a plant culture room (23 ± 1 °C, relative humidity of 40%, 16 h/8 h light/dark cycles). The tissue cultured seedlings were initiated from node bud of runners, which were collected from actively growing plants and disinfected with 70% ethanol (30 s) and 1% NaClO (10 min).

Most of *N. benthamiana* seeds were germinated on MS solid medium with 20% sucrose and grown for 13 days. Then, the seedlings were transplanted to the soil and grown in the culture room. For *N. benthamiana* seeds treated by simulate drought, the seeds were simply spread flat on a filter paper soaked in water or 10% PEG 6000.

4.2. Gene Isolation and Sequence Alignment

Total RNA samples were extracted from collected shoot apicals and leaves using an E.Z.N.A Total RNA Kit (Omega., Norcross, Georgia, USA). HIScript II Reverse Transcriptase (Vazyme, Nanjing, China) was used for cDNA synthesis. The primers of *FaNAC2* were designed according to the GDR Database (Genome Database for Rosaceae) [49]. The CDS (coding sequence) of *FaNAC2* was obtained from the cDNA of "Benihoppe". Phylogenetic analysis was performed using MEGA version 7 (http://www.megasoftware.net/) [50]. Alignments of the *FaNAC2* full-length amino acid sequence with ATAF homologues from other species were performed using BioEdit software (http://www.mbio.ncsu.edu/BioEdit/bioedit.html) and ClustalW for multiple sequence alignments (http://www.ch.embnet.org/software/ClustalW.html).

4.3. Promoter Isolation, Prediction of Cis-Elements and GUS Activity Assay

Genomic DNA was extracted from strawberry "Benihoppe" using a TIANquick Midi Purification Kit (TianGen., Beijing, China). The primers for the promoter of *FaNAC2* were designed using the sequence from the GDR database and the sequence of the promoter was obtained using the DNA of "Benihoppe" strawberry as template, then the fragment was proofread and sequenced. *FaNAC2* promoter was cloned into the pCAMBIA1391 vector using the TrelisfTM SoSoo cloning Kit (TsingKe, Beijing, China) to generate the reporter construct pCAMBIA1391–Pro*FaNAC2*–GUS. The primers used are listed in Table S1. The construct was stably transformed into *N. benthamiana* by *Agrobacterium*-mediated transformation, as described below. Prediction of *cis*-elements was performed using the Plantcare online tool (http://bioinformatics.psb.ugent.be/webtools/plantcare/html/).

For GUS analysis, samples were incubated with GUS staining buffer (including 2 M ferri/ferrocyanide, 0.1% Triton X-100, 0.1 M sodium phosphate buffer, 0.5 mg·mL^{-1} X-gluc, pH 7) at 37 °C for 9 h, then stained samples were decolorized using 75% ethanol.

4.4. Gene Expression Analysis

qRT-PCR was employed to detect the expression of different genes. The qRT-PCR reactions (20 µL volume containing 500 ng cDNA as template) were run using SYBR Premix ExTaq (TAKARA., Beijing, China) as enzyme and ABI QuantStudio™ 6 Flex PCR System (ABI., New York, NY, USA). The $2^{-\Delta\Delta CT}$ method was used for qRT-PCR analysis. *FaACTIN* and *NbACTIN* were used as internal controls for gene expression. The qRT-PCR primers of *NbNCED1/NbNPK1/Nb5CS1/NbproDH2/NbP5CDH* for qRT-PCR were designed according to the *N. benthamiana* database [51]. The template for analyzing the expression of these genes was cDNA of the fifth tobacco leaves from different treatments. PCR was performed in triplicate using RNA samples extracted from three independent plants. Each reaction was performed using three biological replicates and verified by melting curve analysis. The primers are listed in Table S1.

4.5. Stable Transformation of *N. benthamiana*

For overexpression of *FaNAC2* in *N. benthamiana*, the vector pCAMBIA2300 with kanamycin resistance was used for stable transformation. The *FaNAC2* ORF was inserted into pCAMBIA2300 using the TrelisfTM SoSoo cloning Kit (TsingKe) and driven by a CaMV 35S promoter. The *35S::FaHAN* was introduced into *Agrobacterium tumefaciens* strain GV3101 and then transformed into *N. benthamiana* leaf dishes by the method of *Agrobacterium*-mediated transformation [52]. The obtained kanamycin-resistant plants were screened again by RT-PCR. The PCR mix was from TsingKe and the PCR primers used for vector construction are listed in Table S1.

4.6. Abiotic Stress Treatment

The 3-months-old "Benihoppe" strawberry tissue cultured seedlings, which were obtained from the node bud of runners, were used for stress treatment. Three simulated stress conditions of 200 mM NaCl, 20% PEG 6000 and 4 °C were set, and MS liquid medium was used as a control to treat for 0 h, 3 h, 6 h, 12 h and 36 h. The qRT-PCR was performed in biological triplicate using RNA samples extracted from three independent plants. For GUS staining, the abiotic stress treatment for 7 DAG (days after germination) *N. benthamiana* seedlings was 200 mM NaCl, 15% PEG 6000 and 4 °C; the treatment lasted 12 h.

Regarding abiotic stress treatment of *35S::FaNAC2* transgenic *N. benthamiana* lines and WT plants, simulated drought treatment for seeds was used with 10% PEG 6000; drought treatment of 40 DAG seedlings was conducted by stopping the watering for 20 d, until the leaves were all wilted, then re-watering for 5 d to observe the recovery of plants. Cold treatment was performed to put the plants under normal indoor conditions after 4 °C for 2 h, 0 °C for 1 h, −5 °C for 1 h and 4 °C for 1 h, and the status of the plants was observed. Salt treatment for seedings of *35S::FaNAC2* and wild-type lines were MS solid medium with 10 mM NaCl and 100 mM NaCl, respectively. Considering that 40 DAG seedlings were stronger, salt treatment was enhanced to 300 mM NaCl. These experiments were repeated three times and each line had three biological replicates.

4.7. Determination of SOD, POD and CAT Enzyme Activities

Approximately 0.15 g of the fifth expanded leaf from each line (including *35S::FaNAC2* and WT lines) was homogenized in 5 mL pre-cooled Phosphate buffered saline, at 4000 rpm for 10 min at 4 °C, then 2 mL of supernatant was drained and placed on ice for determination of different enzyme activities, according to measurements as follows.

CAT (Catalase) activity was measured spectrophotometrically at 240 nm [53]. The reaction mixture contained 100 mM sodium phosphate buffer (pH 7.0), 30 mM H_2O_2 and 100 μL of crude extract in a total volume of 3 mL. The absorbance was read quickly every 1 min, for a total of 4 min.

The activity of POD (peroxidase) was determined at 420 nm using a spectrophotometer, with callus lignin as substrate. The reaction mixture contained 100 mM sodium phosphate buffer (pH 6.0), 5 mM hydrogen peroxide, 5 mM guaiacol and 100 μL crude extract, with a total volume of 3 mL, at room temperature (±25 °C) [54].

The activity of SOD (superoxide dismutase) was determined by inhibiting the photoreduction of SOD to NBT (nitro-blue tetrazolium) [55]. The reaction mixture contained a final volume of 3 mL of 50 mM sodium phosphate buffer (pH 7.6), 0.1 mM Ethylene Diamine Tetraacetic Acid (EDTA)-Na_2, 50 mM sodium carbonate, 12 mM L-methionine, 50 mM NBT, 10 μM riboflavin and 100 μL of crude extract. Then, the reaction mixture was exposed to white light for 30 min for the SOD reaction. After incubation, the absorbance was recorded at 560 nm with a spectrophotometer.

4.8. Determination of WLR (Water Loss Rate) in *N. benthamiana* Leaves

The fifth leaf was collected from half-month-old tobacco plants which were grown in the soil and dried naturally on filter paper. The temperature was 25 °C, the relative humidity was 38%, and weight measurements were taken at 0 h, 0.5 h, 1 h, 1.5 h, 2 h, 3 h, and 6 h. This experiment was repeated three times to calculate the weight and rate of water loss in each period.

4.9. Proline Measurement

To determine free proline level, 0.1 g of fourth expanded leaf samples from each line (including *35S::FaNAC2* and WT lines) was homogenized in 3% (*w*/*v*) sulphosalycylic acid and then homogenate filtered through filter paper [56]. The mixture was heated at 100 °C for 30 min in a water bath after the addition of acid ninhydrin and glacial acetic acid. The reaction was then stopped by ice bath. The mixture was extracted with toluene and the absorbance of the fraction with toluene aspired from the liquid phase was read at 520 nm. Proline concentration was determined using a calibration curve.

4.10. Statistical Analysis

Microsoft Excel 2019 (Microsoft Corp., Redmond, Washington, USA) and GraphPad (GraphPad Software Inc., San Diego, California, USA) were used for analyzing the experimental data. Data for *p*-values were analyzed by Student's *t* test at a significance level of 0.05 or 0.01. Comparisons between multiple samples were determined using Tukey's multiple comparisons test.

Author Contributions: J.L. and J.Z. designed the experiments and performed the experiments; J.L., J.Z. and H.W. drafted the manuscript; J.L., H.W. and Z.W. revised the manuscript. All authors have read and agreed to the published version of the manuscript.

Acknowledgments: We thank Mingfang Yi (China Agricultural University, Beijing, China) for the generous gift of pCAMBIA2300 and pCAMBIA1391 vectors.

References

1. Choudhury, S.; Panda, P.; Sahoo, L.; Panda, S.K. Reactive oxygen species signaling in plants under abiotic stress. *Plant Signal. Behav.* **2013**, *8*, e23681. [CrossRef] [PubMed]
2. Choudhury, F.K.; Rivero, R.M.; Blumwald, E.; Mittler, R. Reactive oxygen species, abiotic stress and stress combination. *Plant J.* **2017**, *90*, 856–867. [CrossRef]
3. Zhu, J.K. Salt and drought stress signal transduction in plants. *Ann. Rev. Plant Biol.* **2002**, *53*, 247–273. [CrossRef]
4. Zhu, J.K. Abiotic Stress Signaling and Responses in Plants. *Cell* **2016**, *167*, 313–324. [CrossRef] [PubMed]
5. Min, X.; Jin, X.; Zhang, Z.; Wei, X.; Ndayambaza, B.; Wang, Y.; Liu, W. Genome-Wide Identification of NAC Transcription Factor Family and Functional Analysis of the Abiotic Stress-Responsive Genes in Medicago sativa L. *J. Plant Growth Regul.* **2019**, 1–14. [CrossRef]
6. Takada, S.; Hibara, K.; Ishida, T.; Tasaka, M. The CUP-SHAPED COTYLEDON1 gene of Arabidopsis regulates shoot apical meristem formation. *Development* **2001**, *128*, 1127–1135.
7. Mallory, A.C.; Dugas, D.V.; Bartel, D.P.; Bartel, B. MicroRNA Regulation of NAC-Domain Targets Is Required for Proper Formation and Separation of Adjacent Embryonic, Vegetative, and Floral Organs. *Curr. Biol.* **2004**, *14*, 1035–1046. [CrossRef]
8. Olsen, A.N.; Ernst, H.A.; Leggio, L.L.; Skriver, K. NAC transcription factors: Structurally distinct, functionally diverse. *Trends Plant Sci.* **2005**, *10*, 79–87. [CrossRef]
9. Nakashima, K.; Tran, L.S.P.; Van Nguyen, D.; Fujita, M.; Maruyama, K.; Todaka, D.; Ito, Y.; Hayashi, N.; Shinozaki, K.; Yamaguchi-Shinozaki, K. Functional analysis of a NAC-type transcription factor OsNAC6 involved in abiotic and biotic stress-responsive gene expression in rice. *Plant J.* **2007**, *51*, 617–630. [CrossRef]
10. Shao, H.; Wang, H.; Tang, X. NAC transcription factors in plant multiple abiotic stress responses: Progress and prospects. *Front. Plant Sci.* **2015**, *6*, 902. [CrossRef]
11. Fujita, M.; Fujita, Y.; Maruyama, K.; Seki, M.; Hiratsu, K.; Ohme-Takagi, M.; Tran, L.P.; Yamaguchi-Shinozaki, K.; Shinozaki, K. A dehydration-induced NAC protein, RD26, is involved in a novel ABA-dependent stress-signaling pathway. *Plant J.* **2004**, *39*, 863–876. [CrossRef]

12. Tran, L.S.P.; Nakashima, K.; Sakuma, Y.; Simpson, S.D.; Fujita, Y.; Maruyama, K.; Fujita, M.; Seki, M.; Shinozaki, K.; Yamaguchi-Shinozaki, K. Isolation and functional analysis of Arabidopsis stress-inducible NAC transcription factors that bind to a drought-responsive cis-element in the early responsive to dehydration stress 1 promoter. *Plant. Cell* **2004**, *16*, 2481–2498. [CrossRef] [PubMed]
13. Li, X.D.; Zhuang, K.Y.; Liu, Z.M.; Yang, D.Y.; Ma, N.N.; Meng, Q.W. Overexpression of a novel NAC-type tomato transcription factor, SlNAM1, enhances the chilling stress tolerance of transgenic tobacco. *J. Plant. Physiol.* **2016**, *204*, 54–65. [CrossRef] [PubMed]
14. Hu, H.; You, J.; Fang, Y.; Zhu, X.; Qi, Z.; Xiong, L. Characterization of transcription factor gene SNAC2 conferring cold and salt tolerance in rice. *Plant. Mol. Biol.* **2008**, *67*, 169–181. [CrossRef] [PubMed]
15. Peng, H.; Yu, X.; Cheng, H.; Shi, Q.; Zhang, H.; Li, J.; Ma, H. Cloning and characterization of a novel NAC family gene CarNAC1 from chickpea (*Cicer arietinum* L.). *Mol. Biotechnol.* **2010**, *44*, 30. [CrossRef]
16. Xia, N.; Zhang, G.; Liu, X.Y.; Deng, L.; Cai, G.L.; Zhang, Y.; Wang, X.J.; Zhao, J.; Huang, L.L.; Kang, Z.S. Characterization of a novel wheat NAC transcription factor gene involved in defense response against stripe rust pathogen infection and abiotic stresses. *Mol. Biol. Rep.* **2010**, *37*, 3703–3712. [CrossRef]
17. Xia, N.; Zhang, G.; Sun, Y.F.; Zhu, L.; Xu, L.S.; Chen, X.M.; Liu, B.; Yu, Y.T.; Wang, X.J.; Huang, L.L.; et al. TaNAC8, a novel NAC transcription factor gene in wheat, responds to stripe rust pathogen infection and abiotic stresses. *Physiol. Mol. Plant. Pathol.* **2010**, *74*, 394–402. [CrossRef]
18. Collinge, M.; Boller, T. Differential induction of two potato genes, Stprx2 and StNAC, in response to infection by Phytophthora infestans and to wounding. *Plant. Mol. Biol.* **2001**, *46*, 521–529. [CrossRef]
19. Delessert, C.; Kazan, K.; Wilson, I.W.; Straeten, D.V.D.; Manners, J.; Dennis, E.S.; Dolferus, R. The transcription factor ATAF2 represses the expression of pathogenesis-related genes in Arabidopsis. *Plant. J.* **2005**, *43*, 745–757. [CrossRef]
20. Lu, P.L.; Chen, N.Z.; An, R.; Su, Z.; Qi, B.S.; Ren, F.; Chen, J.; Wang, X.C. A novel drought-inducible gene, ATAF1, encodes a NAC family protein that negatively regulates the expression of stress-responsive genes in Arabidopsis. *Plant. Mol. Biol.* **2007**, *63*, 289–305. [CrossRef] [PubMed]
21. Mauch-Mani, B.; Flors, V. The ATAF1 transcription factor: At the convergence point of ABA-dependent plant defense against biotic and abiotic stresses. *Cell Res.* **2009**, *19*, 1322–1323. [CrossRef] [PubMed]
22. Wu, Y.; Deng, Z.; Lai, J.; Zhang, Y.; Yang, C.; Yin, B.; Zhao, Q.; Zhang, L.; Li, Y.; Yang, C.; et al. Dual function of Arabidopsis ATAF1 in abiotic and biotic stress responses. *Cell Res.* **2009**, *19*, 1279–1290. [CrossRef]
23. Gao, F.; Xiong, A.; Peng, R.; Jin, X.; Xu, J.; Zhu, B.; Chen, J.; Yao, Q. OsNAC52, a rice NAC transcription factor, potentially responds to ABA and confers drought tolerance in transgenic plants. *Plant. Cell Tissue Organ. Cult.* **2010**, *100*, 255–262. [CrossRef]
24. Jin, H.; Huang, F.; Cheng, H.; Song, H.; Yu, D. Overexpression of the GmNAC2 gene, an NAC transcription factor, reduces abiotic stress tolerance in tobacco. *Plant. Mol. Biol. Rep.* **2013**, *31*, 435–442. [CrossRef]
25. Yang, X.; Hu, Y.X.; Li, X.L.; Yu, X.D.; Li, Q.L. Molecular characterization and function analysis of SlNAC2 in Suaeda liaotungensis K. *Gene* **2014**, *543*, 190–197. [CrossRef] [PubMed]
26. Wang, L.; Hu, Z.; Zhu, M.; Zhu, Z.; Hu, J.; Qanmber, G.; Chen, G. The abiotic stress-responsive NAC transcription factor SlNAC11 is involved in drought and salt response in tomato (Solanum lycopersicum L.). *Plant. Cell Tissue Organ. Culture* **2017**, *129*, 161–174. [CrossRef]
27. Wang, K.; Zhong, M.; Wu, Y.; Bai, Z.; Liang, Q.; Liu, Q.; Pan, Y.; Zhang, L.; Jiang, B.; Jia, Y.; et al. Overexpression of a chrysanthemum transcription factor gene DgNAC1 improves the salinity tolerance in chrysanthemum. *Plant. Cell Rep.* **2017**, *36*, 571–581. [CrossRef]
28. Wang, J.; Zhang, L.; Cao, Y.; Qi, C.; Li, S.; Liu, L.; Wang, G.; Mao, A.; Ren, S.; Guo, Y. CsATAF1 positively regulates drought stress tolerance by an ABA-dependent pathway and by promoting ROS scavenging in cucumber. *Plant. Cell Physiol.* **2018**, *59*, 930–945. [CrossRef]
29. Brown, J.; Voth, V. Salt damage to strawberries: Types of water, irrigation system, and soil condition found to influence salt accumulation in strawberry plantings. *Calif. Agric.* **1955**, *9*, 11–12.
30. Razavi, F.; Pollet, B.; Steppe, K.; Van, L.M.C. Chlorophyll fluorescence as a tool for evaluation of drought stress in strawberry. *Photosynthetica* **2008**, *46*, 631–633. [CrossRef]
31. Zhang, H.; Kang, H.; Su, C.; Qi, Y.; Liu, X.; Pu, J. Genome-wide identification and expression profile analysis of the NAC transcription factor family during abiotic and biotic stress in woodland strawberry. *PLoS ONE* **2018**, *13*, e0197892. [CrossRef] [PubMed]

32. Souer, E.; van Houwelingen, A.; Kloos, D.; Mol, J.; Koes, R. The no apical meristem gene of Petunia is required for pattern formation in embryos and flowers and is expressed at meristem and primordia boundaries. *Cell* **1996**, *85*, 159–170. [CrossRef]
33. He, X.J.; Mu, R.L.; Cao, W.H.; Zhang, Z.G.; Zhang, J.S.; Chen, S.Y. AtNAC2, a transcription factor downstream of ethylene and auxin signaling pathways, is involved in salt stress response and lateral root development. *Plant. J.* **2005**, *44*, 903–916. [CrossRef] [PubMed]
34. Mitsuda, N.; Seki, M.; Shinozaki, K.; Ohme-Takagi, M. The NAC transcription factors NST1 and NST2 of Arabidopsis regulate secondary wall thickenings and are required for anther dehiscence. *Plant. Cell* **2005**, *17*, 2993–3006. [CrossRef]
35. Uauy, C.; Distelfeld, A.; Fahima, T.; Blechl, A.; Dubcovsky, J. A NAC gene regulating senescence improves grain protein, zinc, and iron content in wheat. *Science* **2006**, *314*, 1298–1301. [CrossRef]
36. Mao, C.; Lu, S.; Lv, B.; Zhang, B.; Shen, J.; He, J.; Luo, L.; Xi, D.; Chen, X.; Ming, F. A rice NAC transcription factor promotes leaf senescence via ABA biosynthesis. *Plant. Physiol.* **2017**, *174*, 1747–1763. [CrossRef]
37. Hetherington, A.M.; Woodward, F.I. The role of stomata in sensing and driving environmental change. *Nature* **2003**, *424*, 901–908. [CrossRef] [PubMed]
38. Roelfsema, M.R.G.; Levchenko, V.; Hedrich, R. ABA depolarizes guard cells in intact plants, through a transient activation of R-and S-type anion channels. *Plant. J.* **2004**, *37*, 578–588. [CrossRef]
39. Zhang, Y.; Xu, W.; Li zhong Deng, X.; Wu, W.; Xue, Y. F-Box Protein DOR Functions as a Novel Inhibitory Factor for Abscisic Acid-Induced Stomatal Closure under Drought Stress in Arabidopsis. *Plant. Physiol.* **2008**, *148*, 4. [CrossRef]
40. Lawson, T.; Blatt, M.R. Stomatal size, speed, and responsiveness impact on photosynthesis and water use efficiency. *Plant. Physiol.* **2014**, *164*, 1556–1570. [CrossRef]
41. Holbrook, N.M.; Shashidhar, V.R.; James, R.A.; Munns, R. Stomatal control in tomato with ABA-deficient roots: Response of grafted plants to soil drying. *J. Exp. Bot.* **2002**, *53*, 1503–1514. [PubMed]
42. Christmann, A.; Weiler, E.W.; Steudle, E.; Grill, E. A hydraulic signal in root-to-shoot signalling of water shortage. *Plant. J.* **2007**, *52*, 167–174. [CrossRef] [PubMed]
43. Mc Adam, S.A.; Sussmilch, F.C.; Brodribb, T.J. Stomatal responses to vapour pressure deficit are regulated by high speed gene expression in angiosperms. *Plant. Cell Environ.* **2016**, *39*, 485–491. [CrossRef]
44. Zhang, F.; Sussmilch, F.; Nichols, D.S.; Cardoso, A.A.; Brodribb, T.J.; McAdam, S.A.M. Leaves, not roots or floral tissue, are the main site of rapid, external pressure-induced ABA biosynthesis in angiosperms. *J. Exp. Bot.* **2018**, *69*, 5. [CrossRef] [PubMed]
45. Saradhi, P.P.; AliaArora, S.; Prasad, K. Proline accumulates in plants exposed to UV radiation and protects them against UV-induced peroxidation. *Biochem. Biophys. Res. Commun.* **1995**, *209*, 1–5. [CrossRef]
46. Hare, P.D.; Cress, W.A. Metabolic implications of stress-induced proline accumulation in plants. *Plant. Growth Regul.* **1997**, *21*, 79–102. [CrossRef]
47. Siripornadulsil, S.; Traina, S.; Verma, D.P.S.; Sayre, R.T. Molecular mechanisms of proline-mediated tolerance to toxic heavy metals in transgenic microalgae. *Plant. Cell* **2002**, *14*, 2837–2847. [CrossRef]
48. Verbruggen, N.; Hermans, C. Proline accumulation in plants: A review. *Amino acids* **2008**, *35*, 753–759. [CrossRef]
49. Jung, S.; Ficklin, S.P.; Lee, T.; Cheng, C.H.; Blenda, A.; Zheng, P.; Yu, J.; Bombarely, A.; Cho, I.; Ru, S.; et al. The Genome Database for Rosaceae (GDR): Year 10 update. *Nucleic Acids Res.* **2014**, *42*, D1237–D1244. [CrossRef]
50. Kumar, S.; Stecher, G.; Tamura, K. MEGA7: Molecular evolutionary genetics analysis version 7.0 for bigger datasets. *Mol. Biol. Evol.* **2016**, *33*, 1870–1874. [CrossRef]
51. Fernandez-Pozo, N.; Menda, N.; Edwards, J.D.; Saha, S.; Tecle, I.Y.; Strickler, S.R.; Bombarely, A.; Fisher-York, T.; Pujar, A.; Foerster, H.; et al. The Sol Genomics Network (SGN)—From genotype to phenotype to breeding. *Nucleic Acids Res.* **2015**, *43*, D1036–D1041. [CrossRef] [PubMed]
52. Wydro, M.; Kozubek, E.; Lehmann, P. Optimization of transient Agrobacterium-mediated gene expression system in leaves of Nicotiana benthamiana. *Acta Biochim. Pol.* **2006**, *53*, 289–298. [CrossRef] [PubMed]
53. Bergmeyer, H.U.; Graßl, M. *Methods of Enzymatic Analysis*, 3rd ed.; Verlag Chemie: Weinheim, Germany, 1983; pp. 267–268.
54. Onsa, G.H.; bin Saari, N.; Selamat, J.; Bakar, J. Purification and characterization of membrane-bound peroxidases from Metroxylon sagu. *Food Chem.* **2004**, *85*, 365–376. [CrossRef]

55. Oktay, M.; Küfreviolu, I.; Kocaçalişkan, I.; Şaklrolu, H. Polyphenoloxidase from Amasya apple. *J. Food Sci.* **1995**, *60*, 494–496. [CrossRef]
56. Bates, L.S.; Waldren, R.P.; Teare, I.D. Rapid determination of free proline for water-stress studies. *Plant. Soil* **1973**, *39*, 205–207. [CrossRef]

3

The Antioxidant Profile Evaluation of Some Tomato Landraces with Soil Salinity Tolerance Correlated with High Nutraceutical and Functional Value

Renata M. Sumalan [1], Sorin I. Ciulca [1], Mariana A. Poiana [2], Diana Moigradean [2], Isidora Radulov [3], Monica Negrea [2], Manuela E. Crisan [4], Lucian Copolovici [5] and Radu L. Sumalan [1,*]

1 Faculty of Horticulture and Forestry, Banat's University of Agricultural Sciences and Veterinary Medicine "King Michael I of Romania" from Timisoara, CaleaAradului 119, 300645 Timisoara, Romania; srenata_maria@yahoo.com (R.M.S); c_i_sorin@yahoo.com (S.I.C.)
2 Faculty of Food Processing, Banat's University of Agricultural Sciences and Veterinary Medicine "King Michael I of Romania"from Timisoara, CaleaAradului 119, 300645 Timisoara, Romania; atenapoiana@yahoo.com (M.A.P.); dimodean@yahoo.com (D.M.); negrea_monica2000@yahoo.com (M.N.)
3 Faculty of Agriculture, Banat's University of Agricultural Sciences and Veterinary Medicine "King Michael I of Romania" from Timisoara, CaleaAradului 119, 300645 Timisoara, Romania; isidoraradulov@yahoo.com
4 "CoriolanDragulescu" Institute of Chemistry, 24 Mihai Viteazul Blvd., 300223 Timisoara, Romania; mdorosencu@yahoo.com
5 Faculty of Food Engineering, Tourism and Environmental Protection, Research Center in Technical and Natural Sciences, "AurelVlaicu" University, Elena Dragoi 2, 310330 Arad, Romania; lucian.copolovici@uav.ro
* Correspondence: sumalanagro@yahoo.com

Abstract: Romania has a wide variety of local landraces and heirloom genotypes. Our study aims to assess the performance of twenty halotolerant tomato landraces, collected from areas with medium and high levels of soil salinity, in terms ofthe accumulation of antioxidant compounds in fruits and to cluster them according to their nutraceutical components. The tomatoes used in the study were harvested once they had attained full ripeness and then analyzed for lycopene (Lyc), ascorbic acid content (AsA), total phenolic content (TPC), and total antioxidant capacity (TAC). The results revealed major differences between genotypes in terms of nutraceutical values. According to principal component analysis, the tomato landraces were grouped into five clusters, characterized by different proportions of compounds with antioxidant activity. The high/moderate nutritional values of Lyc, TAC, TPC, and AsA were obtained from varieties taken from local lands with high soil salinity, over 6.5 dS m^{-1}. These findings support the idea that metabolites and secondary antioxidants are involved in the process of stress adaptation, thereby increasing salinity tolerance in tomatoes. Our results show that there are tomato landraces with a tolerance of adaptation to conditions of high soil salinity and provide information on their ability to synthesize molecules with antioxidant functions that protect plants against oxidative damage.

Keywords: tomato cultivars; salinity tolerance; antioxidant activity; lycopene; ascorbic acid; total polyphenols content

---

1. Introduction

The commercial production of tomato (*Solanum lycopersicum* L.) in the developed regions of the world mostly concerns modern varieties of the fruit, and, more often than not, genetically uniform F1 hybrids that have a high yield, greater tolerance to diseases, and a long shelf life are chosen [1]. Landraces represent an important alternative, as they constitute a reservoir of genetic diversity, with their important abiotic stress tolerance and high fruit quality. Tomato landraces contain valuable alleles uncommon in highlight germplasms [2]; therefore, these local populations represent a valuable resource of genetic traits that can be used in breeding programs for the improvement of the crop [3].

Intensive tomato cultivation technologies require genotypes with good productivity, handling, transport, and storage properties, while the nutraceutical properties are passed on to the secondary level. Nevertheless, consumption of the traditional plant foods that have antioxidant content naturally occurring may be a better strategy to improve the human health status than the consumption of artificial antioxidant products [4]. From this perspective, an appropriate selection of tomato cultivar would help to achieve a higher antioxidant intake with the potential to produce significant health benefits. Considering the growing demand for tasty tomatoes and being rich in phytochemicals, a detailed characterization of the biologically active compounds completed with total antioxidant capacityevaluation should be performed. That is why the tomato landraces and their relatives are of great importance in breeding programs [5–7]. Besides, landraces have low requirements for inputs, which contribute to the development of environmentally-friendly technologies [8–10].

The genetic resources of cultivated plants that come from the soil salinity-affected areas have a major importance due to drought tolerance. The salinity is associated with the physiological drought that may induce the growth of the bioactive compound content with antioxidant properties [11,12]. The information about genetic variability specific to local populations is assumed to be limited [13–15] because traditional locally grown cultivars should not be considered rigorously homogenous.

The tomato taste is correlated with some bioactive components. Some authors consider that AsA (ascorbic acid), TAC (total antioxidant capacity), and TPC (total phenols content) have a direct impact on tomato taste [5], while others assign the taste to the ratio between sugar content and acidity [16–19]. The high content in valuable phytonutrients depends equally on technological, genetic, and storage factors [16,20–25]. Besides dietary fibers and carbohydrates, compounds such as lycopene, β-carotene, ascorbic acid, and polyphenols provide high levels of antioxidants. For this reason, the consumption of raw or processed tomatoes contributes to good activity into an organism by maintaining oxidative stressat a low level [26,27].

Lycopene is the most important and recognized phyto bioactive-compound of tomatoes. It is a carotenoid pigment that is less bio-available compared to β-carotene and lutein [28,29]. In addition to lycopene, the tomato fruits contain vitamins A and C, other carotenoids whose action interacted with those of polyphenols, resulting in an overall benefit on human health [30,31]. The ripening stage of tomato fruits represents a decisive factor regarding the establishment of nutritional values of thereof. From this point of view, the tomatoes harvested at technological ripeness revealed low quantities of lycopene while the content of ascorbic acid was variable, depending on genotype [32].

The aim of this paperis to characterize some tomato local landraces, originating from areas with saline soils located in western Romania (Banat region), concerning the total antioxidantcapacity and the potential of biosynthesis and accumulation of some antioxidant biocomponents, such as phenolic compounds, lycopene and ascorbic acid.

In this work, we test the hypothesis that local tomato landraces originating from areas with soil salinity show a higher biosynthesis capacity of some compounds with antioxidant activity, and this property is maintained even under cultivation on non-saline soil conditions.

2. Materials and Methods

2.1. Plant and Soil Analyses

Twenty halotolerant tomato landraces, collected from local farmers ofthe country-side situated in areas with soils affected by different levels of sodicity from westernRomania (Banat region) were analyzed. Saline-sodic soils are high in exchangeable sodium and low in total soluble salts. The level of ESP (exchangeable sodium percentage) in these soils is 15 or more, which tends to destroy their structure by dispersing the particle, and electrical conductivity (EC) is over 4 dS m^{-1}. Most saline soils in the collecting area have a clayed loamy texture, medium glomerular structure and moderately to high salinization level determined by using an EC-meter (model consort C933, producer De Bruine Instruments bvba, Belgium) determination using the EC 1:2*w/v* method [33] (Table 1).

Seeds and soil samples were collected in the period 2012–2015, and the specimen is available at Plant Physiology Department, Faculty of Horticulture and Forestry, Timisoara. Previously, each local landraces was characterized in morpho-physiologically and genetically manner [34], for each of them was drawn up an identification sheet in which source, specific cultivation technology, productivity, shape, and color of tomato fruits were noticed. Based on productivity traits, high tolerance to salinity and minimum growth requirements proven in the summer of 2015, twenty tomato landraces were selected for assessment of nutraceutical traits (Table 1).

The genotypes were open field cultivated in a plain site located on the northern side of Timisoara (45°78′N; 21°21′E), on cambicchernozem soil [35]. The soil had the following physico-chemical characteristics: clay 402 g kg^{-1}; sand 330 g kg^{-1}; loam 268 g kg^{-1}; organic matter 26.8 g kg^{-1}; pH 6.26; total N_2 g kg^{-1}; available P_2O_5 20.52mg kg^{-1}; exchangeable K_2O 117 mg kg^{-1}; sulfates (mobile in water) 105.6 mg kg^{-1}; sodium (mobile in water) 366.7mg kg^{-1}; calcium (water-soluble) 270.5 mg kg^{-1}; magnesium (soluble in water) 60.8 mg kg^{-1}.

A randomized complete block experimental design with three replicates was used in the field during the spring–summer season. Plants were transplanted at the four-leaf stage on 18 April 2017 in plots of 24 m^2 (6 × 4 m) at 1.66 plants m^{-2}. The average monthly temperatures ranged between 11.6 °C in April to 27.4 °C in July. Plants were spaced 1.2 m between the rows and 0.5 m within the row and watered with a drip irrigation system. Plants were trained with canes and cultivated using a traditional horticultural practice in the area for local tomato varieties. All genotypes have an indeterminate plant growth type.

Table 1. Geographical origin, EC of origin soils, and the main morphological and productive characteristics of tomato landraces.

Genotype Code	Site	GPS Coordinates (lat/long)	Soil EC(dSm^{-1})	Fruit Shape [1]	Tomatoes Weight Average (g) [2]	Full Ripeness Color [3]
CN26	Crai Nou	45°29′17″N/21°0′1″E	6.86	flattened	352.47	light red
PN	Peciu Nou	45°36′54″N/21°01′54″E	5.63	flattened	184.66	light red
Gi	Giera	45°25′21″N/20°57′25″E	5.25	circular	124.18	red
L-189a	Lovrin	45°57′03″N/20°46′32″E	5.02	obovate	73.75	light red
C-102	Cruceni	45°28′23″N/20°52′44″E	7.04	circular	133.55	light red
Pe	Periam	46°01′41″N/20°53′35″E	6.86	flattened	295.76	red
Gr	Gradinari	45°06′16″N/21°34′59″E	4.38	circular	273.00	light red
DV	Dudestii Vechi	46°04′55″N20°26′55″E	5.80	obovate	164.92	red
Ch	Cheglevici	46°6′40″N/20°26′56″E	6.04	flattened	264.54	red
C-60pr12	Cherestur	46°7′60″N/20°22′60″E	5.65	ovate	150.64	light red
Ch-165	Cheglevici	46°6′40″N/20°26′56″E	6.29	circular	105.44	light red
Li	Livezile	45°23′09″N/21°02′43″E	6.44	cilindric	133.62	yellow
L-189b	Lovrin	21°02′43″E/20°46′32″E	6.58	flattened	136.11	light red
Ru	Rudna	45°29′54″N/21°0′31″E	4.50	obovate	81.73	red
SS180	Sanmartinu Sarbesc	45°36′23″N/20°57′38″E	4.47	cordate	303.09	red
T673	Tarnova	45°20′06″N/22°00′08″E	4.11	flattened	309.57	red
T370	Tarnova	45°20′06″N/22°00′08″E	4.23	flattened	348.42	red
IM/pusta	Iecea Mare/Pusta	45°50′51″N/20°54′08″E	4.18	flattened	392.60	red
SS	Sanmartinu Sarbesc	45°36′23″N/20°57′38″E	4.30	flattened	185.59	light red
CN-254	Crai Nou	45°29′17″N/21°0′1″E	7.21	obovate	150.33	red

[1] According to UPOV (International Union for the Protection of New Varieties of Plants) classification in tomatoes fruit shape [36]; [2] data were collected from 15 fruits; [3] USDA (United States Department of Agriculture) tomato ripeness color chart.

2.2. Tomatoes Samples Preparation

From each genotype, samplings were taken at different harvesting times only when tomatoes were at a fully physiological ripening stage. Fifteen fruits were randomly taken from each replication in order to compose the average tomato sample. Evaluations of shape, weight, and colors, as well as chemical analysis, were done. The fruits were stored in polyethylene bags and kept in freezing conditions at −18 °C until performing the chemical analysis. All analyses werecarried out in triplicate.

2.3. Chemical Analysis

The chemical analysis consisted of assessing of TAC, TPC, AsA, and Lyc content from each tomatoes landrace. Prior analysis, the frozen samples were kept in refrigeration condition (4–6 °C) for 6 h and then homogenized in a Bosch Blender (MMB42G0B, 700 W, Germany) for 1 min. Three replicates were prepared from each average sample.

2.3.1. Extract Preparation

Briefly, 10 g of blended tomato sample was mixed with 20 mL ethyl alcohol 70% (*v*/*v*) for 2 h at 25 °C, then, the mixture was filtered and the clear extract was used for the analysis of TAC and TPC.

2.3.2. Reagents and Equipment

All chemicals and reagents were analytical grade or purest quality purchased from Merckand Fluka. Deionized water was used.

TAC Evaluation

TAC of tomatoes was evaluated by ferric reducing antioxidant power (FRAP) assay, according to the method described by Benzie and Strain [37]. This method supposed the reduction of Fe^{3+}-TPTZ (2,4,6-tris(2-pyridyl)-s-triazine) complex to ferrous form at low pH. The ferrous tripyridyltriazine complex has an intense blue color monitored at a wavelength of 593 nm. An aqueous solution of Fe^{2+} with a concentration in the range of 0.1 to 0.8 mM/L was used for the calibration curve preparation. The absorption was measured at $\lambda = 593$ nm after 15 min of incubation at 25 °C using the UV–vis spectrophotometer SPECORD 205 (Analytic Jena, Germany). TAC was expressed as μM Fe^{2+} equivalents·100 g^{-1} FW (fresh weight).

TPC Determination

TPC in tomato samples was evaluated following the Folin–Ciocalteu colorimetric method described by Singleton and Rossi [38]. For analysis, it was used the tomato ethanol extracts diluted 1/10 with bidistilled water. For calibration curve preparation, 0.5 mL aliquot of aqueous gallic acid solution with a concentration in the range 0.2–1.2 μM/mL were mixed with 2.5 mL Folin–Ciocalteu reagent (diluted ten-fold with bidistilled water) and 2.0 mL sodium carbonate (7.5%). The absorption was read at $\lambda = 750$ nm after 2 h of incubation at 20 °C. TPC in mg gallic acid equivalents (GAE)·100 g^{-1} FW was calculated.

AsA Content

The AsA content of tomato samples was measured on the base of the AOAC method [39] by titration with 2,6-dichlorophenolindophenol sodium salt. For this purpose, 10 g of blended tomato sample was mixed with 10 mL bidistilled water for 2 h at 25 °C, then, the mixture was filtered, and the clear extract was used for analysis. Further, 5 mL of extract was diluted with 10 mL bidistilled water, then 1 mL HCl 1N was added and the mixture was titrated with 2,6-dichlorphenolindophenol sodium solution 1 mM in an acid medium (pH = 4). The results are expressed as mg ascorbic acid·100 g^{-1} FW.

Determination of Lyc Content

Lyc was extracted from tomato samples with a hexane–ethanol–acetone (2:1:1) mixture in agreement with the method describes by Sharma and Le Maguer [40]. Briefly, 1 g of tomato blended sample was mixed with 25 mL of the previously mentioned mixture and then placed on a rotary mixer for 30 min. Further, 10 mL of bidistilled water was added, and the mixture was stirred for another 2 min. The obtained mixture solution was separated into two distinct polar and non-polar layers. The absorbance was measured at 502 nm, using hexane as a blank. The Lyc content of tomato samples was calculated on the base of its specific extinction coefficient (E 1%, 1 cm) of 3150 at 502 nm [41]. The Lyc concentration was expressed as mg 100 g^{-1} FW.

2.4. Statistical Analysis

The experimental data were statistically processed using ANOVA, and the means were compared using the multiple range test [42]. The significance of differences was expressed based on letters, being considered as significant the differences between genotypes marked with different letters.

The clustering of genotypes was carried out using the UPGMA (unweighted pair group method with arithmetic mean), with the NEIGHBOR program of PHYLIP package, version 3.5c. [43]. Average intra and inter-cluster distances (D2) were estimated [44], and the percent contribution of each character to the total divergence was calculated by ranking each character on the basis of transformed uncorrelated values. To display the performance of each landrace for each of four nutritional traits in a single graph, the basic principle of the biplot technique was used [45,46].

3. Results

3.1. Fruit Morphological Traits

All the measured fruit morphological traits (shape, weight, and color), showed a large range of phenotypic variation among the 20 tomato landraces.

Fruit shape is one of the most important qualities which can be determined with the naked eye, being used to identify local tomato populations. Globally, there is a huge variety of fruit shapes on tomatoes, specific to different landraces, a fact reported by many studies conducted in different geographic areas and time periods [2,15,36,47–52].

As for the traits related to color and flesh color of fruits, ten genotypes had *red* fruit, nine genotypes had a *light red* color, and one had *yellow* fruits.

Regarding the tomato fruit weight, a number of eight genotypes had large fruits (≥200 g), of which five were very large (≥300 g), ten landraces had average fruits weighing between 100 and 200 g, and only two formed small fruits (below 100 g). The comparative analysis of the form of fruit and their weight shows that six of the local populations with high weight fruit have a flattened shape.

These results are also in line with other studies [53] confirming that, in the process of improving tomatoes, people prefer to increase the size and mass of the fruit, causing and modifying the round shape (wild species) with a flattened one in the most forms cultivated for fresh consumption or elongated for industrialization. On the other hand, some studies [15] found that, in the case of some local Italian and South American tomatoes populations, the flattened form of the fruit has been associated with a small and average weight (50–150 g). Obviously, the fruit mass is a genetically controlled process, but it depends to a large extent on the specific pedo-climatic conditions and applied technology.

Therefore, the study of genetic variability inlocal tomato landraces will be able to provide additional information on the genetic, physiological, and biochemical mechanisms that are based on the correlation of the shape, size, and weight of the fruit, thus contributing to the identification of the alleles and new ecotypes with superior properties in terms of productivity, adaptation, quality, and nutritional value.

3.2. Antioxidants and Nutraceutical Component Analysis

Antioxidants are redox buffers that interact with ROS (reactive oxidative species) and can manifest as a metabolic interface that regulates adaptation responses or programmed cell death [54]. The low values of *F*: 120 and 126, for significant differences at $p < 0.01$, show that there is a lower variability between landraces for TPC and AsA.

3.2.1. Assessment of TAC

Statistical data analysis concerning the performance of TAC of tomato fruit samples reveals significant differences between landraces (Table 2). The quantity and proportion between bioactive compounds with antioxidant capacity depend on the plant's genotype and post-harvest storage conditions [55,56]. The reducing ability recorded for the 20 landraces varies from 561.61 to 240.75 μM Fe^{2+} 100 g^{-1} FW with an amplitude of variation of 320.86 between landraces (Table 2).

Table 2. Total antioxidant capacity (TAC) of tomato fruit samples.

Genotype Code	FRAP μM $Fe^{2\pm}$ 100 g^{-1} FW
CN-26	413.27 ± 3.63 d
PN	305.44 ± 2.99 g
Gi	349.03 ± 3.76 f
L-189a	386.49 ± 3.91 e
C-102	420.97 ± 2.13 d
Pe	451.76 ± 1.36 c
Gr	384.03 ± 3.19 e
DV	411.36 ± 2.02
Ch	274.00 ± 1.53 i
C-60pr12	240.75 ± 1.51 j
Ch-165	304.67 ± 1.65 g
Li	415.14 ± 1.99 d
L-189b	506.51 ± 2.65 b
Ru	300.05 ± 2.42 gh
SS-180	289.21 ± 1.58 h
T-673	287.27 ± 1.60 hi
T-370	407.56 ± 2.31 d
IM/pusta	307.69 ± 1.62 g
SS	387.05 ± 1.92 e
CN-254	561.61 ± 4.37 a
Mean	*370.19 ± 10.48*
Cochran's C Test	*0.145; p = 1.00*
Bartlett's Test	*1.300; p = 0.974*
LSD5%	*13.78*

Data are mean ± SE, $n = 3$. Values within columns with different letters are significantly different ($p = 0.05$).

The high value of TAC was determined for CN-254, 561.61 μM Fe^{2+} 100 g^{-1} FW, followed by L-189b with 506.51 μM Fe^{2+} 100 g^{-1} FW. Both landraces have recorded statistically higher differences related to mean. In the second category fall the local populations Pe (451.76 μM Fe^{2+} 100 g^{-1} FW), C-102 (420.97 μM Fe^{2+} 100 g^{-1} FW), and CN-26 (413.27 μM Fe^{2+} 100 g^{-1} FW). The lowest value of FRAP (240.75 μM Fe^{2+} 100 g^{-1} FW) was determined for genotype C-60pr12. It should be noted that all five cultivars of the first two categories were collected from the areas with high concentrations of soil salinity between 7.21 dS m^{-1} for CN-254 and 6.58 dS m^{-1} at L-189b.

The recorded TAC values are in agreement with those reported by other authors who have used the same method. Thus, mean values of TAC of 506 μM Fe^{2+} 100 g^{-1} FW [57] were reported, while other authors reported limits between 387 and 493 μM Fe^{2+} 100 g^{-1} FW [31] for tomatoes fruits.

Phenolic compounds play an important role in the antioxidant capacity of tomato fruit. In addition, other bioactive compounds such as ascorbic acid and lycopene, the main tomato carotenoid, play a major role in the antioxidant capacity of tomato fruit samples. The ascorbic acid is considered the most important water-soluble antioxidant, with a significant contribution to the antioxidant cellular defense against oxidative stress. FRAP assay also measures the antioxidant capacity of ascorbic acid besides that of phenolic compounds. The phenolic compounds act synergistically with ascorbic acid in order to preserve and regenerate the antioxidant species. Lycopene is one of the most powerful antioxidants among the dietary carotenoids. The total antioxidant capacity of tomato fruit is the result of a combination of different compounds having synergistic and antagonistic effects.

3.2.2. Assessment of TPC, Lyc, and AsAContent

The quantity and quality of phenolic compounds determined in tomato fruit varies in relation to the genotype but also depends on environmental and technology factors [58]. Data from Table 3 reveal that the TPC values range from 51.49 to 123.32 mg GAE·100 g^{-1} FW. The highest value was registered for C-102 (123.32 30 mg GAE·100 g^{-1} FW), followed by L-189b with 117.77 mg GAE·100 g^{-1} FW and CN-254 (114.89 mg GAE·100 g^{-1} FW), but the differences between them have no statistical significance.

The TPC values for the local tomato populations collected from the saline areas are higher compared to other studies [31,56,59,60]. TPC valuesrecorded for industrial processing cultivars(green Ronaldo and red cherry Pera) were between 18.69 to 55.86 mg GAE 100 g^{-1} FW [56] and in the range 19.7–21.1 mg GAE 100 g^{-1} FW [61], while for fresh tomato consumption ranging from 50.86 to 53.88 mg GAE 100 g^{-1} FWwere reported [31].Also, Martinez-Valverde reported TPC values between 25.9 and 49.8 mg GAE 100 g^{-1} FW for determinations of some commercial varieties of tomatoes from supermarkets [59]. The lowest values of TPC in our cultivars were determined for IM/Pusta with 51.49 mg GAE 100 g^{-1} FW, followed by genotype Ch with 53.06 mg GAE 100 g^{-1} FW.

The value of TPC for our landraces (especially C-102, L-189b, and CN-254) are higher than those obtained by other researchers for different hybrids. The highest value of TPC for CN-254 justifies the large TAC value determined for this local population, poly-phenols being well known for their antioxidant activity [30,62]. By correlating the results to the fact that these local populations have adapted over time to high levels of soil salinity, we can assume that they have developed genetic, biochemical, and physiological mechanisms to synthesize some of the antioxidant-related co-agents, to allow survival under specific stress conditions. It seems that the high pressure of the soil salinity influenced the plants which were obliged to adapt by increasing the synthesis capacities of polyphenols and indirectly intensifying of antioxidant processes. These traits appear to be manifested in the conditions of their cultivation outside specific saline areas. Indeed in a recent work, it has been shown that the traditional variety of tomato improved cations homeostasis and increased sucrose content in the fruits as a part of the salt stress tolerance mechanism [63].

Raw tomatoes contain usually Lyc between 3 to 10 mg 100 g^{-1}, but field-grown tomatoes appear to contain higher levels of this compound, ranging from 5.2 to 23.6 mg 100 g^{-1} FW than greenhouse-grown tomatoes (0.1 and 10.8 mg 100 g^{-1} FW) [31].

In our research, the Lyc content recorded a high variability of 13.13 mg 100 g^{-1} FW amplitude of variation (Table 3). The highest value of Lyc was determined for SS-180 (18.43 mg 100 g^{-1} FW) followed by Gr (17.37 mg 100 g^{-1} FW) and T-673 (16.61 mg 100 g^{-1} FW). At the same time, significantly lower mean values of the Lyc content were noticed for landraces Li (5.30 mg 100 g^{-1} FW), Ch-165 (6.79 mg 100 g^{-1} FW), and C-60pr12 (6.81 mg 100 g^{-1} FW) that have recorded statistically assured differences related to mean of experience.

Comparing our results concerning the Lyc content of tomato fruit with other studies [64–67], we found that the local populations collected by the salt-crops cultivated under non-saline conditions recorded higher values. Most previous studies reported values of Lyccontent in fresh tomato fruits ranging from 3 to 10 mg 100 g^{-1} FW. Higher amounts of Lyc (around 25 mg 100 g^{-1} FW) were reported in the cherry tomatoes IIHR-249-1 line [68].

Although most research confirms the growth of Lyccontent in salinity conditions [69–71], there are a few studies that infirm this hypothesis [23,72]. Therefore, the genetic requirements and the specific conditions in which the genotype was cultivated are decisive factors on the capacities of biosynthesis of Lyc, alongside other evident factors such as ripening stage and cultivation technologies (e.g., field/greenhouse, organic/non-organic) [32].

The results obtained through this research do not show a direct connection between the soil salinity level in the harvesting area and the amount of Lyc of the various local populations. Cultivars with high levels of Lyc come from areas with moderate levels of soil salinity (E C $\leq$ 4.5 dS m^{-1}).

The introduction of Lyc as a phyto-compound in the human diet, due to multiple antioxidant properties, led to obtaining products enriched in this biomolecule [73]. From a nutritional point of view and based on recommendations on the consumption of 35 mg Lyc/day [74], this can be achieved by eating 200 g of tomatoes from SS 180 or GR local population for the daily requirement of an adult.

AsA is an antioxidant with an electron donor role in many important reactions [75,76]. It plays a key role in photosynthesis protection during salinity stress. It has been shown that on salinity stress conditions, mutations with deficiency of AsA synthesis accumulate very large amounts of H_2O_2, which coincides with an important decrease of the reduced AsA. The accumulation of H_2O_2 in the foliar apparatus of tomato plants determines the inhibition of photosynthesis by reducing the amount of chlorophyll, CO_2 assimilation, and decrease in PS II activities [54].

Table 3. Lycopene (Lyc),ascorbic acid (AsA),and total phenolic content (TPC)of tomato fruit samples.

Genotype Code	Lycopene mg 100 g^{-1} FW	Ascorbic Acid mg 100 g^{-1} FW	Total Phenols mg GAE 100 g^{-1} FW
CN26	10.72 ± 0.19 hi	16.93 ± 0.20 de	69.15 ± 1.28 f
PN	7.93 ± 0.12 k	15.21 ± 0.14 ij	92.88 ± 1.83 c
Gi	11.43 ± 0.17 g	15.94 ± 0.17 h	93.22 ± 1.71 c
L-189a	11.20 ± 0.13 gh	16.44 ± 0.19 fg	93.68 ± 2.18 c
C-102	12.57 ± 0.12 f	17.37 ± 0.16 bc	123.32 ± 2.10 a
Pe	12.51 ± 0.10 f	17.70 ± 0.18 b	91.76 ± 2.19 cd
Gr	17.37 ± 0.14b	16.56 ± 0.14 ef	86.58 ± 1.67 cd
DV	12.29 ± 0.11 f	16.99 ± 0.16 cde	91.37 ± 1.70 cd
Ch	9.30 ± 0.08 j	13.98 ± 0.16 l	53.06 ± 0.73 g
C-60pr12	6.81 ± 0.06 l	13.17 ± 0.12 m	112.96 ± 1.74 b
Ch-165	6.79 ± 0.07 l	15.07 ± 0.13 ij	75.19 ± 1.53 ef
Li	5.30 ± 0.07 m	17.06 ± 0.11 cd	88.26 ± 2.12 cd
L-189b	10.73 ± 0.12 hi	19.23 ± 0.16 a	117.77 ± 2.53 ab
Ru	12.87 ± 0.17 ef	14.81 ± 0.13 jk	89.29 ± 2.09 cd
SS-180	18.43 ± 0.15 a	14.54 ± 0.11 kl	73.19 ± 1.17 f
T-673	16.61 ± 0.13c	14.31 ± 0.13 l	71.79 ± 1.01 f
T-370	12.83 ± 0.11 ef	16.10 ± 0.18 gh	82.69 ± 1.83 de
IM/pusta	13.23 ± 0.10 e	15.30 ± 0.13 i	51.49 ± 1.19 g
SS	14.00 ± 0.13 d	16.72 ± 0.15 def	93.48 ± 1.46 c
CN-254	10.18 ± 0.08 i	20.15 ± 0.17 a	114.89 ± 2.16 ab
means	*11.65 ± 0.44*	*16.18 ± 0.22*	*88.30 ± 2.49*
Cochran's C Test	*0.128; $p = 1.00$*	*0.086; $p = 1.00$*	*0.102; $p = 1.00$*
Bartlett's Test	*1.191; $p = 0.998$*	*1.057; $p = 1.00$*	*1.162; $p = 0.999$*
LSD5%	*0.65*	*1.56*	*9.49*

Data are mean ±SE, $n = 3$. Values within columns with different letters are significantly different ($p = 0.05$).

Consequently, our results show that AsA content of tomato fruit has not a large amplitude, the determined quantities ranging from 13.17 to 20.15 mg 100 g^{-1} FW. Particularly notable are CN-254 and L-189b landraces, having the highest values of AsA content of 20.15 and 19.23 mg 100 g^{-1} FW, respectively, followed by Pe (17.70 mg 100 g^{-1} FW) and C-102 (17.37 100 g^{-1} FW).

The analysis of the results confirms that the above-presented genotypes with high synthesis capacity of the AsA have manifested very good qualities of TAC and TPC (Table 2).The results prove that the AsA content is conditioned by genotype to a minimal extent. High AsAcontent was reported in F1 tomato hybrid for which heterosis effect was manifested, the maximum content determined of 36.3 mg 100 g^{-1} FW being over mid parental values [20,77].

It was demonstrated that AsA increase is linked to the adopted cultivation system. Studies on tomato growing technologies have highlighted the importance of nutrient availability for tomato plant requirements for AsA accumulation [78].Therefore, there are numerous reports of increasing the content of AsA in tomato fruit subject to various types of stress [79–84], as well as some results reported in which stress resulted in the reduction of AsA [85,86]. The contradictory results can be attributed to the genetic differences regarding the sensitivity of the different genotypes to the oxidative stress manifested by salinity.

It is known that there are universal mechanisms as responses to the action of stress factors, but their relative impacts can vary from one species to another and within the same species from one genotype to another, depending, actually, on specific metabolic background.

3.3. Comparison of Tomato Landraces for Nutraceutical Traits

Using the UPGMA method for 20 variables of tomato landraces, a dendrogram was designed that has identified five groups (clusters) on the basis of coefficients similarity for bioactive compounds (Figure 1). The first cluster brings together nine landraces: CN26, Gi, L-189a, DV, SS, T370, Pe, Gr, and C-102. These are characterized by higher TAC and AsA values, while TPC and Lyc have recorded values above landrace average. Three landraces are grouped in the second cluster, namely: PN, Ch-165, and Li; these recorded low values of Lyc but medium values of TAC, TPC, and AsA content.

In Cluster III, there are grouped five landraces characterized by a high Lyc, while TAC and AsA have lower values than average, and TPC has the lowest values. In Cluster IV, one single landraceis noticed; C-60pr12 has a high value of TPC but a low one of Lyc, while TAC and AsA contents are lower than mean values of landraces.

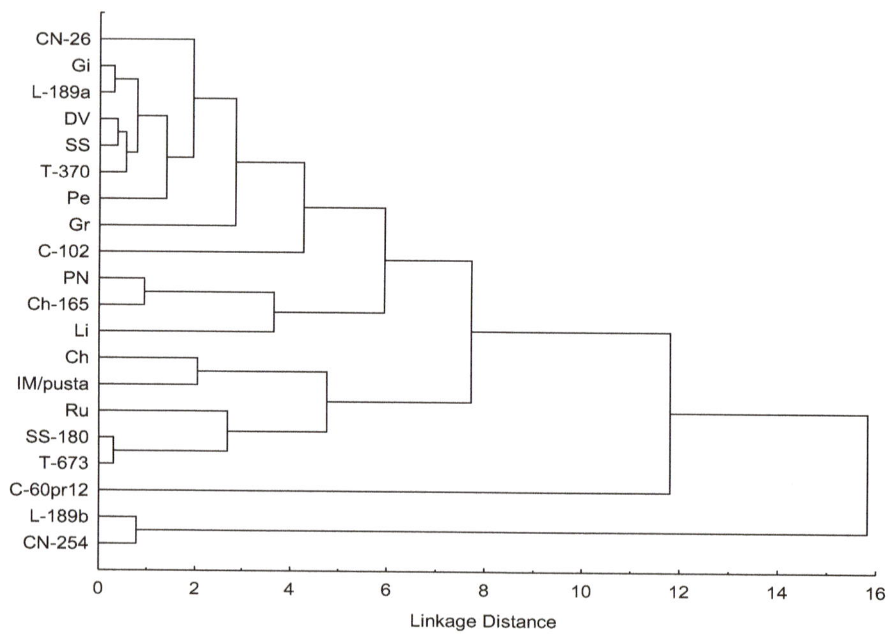

Figure 1. Dendrogram of tomato landraces for quality traits.

The landraces L-189b and CN-254 are grouped in Cluster V, which has recorded higher TAC, TPC, and AsA values, but medium Lyc content. Regarding the contribution of different traits to diversity between clusters, it was found that TPC is the most important by 45%, contributing to divergence; meanwhile, AsA has the lowest contribution, respectively, 13.5% (Table 4). Therefore, studied tomato landraces can be distinguished between each other to a much lesser extent in terms of AsA content.

Table 4. Cluster's mean for five traits of tomato landraces and the contribution of each trait to the total divergence.

Traits	Clusters					Times Ranked First	Contribution to Divergence (%)
	I	II	III	IV	V		
TAC	401.28	341.75	291.77	240.75	534.06	35	17.5
TPC	91.79	85.44	67.76	112.96	116.33	90	45.0
Lyc	12.77	6.67	14.09	6.81	10.46	48	24.0
AsA	16.75	15.78	14.59	13.17	19.69	27	13.5

It seems that this bioactive compound might not be important in landraces recognition and might oscillate depending on the tomato growing technology. Our observation is also supported by other researches regarding the accumulation of AsA in tomatoes; major differences have been found among the individual samples but not between tomato varieties. It seems that light exposure of tomato fruits directly affects the accumulation of phyto-compound [87]. Analyzing the contribution of different traits to inside cluster diversity, it was found that the highest diversity exists between the landraces Ch, IM/pusta, Ru, SS180, and T-673 grouped in Cluster III (D^2 = 3.62), while between landraces L-189b and CN 254 from Cluster V, a high similarity (D^2 = 0.78) was recorded to all nutritional components (Table 5).

According to the inter-cluster distances, it was observed that the landraces L-189b and CN254 from Cluster V differ significantly to the landraces from other clusters except for the nine landraces of Cluster I. Also, the landrace Ch-60pr12, characterized by low values of these quality traits, differs significantly to the landraces from Clusters I and III.

Table 5. Average intra- (bold diagonal) and inter-cluster (off diagonal) D2 values.

Cluster	Landraces	I	II	III	IV	V
I	CN-26; Gi; L-189a; DV; SS; T-370; Pe; Gr; C102	2.30	5.94	7.42	13.22 *	8.71
II	PN; Ch-165; Li		2.76	8.63	6.70	15.42 **
III	Ch; IM/pusta; Ru; SS180; T-673			3.62	12.31 *	26.36 ***
IV	Ch-60pr12				0.00	28.31 ***
V	L-189b; CN254					0.78

$\chi2 = 9.49$ ($p = 0.05$); $\chi2 = 13.28$ ($p = 0.01$); $\chi2 = 18.47$ ($p = 0.001$). The data show correlation index values; * $p < 0.05$, ** $p < 0.01$, *** $p < 0.001$.

The four principal components account for the whole variability among the studied tomato landraces for the analyzed quality traits (Table 6). The first principal component (PC1) has a major contribution of 59.84% to the total variation. Only Lyc (0.144) contributed positively to PC1, while the other traits contributed negatively to this principal component.

Table 6. Eigen vectors and eigen values of the first four principal components for quality traits of tomato landraces.

Traits	PC1	PC2	PC3	PC4
TAC	−0.956	0.171	0.225	−0.081
TPC	−0.737	−0.253	−0.627	0.000
Lyc	0.144	0.959	−0.244	0.000
AsA	−0.957	0.169	0.221	0.082
Eigen value	2.394	1.041	0.552	0.013
Cumulative eigen value	2.394	3.435	3.987	4.000
Proportion variance	59.84	26.02	13.80	0.33
Cumulative variance	59.84	85.86	99.67	100.00

The second principal component (PC2) accounted for 26.02% of the total variation, with positive support of Lyc (0.959), TAC (0.171), and AsA (0.169), while TPC (−0.253) has a negative involvement. The third principal component (PC3) showed 13.80% of the overall variation and was positively associated with TAC (0.225), and AsA, as well as TPC (0.627) and Lyc (−0.244), were negatively associated with PC3. The fourth principal component (PC4) depicted a low proportion of the whole variability (0.33%), indicating the strongest discriminatory power of these two principal components (Figure 2). The biplot reveals a broad dispersion of the landraces and explained 85.86% of the variability. Negative values at PC1 indicate landraces with high TAC, ascorbic acid, and TPC.In this regard were highlighted the landraces from Cluster V, L-189a, C-102, and CN-245 have the highest values for TAC, AsA, and TPC, but a medium Lyc content. Positive values for PC2 belong to landraces having a high Lyc amount, SS-180, T-673, IM/pusta, and Ru, from Cluster III. The negative values of PC2 associated with positive values of PC1 are characteristic of the landraces with low levels of these traits like PN, Ch-165, and Ch-60pr12, grouped in Clusters II and IV, respectively. The landrace Li shows a higher TAC compared to PN and Ch-165 from Cluster II, thus being the main contributor to intra-cluster diversity. According to the dendrogram (Figure 1), it was noticed that there are different landraces groups inside Cluster I. Thus, landrace Gr with a high Lyc content exceeds the average values for TAC and ascorbic acid content, while SS-180 and T-673 with a high content of Lyc are associated with low values for the other qualitative traits.

The nutritional value of tomatoes proven by the content in bioactive compounds with high value is influenced by several factors. Many studies on tomato cultivars highlighted that variation of both abiotic [23,29,60] and technological [10,25,78,88] factors have a decisive effect on increasing the nutritional value of tomatoes.

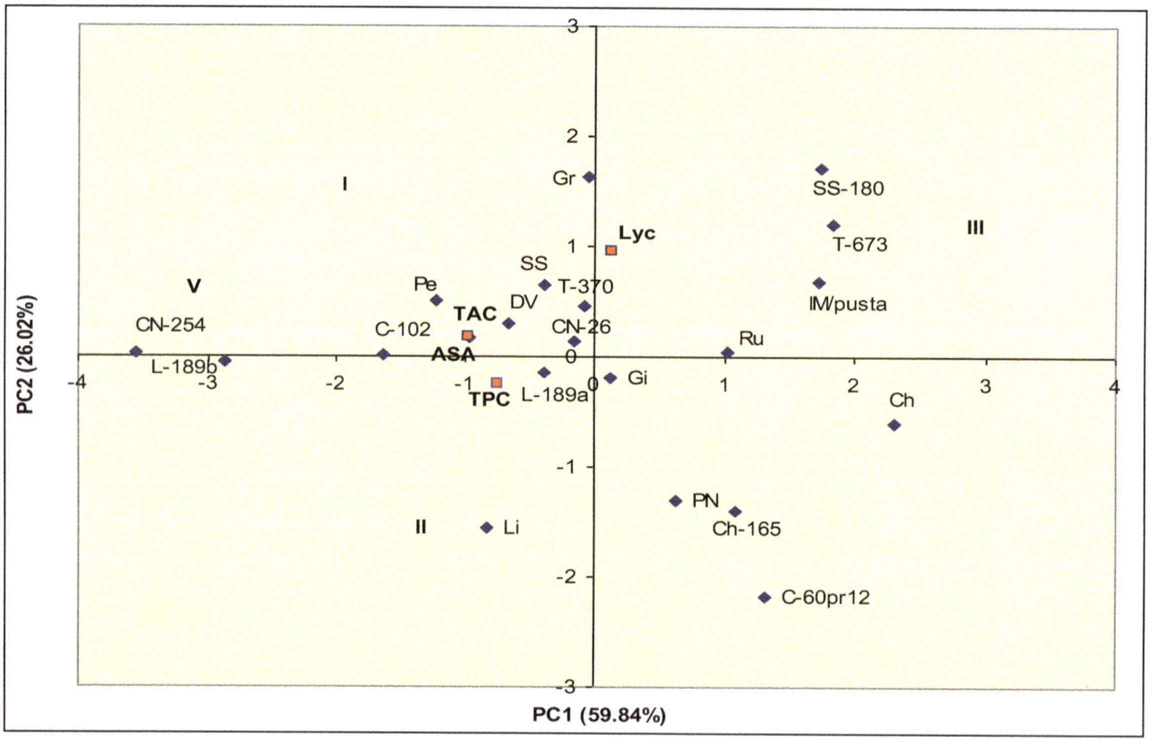

Figure 2. Biplot of the first two principal components for 20 tomato landraces and four quality traits.

4. Conclusions

The results reveal that the 20 tomato landraces with tolerance to salinity have high potential in phyto-compound accumulation with high antioxidant levels. The ratio between these is different. Even if it is widely accepted nowadays that the idea ofphyto-chemicals with high nutraceutical value depend on plants' genetic information, environmental factors may alter the expression of these genes.

The research confirms the hypothesis that tomato landraces with tolerance to soil salinity have a higher ability to accumulate in ripe fruits large amounts of antioxidants such as phenolic compounds and carotenoids. The largest amounts of antioxidants were recorded in that local populations originating from the areas with a high level of soil salinity, whose electrical conductance was over 6.5 dS m^{-1}.

Correlating the results to the fact that these tomato landraces have adapted over time to high levels of soil salinity, we can assume that they have developed genetic, biochemical, and physiological mechanisms to synthesize some of the antioxidant-related co-agents, to allow survival under specific stress conditions. Under the higher soil-salinity conditions, the plants were more obliged to adapt by increasing the synthesis capacities of polyphenols and indirectly intensifying of antioxidant processes. These traits seem to be manifested in the conditions of their cultivation outside specific saline areas. The compositional evaluation highlighted that tomato halo-tolerant landraces are an inexhaustible resource of variability with nutraceutical properties that have been proven. These resources can be exploited in breeding programs or could be cultivated in traditional farms that adopt ecological technology.

The results of the TAC, TPC, Lyc, and AsA determinations can be considered the cumulative response of the genetic fund interaction with all the interactive effects occurring during the maturation phases of the fruit. This approach can provide meaningful information on the modeling of the nutritional quality of tomato fruit and also provides interesting insights into the metabolic capacities of the old local populations that have adapted to the conditions of high soil salinity. However, additional functional research is still needed to link the direct determinations with genetic and metabolic analyses.

Author Contributions: Conceptualization, R.M.S. and R.L.S.; methodology, M.A.P., R.L.S., R.M.S., and L.C.; software, S.I.C.; validation, M.E.C., L.C., and I.R.; formal analysis, L.C.; investigation, R.L.S., S.I.C., R.M.S., D.M., and M.N., resources, R.L.S.; data curation, D.M., M.N., and S.I.C.; writing—original draft preparation, R.M.S. and R.L.S.; writing—review and editing, M.E.C. and L.C.; project administration, R.L.S. and I.R.; funding acquisition, I.R. All authors have read and agreed to the published version of the manuscript.

Acknowledgments: The authors thanks for the financial support for publishing throught the project "Ensuring excellence in RDI activities within USAMVBT", code 35PFE/2018 financed by the Ministry of Research and Innovation (MCI) through Program 1–Development of the national research and development system, Subprogram 1.2–Institutional performance, Institutional development projects–Projects to fund excellence in RDI; Special thanks to all persons from the villages who were very kind to let us visit their gardens and supplied us with the necessary plant material.

References

1. Díez, M.J.; Nuez, F. *Vegetables II Fabaceae, Liliaceae, Solanaceae, and Umbelliferae*; Prohens, J., Nuez, F., Eds.; Springer: New York, NY, USA, 2008; pp. 249–323. [CrossRef]
2. Sardaro, M.L.S.; Marmiroli, M.; Maestri, E.; Marmiroli, N. Genetic characterization of Italian tomato varieties and their traceability in tomatofood products. *Food Sci. Nutr.* **2013**, *1*, 54–62. [CrossRef] [PubMed]
3. Shaye, N.A.; Migdadi, H.; Charbaji, A.; Alsayegh, S.; Daoud, S.; Al-Anazi, W.; Alghamdi, S. Genetic variation among Saudi tomato (*Solanum lycopersicum* L.) landraces studied using SDS-PAGE and SRAP markers. *Saudi J. Biol. Sci.* **2018**, *25*, 1007–1015. [CrossRef] [PubMed]
4. Abete, I.; Perez-Cornago, A.; Navas-Carretero, S.; Bondia-Pons, I.; Zulet, M.A.; Martinez, J.A. A regular lycopene enriched tomato sauce consumption influences antioxidant status of healthy young-subjects: A crossover study. *J. Funct. Foods* **2013**, *5*, 28–35. [CrossRef]
5. Figàs, M.R.; Prohens, J.; Raigón, M.D.; Fita, A.; García-Martínez, M.D.; Casanova, C.; Borràs, D.; Plazas, M.; Andújar, I.; Soler, S. Characterization of composition traits related to organoleptic and functional quality for the differentiation, selection and enhancement of local varieties of tomato from different cultivar groups. *Food Chem.* **2015**, *187*, 517–524. [CrossRef]
6. Scalzo, J.; Politi, A.; Pellegrini, N.; Mezzetti, B.; Battino, M. Plant genotype affects total antioxidant capacity and phenolic contents in fruit. *Nutrition* **2005**, *21*, 207–213. [CrossRef]

7. Figàs, M.R.; Prohens, J.; Raigón, M.D.; Fernández de Córdova, P.; Fita, A.; Soler, S. Characterization of a collection of local varieties of tomato (*Solanum lycopersicum* L.) using conventional descriptors and the high-throughput phenomics tool Tomato Analyzer. *Genet. Resour. Crop Evol.* **2015**, *62*, 189–204. [CrossRef]
8. Schreiner, M. Vegetable crop management strategies to increase the quantity of phytochemicals. *Eur. J. Nutr.* **2005**, *44*, 85–94. [CrossRef]
9. Ceccarelli, S. Landraces: Importance and use in breeding and environmentally friendly agronomic systems. In *Agrobiodiversity Conservation: Securing the Diversity of Crop Wild Relatives and Landraces*; Maxted, N., EhsanDulloo, M., Ford-Lloyd, B.V., Frese, L., Iriondo, J.M., Pinheiro de Carvalho, M.A.A., Eds.; CAB International: Wallingford, UK, 2012; Chapter 15; pp. 103–117. [CrossRef]
10. Zoran, I.S.; Nikolaos, K.; Ljubomir, Š. Tomato Fruit Quality from Organic and Conventional Production. In *Organic Agriculture towards Sustainability*; Pilipavicius, V., Ed.; In Tech: Dubrovnik, Croatia, 2014; Chapter 7; pp. 17–169. [CrossRef]
11. Jacobo-Velázquez, D.A.; Cisneros-Zevallos, L. An alternative use of horticultural crops: Stressed plants as biofactories of bioactive phenolic compounds. *Agriculture* **2012**, *2*, 259–271. [CrossRef]
12. Rozema, J.; Schat, H. Salt tolerance of halophytes, research questions reviewed in the perspective of saline agriculture. *Environ. Exp. Bot.* **2013**, *92*, 83–95. [CrossRef]
13. Mazzucato, A.; Ficcadenti, N.; Caioni, M.; Mosconi, P.; Piccinini, E.; Rami, V.; Sanampudi, R.; Sestili, S.; Ferrari, V. Genetic diversity and distinctiveness in tomato (*Solanum lycopersicum* L.) landraces: The Italian case study of 'A peraAbruzzese'. *Sci. Hortic.* **2010**, *125*, 55–62. [CrossRef]
14. Terzopoulos, P.J.; Bebeli, P.J. Phenotypic diversity in Greek tomato (*Solanum lycopersicum* L.) landraces. *Sci. Hortic.* **2010**, *126*, 138–144. [CrossRef]
15. Sacco, A.; Ruggieri, V.; Parisi, M.; Festa, G.; Rigano, M.M.; Picarella, M.E.; Mazzucato, A.; Barone, A. Exploring a Tomato Landraces Collection for Fruit-Related Traits by the Aid of a High-Throughput Genomic Platform. *PLoS ONE* **2015**, *10*, e0137139. [CrossRef] [PubMed]
16. Beckles, D.M. Factors affecting the postharvest soluble solids and sugar content of tomato (*Solanum lycopersicum* L.) fruit. *Postharvest Biol. Technol.* **2012**, *63*, 129–140. [CrossRef]
17. Oliveira, S.M.; Brandao, T.R.S.; Silva, C.L.M. Influence of drying processes and pretreatments on nutritional and bioactive characteristics of dried vegetables: A review. *Food Eng. Rev.* **2016**, *8*, 134–163. [CrossRef]
18. Causse, M.; Friguet, C.; Coiret, C.; Lépicier, M.; Navez, B.; Lee, M.; Holthuysen, L.; Sinesio, F.; Moneta, E.; Grandillo, S. Consumer preferences for fresh tomato at the European scale: A common segmentation on taste and firmness. *J. Food Sci.* **2010**, *75*, S531–S541. [CrossRef]
19. Siddiqui, M.W.; Ayala-Zavala, J.F.; Dhua, R.S. Genotypic variation intomatoes affecting processing and antioxidant properties. *Crit. Rev. Food Sci.Nutr.* **2015**, *55*, 1819–1835. [CrossRef]
20. Garg, N.; Cheema, D.S. Assessment of fruit quality attributes of tomato hybrids involving ripening mutants under high temperature conditions. *Sci. Hortic.* **2011**, *131*, 29–38. [CrossRef]
21. Fernandez-Ruiz, V.; Olives, A.I.; Camara, M.; Sanchez-Mata, M.C.; Torija, M.E. Mineral and trace elements content in 30 accessions of tomato fruits (*Solanum lycopersicum* L.) and wild relatives (*Solanum pimpinellifolium* L., *Solanum cheesmaniae* L. Riley, and *Solanum habrochaites* S. Knapp & D.M. Spooner). *Biol. Trace Elem. Res.* **2011**, *141*, 329–339. [CrossRef]
22. Antunes, M.D.C.; Rodrigues, D.; Pantazis, V.; Cavaco, A.M.; Siomos, A.; Miguel, G. Nutritional quality changes of fresh-cut tomato during shelf life. *Food Sci. Biotechnol.* **2013**, *22*, 1–8. [CrossRef]
23. Hala, E.M.A.; Ghada, S.M.I. Tomato fruit quality as influenced by salinity and nitric oxide. *Turk. J. Bot.* **2014**, *38*, 122–129. [CrossRef]
24. Barros, L.; Duenas, M.; Pinela, J.; Carvalho, A.M.; Buelga, C.S.; Ferreira, I.C. Characterization and quantification of phenolic compounds in four tomato (*Lycopersicon esculentum* L.) farmers' varieties in northeastern Portugal homegardens. *Plant Foods Hum. Nutr.* **2012**, *67*, 229–234. [CrossRef] [PubMed]
25. Doncean, A.; Sumalan, R.M.; Beinsan, C.; Gergen, I.; Sumalan, R.L. Influence of different types and mixtures of composts on quality of tomatoes fruits (*Solanum lycopersicum* L.). *J. Hyg. Eng. Des.* **2015**, *8*, 40–47.
26. Gitenay, D.; Lyan, B.; Rambeau, M.; Mazur, A.; Rock, E. Comparison of lycopene and tomato effects on biomarkers of oxidative stress in vitamin E deficient rats. *Eur. J. Nutr.* **2007**, *46*, 468–475. [CrossRef] [PubMed]
27. Pinela, J.; Barros, L.; Carvalho, A.M.; Ferreira, I.C. Nutritional composition and antioxidant activity of four tomato (*Lycopersicon esculentum* L.) farmer' varieties in Northeastern Portugal homegardens. *Food Chem. Toxicol.* **2012**, *50*, 829–834. [CrossRef]

28. Khoo, H.E.; Prasad, K.N.; Kong, K.W.; Jiang, Y.; Ismail, A. Carotenoids and their isomers: Color pigments in fruits and vegetables. *Molecules* **2011**, *16*, 1710–1738. [CrossRef]
29. Srivastava, S.; Srivastava, A.K. Lycopene; chemistry, biosynthesis, metabolism and degradation under various abiotic parameters. *J. Food Sci. Technol.* **2015**, *52*, 41–53. [CrossRef]
30. Chiva-Blanch, G.; Visioli, F. Polyphenols and health: Moving beyond antioxidants. *J. Berry Res.* **2012**, *2*, 63–71. [CrossRef]
31. Raiola, A.; Del Giudice, R.; Monti, D.M.; Tenore, G.C.; Barone, A.; Rigano, M.M. Bioactive compound content and cytotoxic effect on human cancer cells of fresh and processed yellow tomatoes. *Molecules* **2015**, *21*, 33–47. [CrossRef]
32. Opara, U.L.; Al-Ani, M.R.; Al-Rahbi, M. Effect of fruit ripening stage on physico-chemical properties, nutritional composition and antioxidant components of tomato (*Lycopersicum esculentum*) cultivars. *Food Bioprocess Technol.* **2012**, *5*, 3236–3243. [CrossRef]
33. Corwin, D.L.; Lesch, S.M. Application of soil electrical conductivity to precision agriculture: Theory principles and guideline. *Agron. J.* **2003**, *95*, 455–471. [CrossRef]
34. Sumalan, R.L.; Popescu, I.; Schmidt, B.; Sumalan, R.M.; Popescu, C.; Gaspar, S. Salt tolerant tomatoes local landraces from Romania–Preserving the genetic resources for future sustainable agriculture. *J. Biotehnol.* **2015**, *208*, S18. [CrossRef]
35. IUSS Working Group WRB. World Reference Base for Soil Resources 2014, update 2015. In *International Soil Classification System for Naming Soils and Creating Legends for Soil Maps*; World Soil Resources Reports No. 106; FAO: Rome, Italy, 2015.
36. UPOV—International Union for the Protection of New Varieties of Plants–Geneva. TOMATO UPOV Code: SOLAN_LYC *Solanum lycopersicum* L. Guidelines for the Conduct of Tests for Distinctness, Uniformity and Stability. 2018. Available online: https://www.upov.int/edocs/tgdocs/en/tg044.pdf (accessed on 27 March 2020).
37. Benzie, I.F.; Strain, J.J. Ferric reducing ability of plasma (FRAP) as a measure of "antioxidant power": The FRAP assay. *Anal. Biochem.* **1996**, *239*, 70–76. [CrossRef] [PubMed]
38. Singleton, V.L.; Orthofer, R.; Lamuela-Raventos, R.M. Analysis of total phenols and other oxidation substrates and antioxidants by means of Folin-Ciocalteu reagent. *Methods Enzymol.* **1999**, *299*, 152–178. [CrossRef]
39. AOAC. Vitamin C (ascorbic acid) in vitamin preparations and juices. In *AOAC: Official Methods of Analysis*, 15th ed.; Helrich, K., Ed.; Association of Official Analytical Chemists, Inc.: Arlington County, VA, USA, 1990; Volume 1, pp. 1058–1059.
40. Sharma, S.K.; Le Maguer, M. Lycopene in tomatoes and tomato pulp fractions. *Ital. J. Food Sci.* **1996**, *8*, 107–113.
41. Toor, R.K. Influence of different types of fertilizers on the major antioxidant components of tomatoes. *J. Food Compos. Anal.* **2006**, *19*, 20–27. [CrossRef]
42. Ciulca, S. *Metodologii de Experimentare în Agricultura şi Biologie*; Editura Agroprint: Timisoara, Romania, 2006; pp. 33–53.
43. Felsenstein, J. *PHYLIP (Phylogeny Inference Package) Version 3.5c*; Distributed by the author; Department of Genetics, University of Washington: Seattle, WA, USA, 1993.
44. Singh, R.K.; Chaudhary, B.D. *Biometrical Methods in Quantitative Genetic Analysis*; Kalyani Publishers: Ludhiana, India, 1979; pp. 191–200.
45. Yan, W.; Hunt, L.A.; Sheng, Q.; Szlavnics, Z. Cultivar evaluation and mega-environment investigation based on the GGE biplot. *Crop Sci.* **2000**, *40*, 597–605. [CrossRef]
46. Yan, W.; Kang, M.S. *GGE Biplot Analysis: A Graphical Tool for Breeders, Geneticists, and Agronomists*, 1st ed.; CRC Press: Boca Raton, FL, USA, 2002; p. 144.
47. Corrado, G.; Piffanelli, P.; Caramante, M.; Coppola, M.; Rao, R. SNP genotyping reveals genetic diversity between cultivated landraces and contemporary varieties of tomato. *BMC Genom.* **2013**, *14*, 835–848. [CrossRef]
48. Cebola-Cornejo, J.; Rosello, S.; Nuez, F. Phenotypic and genetic diversity of Spanish tomato landraces. *Sci. Hortic.* **2013**, *162*, 150–164. [CrossRef]
49. Knapp, S.; Peralta, I.E. The tomato (*Solanum lycopersicum* L. Solanaceae) and its botanical relatives. In *The Tomato Genome (Compendium of Plant Genomes)*; Causse, M., Giovannoni, J., Bouzayen, M., Zouine, M., Eds.; Springer: Berlin/Heidelberg, Germany, 2016; pp. 7–21. [CrossRef]

50. Corrado, G.; Caramante, M.; Piffanelli, P.; Rao, R. Genetic diversity in Italian tomato landraces: Implications for the development of a core collection. *Sci. Hortic.* **2014**, *168*, 138–144. [CrossRef]
51. Tomescu, D.; Sumalan, R.L.; Copolovici, L.; Copolovici, D. The influence of soil salinity on volatile organic compounds emission and photosynthetic parameters of *Solanum lycopersicum* L. varieties. *Open Life Sci.* **2017**, *12*, 135–142. [CrossRef]
52. Salim, M.M.R.; Rashid, M.H.; Hossain, M.M.; Zakaria, M. Morphological characterization of tomato (*Solanum lycopersicum* L.) genotypes. *J. Saudi Soc. Agric. Sci.* **2018**, in press. [CrossRef]
53. Tanksley, S.D. The genetic, developmental, and molecular bases of fruit size and shape variation in tomato. *Plant Cell* **2004**, *16*, S181–S189. [CrossRef]
54. Miller, G.; Suzuki, N.; Ciftci-Yilmaz, S.; Mittler, R. Reactive oxygen species homeostasis and signaling during drought and salinity stresses. *Plant Cell Environ.* **2010**, *33*, 453–467. [CrossRef]
55. Passam, H.C.; Karapanos, I.C.; Bebeli, P.J.; Savvas, D. A review of recent research on tomato nutrition, breeding and post-harvest technology with reference to fruit quality. *Eur. J. Plant Sci. Biotechnol.* **2007**, *1*, 1–21.
56. García-Valverde, V.; Navarro-González, I.; García-Alonso, J.; Periago, M.J. Antioxidant bioactive compounds in selected industrial processing and fresh consumption tomato cultivars. *Food Bioprocess Technol.* **2013**, *6*, 391–402. [CrossRef]
57. Deng, G.-F.; Lin, X.; Xu, X.-R.; Gao, L.-L.; Xie, J.-F.; Li, H.-B. Antioxidant capacities and total phenolic contents of 56 vegetables. *J. Funct. Foods* **2013**, *5*, 260–266. [CrossRef]
58. Ilahy, R.; Hdider, C.; Lenucci, M.S.; Tlili, I.; Dalessandro, G. Antioxidant activity and bioactive compound changes during fruit ripening of high-lycopene tomato cultivars. *J. Food Compos. Anal.* **2011**, *24*, 588–595. [CrossRef]
59. Martinez-Valverde, I.; Periago, M.J.; Provan, G.; Chesson, A. Phenolic compounds, lycopene and antioxidant activity in commercial varieties of tomato (*Lycopersicum esculentum*). *J. Sci. Food Agric.* **2002**, *82*, 323–330. [CrossRef]
60. Drakou, M.; Birmpa, A.; Koutelidakis, A.E.; Komaitis, M.; Panagou, E.Z.; Kapsokefalou, M. Total antioxidant capacity, total phenolic content and iron and zinc dialyzability in selected Greek varieties of table olives, tomatoes and legumes from conventional and organic farming. *Int. J. Food Sci. Nutr.* **2015**, *66*, 197–202. [CrossRef] [PubMed]
61. Hernández, V.; Hellín, P.; Fenoll, J.; Flores, P.J. Increased temperature produces changes in the bioactive composition of tomato, depending on its developmental stage. *J. Agric. Food Chem.* **2015**, *63*, 2378–2382. [CrossRef]
62. Lairon, D. Nutritional quality and safety of organic food. A review. *Agron. Sustain. Dev.* **2010**, *30*, 33–41. [CrossRef]
63. Massaretto, I.L.; Albaladejo, I.; Purgatto, E.; Flores, F.B.; Plasencia, F.; Egea-Fernanadez, J.M.; Bolarin, M.C.; Egea, I. Recovering tomato landraces to simultaneously fruit yield and nutritional quality against salt stress. *Front. Plant Sci.* **2018**, *9*, 1–17. [CrossRef] [PubMed]
64. Alba, R.; Cordonnier-Pratt, M.M.; Pratt, L.H. Fruit-localized phytochromes regulate lycopene accumulation independently of ethylene production in tomato. *Plant Physiol.* **2000**, *123*, 363–370. [CrossRef] [PubMed]
65. Rosati, C.; Aquilani, R.; Dharmapuri, S.; Pallara, P.; Marusic, C.; Tavazza, R.; Bouvier, F.; Camara, B.; Giuliano, G. Metabolic engineering of beta-carotene and lycopene content in tomato fruit. *Plant J.* **2000**, *24*, 413–419. [CrossRef] [PubMed]
66. Giovannetti, M.; Avio, L.; Barale, L.; Ceccarelli, N.; Cristofani, R.; Iezzi, A.; Mignolli, F.; Picciarelli, P.; Pinto, B.; Reali, D.; et al. Nutraceutical value and safety of tomato fruits produced by mycorrhizal plants. *Br. J. Nutr.* **2012**, *107*, 242–251. [CrossRef]
67. Perveen, R.; Rasul-Suleria, H.A.; Anjum, F.M.; Butt, M.S.; Pasha, I.; Sarfraz, A. Tomato (*Solanumlycopersicum*) carotenoids and lycopenes chemistry; metabolism, absorption, nutrition, and allied health claims—A comprehensive review. *Crit. Rev. Food Sci. Nutr.* **2015**, *55*, 919–929. [CrossRef]
68. Kavitha, P.; Shivashankara, K.S.; Rao, V.K.; Sadashiva, A.T.; Ravishankar, K.V.; Sathish, G.J. Genotypic variability for antioxidant and quality parameters among tomato cultivars, hybrids, cherry tomatoes and wild species. *J. Sci. Food Agric.* **2014**, *94*, 993–999. [CrossRef]

69. Borghesi, E.; González-Miret, M.L.; Escudero-Gilete, M.L.; Malorgio, F.; Heredia, F.J.; Meléndez-Martínez, A.J. Effects of Salinity Stress on Carotenoids, Anthocyanins, and Color of Diverse Tomato Genotypes. *J. Agric. Food Chem.* **2011**, *59*, 11676–11682. [CrossRef]
70. Giannakoula, A.E.; Ilias, A.F. The effect of water stress and salinity on growth and physiology of tomato (*Lycopersicon Esculentum* Mill.). *Arch. Biol. Sci.* **2013**, *65*, 611–620. [CrossRef]
71. Ehret, D.L.; Usher, K.; Helmer, T.; Block, G.; Steinke, D.; Frey, B.; Kuang, T.; Diarra, M. Tomato Fruit Antioxidants in Relation to Salinity and Greenhouse Climate. *J. Agric. Food Chem.* **2013**, *61*, 1138–1145. [CrossRef]
72. Van Meulebroek, L.; Hanssens, J.; Steppe, K.; Vanhaecke, L. Metabolic Fingerprinting to Assess the Impact of Salinity on Carotenoid Content in Developing Tomato Fruits. *Int. J. Mol. Sci.* **2016**, *17*, 821. [CrossRef] [PubMed]
73. Kaur, S.; Das, M. Functional foods: An overview. *Food Sci. Biotechnol.* **2011**, *20*, 861–875. [CrossRef]
74. Rao, A.V.; Agarwal, S. Role of antioxidant lycopene in cancer and heart disease. *J. Am. Coll. Nutr.* **2000**, *19*, 563–569. [CrossRef] [PubMed]
75. Ivanov, B.; Asada, K.; Kramer, D.M.; Edwards, G. Characterization of photosynthetic electron transport in bundle sheath cells of maize. I. Ascorbate effectively stimulates cyclic electron flow around PSI. *Planta* **2005**, *220*, 572–581. [CrossRef]
76. Shao, H.B.; Chu, L.Y.; Shao, M.A.; Jaleel, C.A.; Mi, H.M. Higher plant antioxidants and redox signaling under environmental stresses. *Comptes Rendus Biol.* **2008**, *331*, 433–441. [CrossRef]
77. Bodnarescu, F.; Sumalan, R.M.; Ciulca, S.; Copolovici, L.; Sumalan, R.L. The influence of parental lines on lycopene and ß-carotene content in tomato F1 hybrids (*Solanum lycopersicum* L.). *Res. J. Agric. Sci.* **2018**, *50*, 90–97.
78. Rossi, F.; Godani, F.; Bertuzzi, T.; Trevisan, M.; Ferrari, F.; Gatti, S. Health-promoting substances and heavy metal content in tomatoes grown with different farming techniques. *Eur. J. Nutr.* **2008**, *47*, 266–272. [CrossRef]
79. Carnovale, E. La qualitanutrizionaledeiprodottidell'agricolturabiologica. *Italus Hortus* **1999**, *6*, 41–44.
80. Lucarini, M.; Carbonaro, M.; Nicoli, S.; Aguzzi, A.; Cappelloni, M.; Ruggeri, S.; DiLullo, G.; Gambelli, L.; Carnovale, E. Endogenous markers for organic versus conventional plant products. In *Agrifood Quality II: Quality Managements of Fruits and Vegetables, Session VI: Quality Assessment*; Hagg, M., Ahvenainen, R., Evers, A.M., Tiilikkala, K., Eds.; Royal Society of Chemistry: Cambridge, UK, 1999; pp. 306–310.
81. Asami, D.K.; Hong, Y.J.; Barrett, D.M.; Mitchell, A.E. Comparison of the total phenolic and ascorbic acid content of freeze-dried and air-dried marionberry, strawberry and corn grown using conventional, organic and sustainable agricultural practices. *J. Agric. Food Chem.* **2003**, *51*, 1237–1241. [CrossRef]
82. Caris-Veyrat, C.; Amiot, M.J.; Tyssandier, V.; Grasselly, D.; Buret, M.; Mikolajczak, M.; Guilland, J.C.; Bouteloup-Demange, C.; Borel, P. Influence of organic versus conventional agricultural practice on the antioxidant microconstituent content of tomatoes and derived pureed; consequences on antioxidant plasma status in humans. *J. Agric. Food Chem.* **2004**, *52*, 6503–6509. [CrossRef]
83. Auclair, L.; Zee, J.A.; Karam, A.; Rochat, E. Valeur nutritive qualitéorganoleptique et productivite des tomatoes de serre en function de leur mode de production: Biologique-conventionel-hydroponique. *Sci. Aliment.* **1995**, *15*, 511–528.
84. Dumas, Y.; Dadomo, A.; Di Lucca, G.; Grolier, P. Effects of environmental factors and agricultural techniques on antioxidant content of tomatoes. *J. Sci. Food Agric.* **2003**, *83*, 369–382. [CrossRef]
85. Dorais, M.; Ehret, D.L.; Papadopoulos, A.P. Tomato (*Solanum lycopersicum*) health components: From the seed to the consumer. *Phytochem. Rev.* **2008**, *7*, 231–250. [CrossRef]
86. Fanasca, S.; Martino, A.; Heuvelink, E.; Stanghellini, C. Effect of electrical conductivity, fruit pruning, and truss position on quality in greenhouse tomato fruit. *J. Hort. Sci. Biotechnol.* **2007**, *82*, 488–494. [CrossRef]
87. Helyes, L.; Pék, Z.; Lugasi, A. Function of the variety technological traits and growing conditions on fruit components of tomato (*Lycopersicon Lycopersicum* L Karsten). *Acta Aliment.* **2008**, *37*, 427–436. [CrossRef]
88. Hallmann, E.; Lipowski, J.; Marszałek, K.; Rembiałkowska, E. The seasonal variation in bioactive compounds content in juice from organic and non-organic tomatoes. *Plant Foods Hum. Nutr.* **2013**, *68*, 171–176. [CrossRef]

Assessing the Potential of Extra-Early Maturing Landraces for Improving Tolerance to Drought, Heat and both Combined Stresses in Maize

Charles Nelimor [1,2,3], Baffour Badu-Apraku [2,*], Antonia Yarney Tetteh [4], Ana Luísa Garcia-Oliveira [2] and Assanvo Simon-Pierre N'guetta [3]

1 West African Science Service Centre on Climate Change and Adapted Land-use (WASCAL), Graduate Research Program on Climate Change and Biodiversity, Université Felix Houphouët Boigny, Abidjan 31 B.P. 165, Cote d'Ivoire; nelimor.c@edu.wascal.org
2 International Institute of Tropical Agriculture, Ibadan 200001, Nigeria; o.oliveira@cgiar.org
3 Department of Bioscience, Université Felix Houphouët Boigny, Abidjan 22 B.P. 461, Cote d'Ivoire; nguettaewatty@gmail.com
4 Department of Biochemistry and Biotechnology, Kwame Nkrumah University of Science and Technology, University Post Office Box PMP, Kumasi 00233, Ghana; aytetteh@gmail.com
* Correspondence: b.badu-apraku@cgiar.com

Abstract: Maize landrace accessions constitute an invaluable gene pool of unexplored alleles that can be harnessed to mitigate the challenges of the narrowing genetic base, declined genetic gains, and reduced resilience to abiotic stress in modern varieties developed from repeated recycling of few superior breeding lines. The objective of this study was to identify extra-early maize landraces that express tolerance to drought and/or heat stress and maintain high grain yield (GY) with other desirable agronomic/morpho-physiological traits. Field experiments were carried out over two years on 66 extra-early maturing maize landraces and six drought and/or heat-tolerant populations under drought stress (DS), heat stress (HS), combined both stresses (DSHS), and non-stress (NS) conditions as a control. Wide variations were observed across the accessions for measured traits under each stress, demonstrating the existence of substantial natural variation for tolerance to the abiotic stresses in the maize accessions. Performance under DS was predictive of yield potential under DSHS, but tolerance to HS was independent of tolerance to DS and DSHS. The accessions displayed greater tolerance to HS (23% yield loss) relative to DS (49% yield loss) and DSHS (yield loss = 58%). Accessions TZm-1162, TZm-1167, TZm-1472, and TZm-1508 showed particularly good adaptation to the three stresses. These landrace accessions should be further explored to identify the genes underlying their high tolerance and they could be exploited in maize breeding as a resource for broadening the genetic base and increasing the abiotic stress resilience of elite maize varieties.

Keywords: Abiotic stress; climate change; combined drought and heat stress; drought; heat stress; genetic resources; landrace accessions; maize

1. Introduction

Cultivation of maize (*Zea mays* L.) across the different agro-ecological zones of Africa dates to precolonial times after its introduction by Portuguese sailors in the late fifteenth century [1]. Hybridization between different populations, natural and artificial selection, and cultivation in diverse edaphic and climatic conditions, led to a plethora landraces/local varieties adapted to different agro-ecological zones, cultivation practices, and uses [2]. Whereas about 45% of these landraces/local varieties are still being grown across sub-Saharan Africa (SSA), many of these genotypes were collected and preserved in germplasm banks for avoiding the threat of extinction due to adoption of modern varieties or hybrids [3]. Historically, maize breeders identified and composited the most productive of these landraces to generate genetically diverse populations, constituting the foundation of hybrid maize breeding and developing open-pollinated varieties (OPVs) that displayed high yielding with tolerance to biotic and abiotic stresses [4,5].

Nonetheless, nowadays breeders are generally reluctant to use landraces because of the long-term commitment required to identify useful, novel diversity and introgress it into well-adapted elite cultivars while reducing the effects of undesired linked genes [6]. Breeders often resort to their own working collection consisting of elite breeding lines and some germplasm lines as parents in crossing, resulting in recirculating of the same germplasm. As a direct consequence, some newly developed maize varieties, including hybrids exhibit reduced genetic diversity [7], which may limit genetic gains and resilience to abiotic stresses. For example, by assessing the genetic diversity among selected elite CIMMYT Maize Hybrids in East and Southern Africa (ESA), Masuka et al. [7] observed that repeated use of four inbred parents resulted in narrowing of the genetic base of 29 to 58% of the hybrids, which could pose major risk in case of pest or disease outbreaks that are most likely under the prevailing climate changes.

The global climate over the past decade has changed rapidly, and temperatures are predicted to increase [8]. Similarly, precipitation patterns are expected to change significantly, which could adversely affect crop productivity either through drought or waterlogging [8]. Drought stress (DS) and heat (HS) stress are the two most critical and frequently co-occurring abiotic factors on farmers' fields. The tropical location, socio-economic, demographic, policy and farming characteristics of SSA heighten the risk of these stresses [9,10]. Thus, climate change represents a major impediment to African economy and subsistence.

Predicted impacts of climate change on major staple crops including maize in SSA are significant and primarily negative [11,12]. Drought and HS impede maize performance at all stages of plant growth and development. At flowering and early grain-filling stages of maize, DS and HS result in longer anthesis-silking interval (ASI) owing to the delay of silk extrusion, premature lodging, and reduced rates of net photosynthesis arising from oxidative damage to chloroplasts. Damage to chloroplasts is initiated by other detrimental effects of stomatal closure, leaf firing, tassel blasting, and senescence [13–18]. Consequently, reduced pollination efficiency, abnormal development of the embryo sac, as well as zygotic and early kernel abortion occur [14,15]. For these reasons, DS and HS occurring just before and shortly after pollination have the most profound negative effects on maize grain yield (GY) [16]. Yield losses attributable to DS and HS at flowering and the early grain-filling stages of maize were estimated at between 46–90% [17–19] and 45–55%, respectively [19,20]. Worse of all, DS and HS often co-occur under natural field conditions and their combined effects on growth, GY and related traits can either be synergistic, antagonistic or hypo-additive [21–24]. Recently, we screened 36 early maturing maize accessions for tolerance to abiotic stresses and found that on average, GY was reduced by 46%, 55% and 66% under DS, HS and DSHS conditions, respectively [19]. The high GY loss under DSHS reflects hypo-additive effect of DS and HS, which under extreme conditions could force farmers to abandon their farmlands [18,25].

To meet the food and feed needs of the projected population by 2050, it is essential for maize improvement programs in SSA to focus on the development of maize varieties with climate-adaptive traits, specifically, enhanced DS and/or HS tolerance either through conventional or biotechnology means. Genetic resources of maize are crucial to this goal. In particular, the landraces of maize, given their millennia evolutionary history and adaptation to low-input agricultural systems, harbor wide genetic diversity for improved productivity, climate adaptation, nutritional value, and quality attributes [26]. Despite their potential, most genetic diversity studies on maize landraces have focused on the analysis of variation in genetic parameters without much consideration for traits that confer tolerance to abiotic stress conditions. Only few studies have assessed the amount and nature of variation in landrace accessions of maize under individual and combined DS and HS [19,27].

Besides high yield potential, an appropriate cultivar for large areas of West and Central Africa (WCA), especially the Sudan savanna agro-ecological zone where growing cycles are short, must be DS and/or HS-tolerant at an extra-early stage (requiring 80–85 days to maturity) or early stage (90–95 days to maturity) [28]. The success of maize improvement for DS tolerance in tropical maize is well acknowledged [29]. Nonetheless, new DS tolerant source populations are needed to sustain increased genetic gains for food security [29]. Moreover, in comparison to DS, research on HS tolerance in tropical maize is still at the infant stage [29,30], and more studies are needed to complement these breeding efforts. Therefore, performing phenotypic screens to capture the genetic diversity that exist in extra-early landrace accessions of maize resting idle in germplasm banks under DS, HS and DSHS conditions is a promising strategy to uncover new genetic sources, which when introgressed into breeding stocks could contribute to the broadening of the genetic base and development of 'next generation' maize varieties with enhanced quality and other end-user preferred traits.

The present study aimed at identifying DS, HS and DSHS tolerant extra-early maturing maize landraces by exploring agronomic and morpho-physiological traits along with grain yield.

2. Materials and Methods

2.1. Plant Materials

One hundred and ninety-six (196) landrace accessions of maize from Burkina Faso, Ghana and Togo, were randomly sampled from gene banks at the International Institute of Tropical Agriculture (IITA), Nigeria and the Plant Genetics Resources Institute of Ghana in January 2017. Records on collection dates and geographical co-ordinates of the collection sites of these landraces were not available. However, the agro-ecologies/environments in these countries differ in a number of ways in terms of temperature and precipitation, and the adaptation of the accessions to these contrasting agro-ecologies may have been different. We characterized the maize collections under non-stress (NS) conditions for two years, and identified traits related to adaptation to local environmental conditions [31]. A total of 66 landraces were selected for the present study based on variation in standability, plant architecture, agronomic traits (lodging, earlier flowering, anthesis-silking interval, ear husk cover) and grain yield potential under optimal conditions. Six extra-early maturing drought and /or heat-tolerant populations developed by the Maize Improvement Program at the International Institute of Tropical Agriculture (IITA-MIP), Ibadan, Nigeria were included as checks. Details on the 72 maize accessions assessed in this study are available in Supplementary Table S1.

2.2. Agronomic Management

In the present study, each location-year combination was regarded as an environment. At all environments, the experiment was arranged in a 9 × 8 alpha lattice design with two replications. Each plot consisted of one-row plots each 3-m long with plants within and between rows spaced 0.40 m and 0.75 m apart, respectively. Three seeds were sown per hill and thinned to two, two weeks after emergence, resulting in a final plant population density of 66,666 plants per hectare. All stress trials were conducted during the dry season to allow DS, HS and DSHS to be imposed at the period considered most critical for maize growth and development [16].

As control, trials conducted under NS conditions were planted between June and July at the IITA experimental stations at Ikenne (6°53′ N, 3°42′ E, 60 m altitude, 1200 mm annual rainfall) and Mokwa (9°18′ N, 5°185 4′ E, altitude 457 m, 1100 mm annual rainfall), all in Nigeria, in 2017 and 2018. At 2 weeks after planting (WAP), 60 kg ha^{-1} each of nitrogen (N), phosphorus (P) and potassium (K) was applied. Four weeks later, the trials were top-dressed with 30 kg ha^{-1} of urea. In the stress trials, all plots received 60 kg N ha^{-1}, 60 kg P ha^{-1} and 60 kg K ha^{-1} as NPK 15–15–15 at sowing. A second application of N (30 kg N ha^{-1}) was applied as urea two weeks later. Weeds were controlled with herbicides and/or manually.

2.2.1. Drought Stress Trials

Drought experiments were planted at Ikenne during the last week of November in 2017 and 2018, so that flowering and early grain filling stages occurred in January when the incidence of rainfall was negligible. In both years, irrigation was applied using a sprinkler system (flow rate 17 mm day^{-1}) for the first 25 days after planting (DAP). Irrigation was suspended two weeks before anthesis until physiological maturity, so that the maize plants depended on stored water in the soil for growth and development.

2.2.2. Heat and Combined Drought and Heat Stress Trials

HS and DSHS trials were carried out at Kadawa (11°39′ N, 8°27′ E, 500 m altitude), Nigeria, where extreme DS at elevated temperatures occur between February and June each year. During this period, air temperature often varies between 33 and 45 °C [18], allowing for the establishment of trials under HS and DSHS. The HS and DSHS trials were laid in adjacent blocks, 15 m apart and planting was done on the same day in mid-February 2018 and 2019. Temperature and rainfall patterns during the experiment were measured by an automatic weather station installed at Kano, Nigeria. The average temperature at the experimental sites during the study period ranged from 36 °C/17 °C (day/night) to 40 °C/24 °C (day/night) in 2018, and from 31 °C/18 °C (day/night) to 39 °C/26 °C (day/night) in 2019 (Figure 1). There were minor incidences of rainfall after grain filling in May and June in both years (Figure 1). Flowering and grain-filling stages occurred in April, coinciding with extreme DS at elevated temperatures, which resulted in irreversible tissue injuries (leaf firing and tassel blasting) to susceptible genotypes (Supplementary Figure S1). During the reproductive stages in April, the minimum temperature was 36 °C/18 °C (day/night) in 2018, and 36 °C/23 °C (day/night) in 2019. The maximum temperature observed was 41 °C/27 °C (day/night) in 2018, and 41 °C/29 °C (day/night) in 2019, with a mean of 39 °C/23 °C (day/night) and 39 °C/26 °C (day/night) in 2018 and 2019, respectively (Figure 2). Irrigation was applied twice weekly on both HS and DSHS blocks using a furrow irrigation system. Irrigation was suspended on the DSHS block at 32 DAP but was resumed after grain-filling and applied once in order to avoid complete loss of trials. The HS block on the hand continued to receive irrigation every four days until physiological maturity.

2.3. Trait Measurement

At flowering, days to anthesis (AD) and silking (SD) were recorded when 50% of the plants in a plot had shed pollen and extruded silks, respectively. ASI was computed as SD minus AD. Plants with leaf firing (LF) and tassel blasting (TB) were counted on HS and DSHS plots and converted to percentages. At physiological maturity, plant and ear heights (PLHT and EHT) were measured on ten representative plants per plot as the length from the base of the plant to the height of the first tassel branch and the node bearing the upper ear, respectively. Plant aspect (PASP) was visually scored based on the general appeal of plants in a plot (standability, vigour, plant, and ear height, uniformity of plants, ear placement and size, as well as disease damage and lodging) using a scale of 1 to 9, where 1 = excellent overall phenotypic appeal; 2 = very good overall phenotypic appeal; 3 = good overall phenotypic appeal; 4 = satisfactory overall phenotypic appeal; 5 = acceptable phenotypic appeal; 6 = undesirable phenotypic appeal, 7 = poor overall phenotypic appeal, 8 = very poor phenotypic appeal and 9 = completely undesirable phenotypic appeal. Similarly, husk cover (HC) was rated on a scale of 1 to 9, where 1 = husks tightly arranged and extended beyond the ear tip and 9 = exposed ears. Stay green characteristic (SG) was recorded on all stressed plots at 70 DAP using a scale of1 to 9, where 1 = 10% dead leaf area; 2 = 20% dead leaf area; 3 = 30% dead leaf area, 4 = 40% dead leaf area; 5 = 50% dead leaf area; 6 = 60% dead leaf area; 7 = 70% dead leaf area; 8 = 80% dead leaf area and 9 represented 90–100% dead leaf area. Few days before harvest, root and stalk lodging (RL and SL) were recorded as the percentage of plants leaning more than 30° from the vertical, and percentage plants broken at or below the highest ear node, respectively then all plants were hand harvested. At harvest, ear aspect (EASP) was rated based on the general appeal of the ears without the husks (ear size and number; uniformity of size, colour, and texture; extent of grain filling, insect and disease damage) using a scale of 1 to 9, where 1 = excellent (clean, uniform, large, and well-filled and disease-free ears); 2 = very good ears with no disease damage and fully filled grains; 3 = good ears with no disease damage and fully filled grains; 4 = no disease, fully filled grains, one or two irregularity in cob size poor; 5 = mild disease damage and fully filled grains, one or two irregularity in cob size, 6 = severe disease damage and fully filled grains, smaller cobs, non-uniform cob size; 7 = severe disease, scanty grain filling, few ears, non-uniformity of cobs; 8 = severe disease damage, very scanty grain filling, few ears, and 9 = only one or no ears produced (Supplementary Figure S2). Ears with rot (EAROT) were counted on plot basis and converted to percentage. Grain yield (kg ha^{-1}) was estimated based on 80% shelling percentage for NS plots. In contrast, total shelled grain weight was obtained for stress plots and GY was adjusted to 15% moisture content.

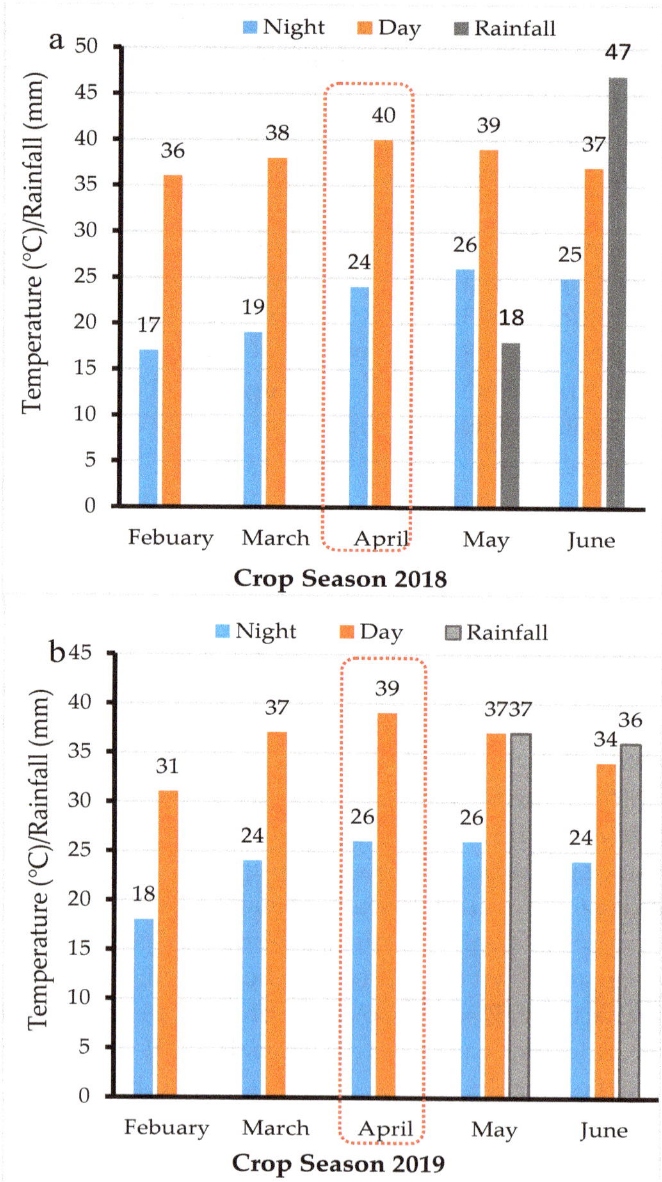

Figure 1. Monthly average temperature and rainfall during the experimental period at Kadawa, Nigeria. Note: Dotted rectangles represent flowering and early grain filling stages. (**a**) Trial period in 2018 (**b**) Trial period in 2019.

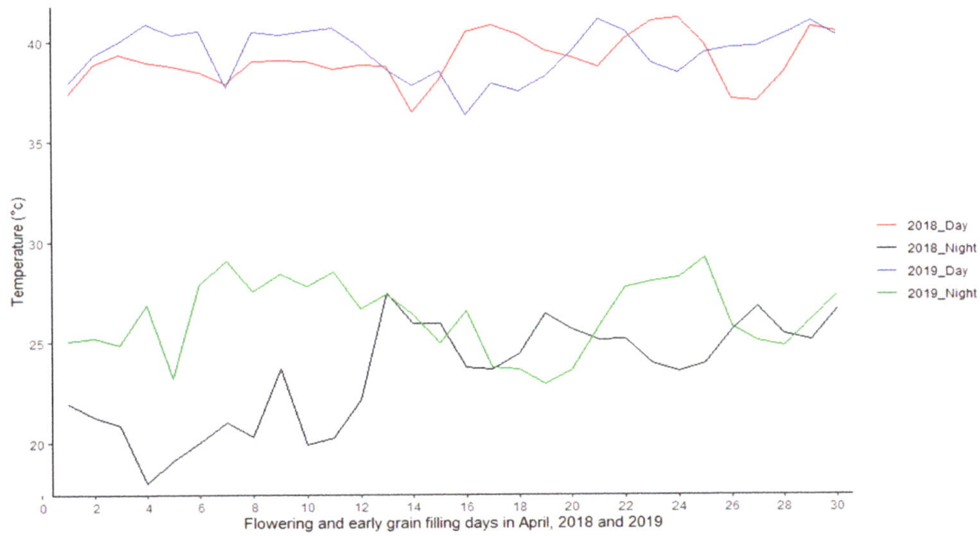

Figure 2. Average temperature (night and day) at Kadawa during flowering and early grain filling in 2018 and 2019.

2.4. Statistical Analysis

Variance components were estimated for each trial/environment by restricted maximum likelihood (Reml) using PROC Mixed of SAS version 9.4 [32]. The corresponding linear mixed model for the response variable was represented by:

$$Y_{ikl} = \mu + R_k + IB(R)_{lk} + G_i + e_{ikl} \quad (1)$$

where Y is the trait of interest, μ is the overall mean, R_k is the effect of the kth replicate, $IB(R)_{lk}$ is the effect of the incomplete block within the kth replicate, G_i is the effect of the ith genotype, and e_{ikl} is the experimental error. All effects were considered random except replicates. Broad-sense heritability (H) of grain yield was estimated for each environment as:

$$H = \frac{\sigma_g^2}{\left(\sigma_g^2 + \frac{\sigma_e^2}{r}\right)} \quad (2)$$

where σ_g^2 and σ_e^2 represented genotypic and residual variance, respectively and r denote the number of replicates. Trials with H of GY less than 0.30 were highly influenced by the environment and were removed from the analyses.

Combined analysis of variance (ANOVA) across environments were performed for each treatment (NS, HS, DS, and DSHS) with PROC GLM in SAS 9.4 using a RANDOM statement with the TEST option [32]. In the ANOVA, test environments, replicates, blocks and their interactions were considered as random factors while accessions were regarded as fixed effects.

Repeatability of the traits [33] under each treatment were computed on accession-mean basis using the following formula:

$$R = \frac{\sigma_g^2}{\sigma_g^2 + \frac{\sigma_{ge}^2}{e} + \frac{\sigma_e}{re}} \quad (3)$$

where σ_g^2 is the genotypic variance, σ_{ge}^2 is the variance of genotype × environment, σ_e is the residual variance; e is the number of environments, and r is the number of replicates per environment.

Phenotypic correlations among grain yield of the treatments were computed to determine the mechanism of abiotic stress tolerance using R library Performance Analytics [34]. Genetic correlations between phenology (AD and SD) and other secondary traits were estimated following the procedure of Cooper et al. [35] using Multi Environment Trial Analysis (META) [36].

Sequential path coefficient analysis [37] was used to identify traits with significant contributions to GY of the maize accessions under the evaluation conditions. A stepwise regression analysis was used to categorize the predictor traits into first, second and third order based on their individual contributions to total variation in GY with minimized multicollinearity [38]. The first step involved the regression of all the traits on GY and those with significant contributions to GY at $p < 0.05$ were identified as first order traits. Subsequently, traits that were not identified as first order traits were regressed on each of the first order traits to identify those with significant contributions to GY through each of the first order traits and were categorized as second-order traits. The procedure was repeated to identify traits in subsequent orders. The path coefficients were represented by the standardized b-values obtained from the regression analysis [38–40]. The sequential path coefficient including the stepwise regression analyses was done using the Statistical Package for Social Sciences, SPSS version 17.0 [41]

A base index (BI) that integrated superior grain yield, EPP, ASI, PASP, EASP, and SG was used to select the best and worst performing genotypes under each treatment [40]. Each trait was first standardized with standard deviation of 1 and a mean of zero to minimize the effect of the different scales prior to integrating into the BI. The BI was computed using the equation:

$$BI = [(2 \times YLD_S) + EPP - ASI - PASP - EASP - SG] \quad (4)$$

where YLD_S is GY under stress, $PASP$ is plant aspect, $EASP$ is ear aspect, EPP is ears per plant, ASI is anthesis-silking interval and SG is the stay-green characteristic. A positive BI value indicated tolerance to the applied stress while negative values indicated susceptibility [40].

Principal components analysis (PCA) was performed and the results were graphically visualized in a biplot that displayed different genetic groups (highlighted with different colours) and the association between accessions and measured traits. Grain yield was regressed on TB scores, plotted on a graph to show the effect of TB on GY of the maize accessions under HS and DSHS. Standardized data of the traits included in the BI selection were subjected to cluster analysis, in which phylogenetic constellation plots were generated, depicting the genetic relationships among the accessions under each stress treatment. The PCA, regression and cluster analysis were performed using JMP pro 14.10 [32].

3. Results

3.1. Analysis of Variance and Broad-Sense Heritability

Detailed values for variance components are presented in Tables 1 and 2. There was a large variation among the accessions for GY and other traits under NS, DS, HS and DSHS, which facilitated the grouping of abiotic stress-tolerant genotypes from their susceptible counterparts, as well as identification of traits that maximize variance in tolerance to the imposed stresses. Broad-sense heritability estimates of GY of individual trials ranged from 0.88 to 0.96 under NS, 0.51 to 0.61 under DS, 0.48 to 0.55 under HS, and 0.51 to 0.68 under DSHS (Supplementary Table S1). Hence, no trial was eliminated from the combined analysis. Within-treatments, repeatability measured across environmental conditions ranged from 0.41 for HC to 0.94 for GY under NS, 0.30 for EPP to 0.95 for EHT under DS, 0.21 for TB to 0.94 for AD under HS, and from 0.25 for EPP to 0.95 for AD under DSHS (Tables 1 and 2).

Table 1. Mean squares, and repeatability of grain yield and other traits of 72 extra-early maturing maize accessions evaluated under non-stress and drought stress conditions between 2017 and 2019 at Kadawa, Nigeria.

SV	df	GY	AD	SD	ASI	PLHT	EHT	HC	EPP	PASP	EASP	SG	RL	SL	EAROT
Non-Stress Conditions (NS)															
Env	2	5,746,960.3**	1.3	108.8*	133.4**	83913.4**	32981.7**	86.7**	0.5**	24.5**	3.3*	-	-	-	-
Rep (Env)	3	169,368.8	9.5*	0.8	5.4	1485.9*	1260.4**	7.6**	0	1.4*	0	-	-	-	-
Block (Env × Rep)	42	530,887.1*	3.4	8	5.2	690.1**	139	1.3**	0	0.9*	1.2*	-	-	-	-
Genotype	71	2,466,309.9**	47.7**	75.6**	13.2**	686.7**	344.2**	0.8*	0.1**	1.7**	3.8**	-	-	-	-
Env × Genotype	142	84,207.7	5.5**	13.9*	6.8*	180.3	85.9	0.5	0	0.4	0.4	-	-	-	-
Error	171	204,531.2	2.5	7.9	5.1	214.4	95	0.4	0	0.4	0.7	-	-	-	-
Repeatability		0.94	0.89	0.81	0.5	0.75	0.77	0.41	0.7	0.82	0.86	-	-	-	-
Grand Mean		2992.38	47	50	3	152.72	68.81	4	0.92	5	5	-	-	-	-
Drought Stress (DS)															
Env	1	3,976,484	75.4**	130.7**	19.4**	10811.4**	852.3**	54.2**	1.6**	24.3**	34.9**	9.4**	0.9**	0.5**	-
Rep (Env)	2	2,258,791	4.9*	6.5*	0.5	280.5	206.4*	0.9	0	0.6	0.7	4.8**	0.2**	0	-
Block (Env × Rep)	28	1,039,093	3.2*	3.4*	0.6	402.9**	177.5**	0.7	0	0.5	0.5	0.94*	0	0	-
Genotype	71	6,056,025.5 **	72.4**	76.4**	2.6**	1531.6**	946.2**	3.0**	0.1**	4.2**	4.2**	1.6**	0*	0**	-
Env × Genotype	71	1,840,190	3.9**	5.2**	1.5**	105.2	43	1.0*	0.0**	0.6*	0.8**	0.37	0	0	-
Error	114	1,638,119	1.6	2	0.8	109.5	60	0.7	0	0.4	0.4	0.39	0	0	-
Repeatability		0.69	0.94	0.93	0.45	0.94	0.95	0.68	0.3	0.86	0.82	0.79	0.33	0.68	-
Grand Mean		1523.92	46	50	4	116.73	54.95	4	0.68	5	5	4	0.1	0.22	-
Reduction (%)		0.49	1	-	1	0.24	0.20	-	0.26	-	-	-	-	-	-

*, ** Significance at 0.01 and 0.001, respectively; GY: Grain yield (kg/ha); AD: Days to 50% anthesis; SD: Days to 50% silking; ASI: Anthesis-silking interval; PLHT: Plant height (cm); EHT: Ear height (cm); HC: Husk cover (scale:1-9); EPP: Ears per plant; PASP: Plant aspect (scale:1-9); EASP: Ear aspect (scale:1-9); SG: Stay green characteristic (scale:1-9); RL: Root lodging (%); SL: Stalk lodging (%); EAROT: Ear rot.

Table 2. Mean squares and repeatability of grain yield and other traits of 72 extra-early maturing maize accessions evaluated under heat stress and combined drought and heat stress conditions between 2017 and 2019 at Kadawa, Nigeria.

SV	df	GY	AD	SD	ASI	PLHT	EHT	HC	EPP	PASP	EASP	SG	RL	SL	TB	LF	EAROT
Heat Stress (HS)																	
Env	1	16,037,142.5**	9.8	20.1*	1.8	40110.6**	23562.6**	103.9**	1.8**	0.2	8.7*	48.3**	0.2**	0.1*	0.2**	0.0*	-
Rep (Env)	2	8,533,488.5**	11.9	11.6	1	287.9	652.8*	1.7	0.6**	0.8	2.9*	3.5*	0.1**	0	0.0*	0	-
Block (Env × Rep)	28	1,970,909.7*	9.1*	9.2*	1.1	1363.7**	501.8**	1	0.1	0.6	1.3*	0.7	0	0	0	0	-
Genotype	71	2,330,918.4**	59.9**	51. 0**	3.4**	1181.9**	716.6**	1.5*	0.1**	2.0**	2.0**	1.4**	0	0.0**	0.0*	0.0*	-
Env × Genotype	71	753,004.1	4.5	3.2	1.3	286.3	168.4	0.9	0	0.3	0.6	0.6	0.0*	0	0.0*	0	-
Error	114	841,366	4.3	4.6	1.5	336.8	135.6	0.7	0	0.4	0.7	0.5	0	0	0	0	-
Repeatability		0.67	0.94	0.93	0.63	0.76	0.79	0.34	0.63	0.85	0.69	0.56	0.37	0.64	0.21	0.38	-
Mean		2301.39	57	60	3	144.12	59.65	4	0.82	5	5	3	7.33	12.71	3.91	4.19	-
Reduction (%)		0.23	−10	−10	-	0.06	0.13	-	0.11	-	-	-	-	-	-	-	
Combined Drought and Heat Stress (DSHS)																	
Env	1	613,509,9.7*	1.9	0.1	2.5	155177.3**	58356.3**	38.9**	2.3*	0.1	5.7*	23.0**	0.1*	0	0	0.1*	-
Rep (Env)	2	843,164.7	14.4*	17.2*	0.3	522.3*	21.3	1.4	0.1	0.5	2.9*	0.4	0	0	0	0.1*	-
Block (Env × Rep)	28	1,522,778.5*	7.7*	13.0**	2.8*	913.3**	263.3**	2.0**	0.3	0.7	2.8**	0.9	0	0.0*	0	0.0*	-
Genotype	71	2,575,212.4**	47.3**	46.6**	4.1**	1010.2**	719.7**	1.6**	0.2*	1.8**	3.5**	2.1**	0.0*	0.0**	0.0**	0.0*	-
Env × Genotype	71	4,923,88.2	2.3	3.1	0.7	254.1*	113.4**	0.6	0.2	0.3	0.8	0.6	0	0	0	0	-
Error	114	595,224.5	3.2	4.7	1.7	161.7	46.8	0.7	0.2	0.5	0.8	0.6	0	0	0	0	-
Repeatability		0.79	0.95	0.93	0.66	0.77	0.86	0.62	0.25	0.78	0.77	0.69	0.57	0.62	0.64	0.55	-
Grand Mean		1258.51	56	58	3	142.68	58.59	4	0.68	5	5	4	9.75	15.71	9.94	11.97	-
Reduction (%)		0.58	−9	−8	-	0.07	0.15	-	0.26	-	-	-	-	-	-	-	-

*, ** Significance at 0.01 and 0.001, respectively; GY: Grain yield (kg/ha); AD: Days to 50% anthesis; SD: Days to 50% silking; ASI: Anthesis-silking interval; PLHT: Plant height (cm); EHT: Ear height (cm); HC: Husk cover (scale:1-9); EPP: Ears per plant; PASP: Plant aspect (scale:1-9); EASP: Ear aspect (scale:1-9); SG: Stay green characteristic (scale:1-9); RL: Root lodging (%); SL: Stalk lodging (%); TB: Tassel blast (scale:1-9); LF: Leaf firing (scale:1-9); EAROT: Ear rot.

Mean GY under NS was ~2992 kg/ha, ~2301 kg/ha under HS, ~1524 kg/ha under DS, and ~1259 kg/ha under DSHS (Tables 1 and 2). Compared with the NS environment, DS reduced GY by between 2.5% to ~85% (mean = 49%), 1.1% to 68.4% with an average of 23% under HS and, between 10% to 94% with a mean of 58% under DSHS. Anthesis was, on average, reached after 47 and 46 days under NS and DS conditions, respectively. Under HS, anthesis was delayed by 10 days and by 9 days under DSHS. Similarly, silking was on the average reached after 50 days under both NS and DS conditions but delayed by 10 days under HS and 8 days under DSHS. Despite the stress, ASI remained largely unchanged under all the treatments (averaging 3 days under NS, HS and DSHS, and 4 days under DS). Plant height was reduced by ~24% under DS and by 6% and 7% under HS and DSHS, respectively. Reduction in EPP was highest under DS (26%) and DSHS (26%) and lowest under HS (11%). Traits such as HC, and PASP and EASP were unaffected by the stresses as indicated by the average score of 4 and 5 under all the contrasting conditions. Leaf senescence, RL, SL, TB, and LF were on average, higher under DSHS relative to HS conditions.

3.2. *Genetic Correlations and Sequential Regression Analysis*

Phenotypic correlations between grain yields under NS, DS, HS and DSHS are presented in Table 3. Grain yield observed under NS was strongly and positively correlated with GY under HS ($r = 0.75$; $p < 0.0001$) while the correlation between GY under NS and those under DS, and DSHS were moderate and positive. Similarly, the correlation between GY under DS and DSHS was positive and moderate ($r = 0.60$; $p < 0.0001$). However, weak and positive phenotypic correlations were observed between GY under DS and HS ($r = 0.48$; $p < 0.001$) as well as HS and DSHS ($r = 0.37$; $p < 0.001$). Strong and positive correlations ($r \geq 0.73$; $p < 0.0001$) were observed for both AD and SD under NS and those under DS, and DSHS as well as between DS and DSHS. No significant correlation was observed between EPP under the different treatments except between NS and DS ($r = 0.37$; $p < 0.001$). Weak to moderate correlations were recorded for PASP under the different treatments except those between HS and DS ($r = 0.19$) and, HS and DSHS ($r = 0.05$). A similar trend was observed for EASP under the treatments. Negative genetic correlations were observed between phenology (AD and SD) and GY, HC, PASP, EASP, SG, LF, and TB under the applied stresses whereas, genetic correlations between flowering traits (AD and SD) and ASI, PLHT and EHT were positive (Table 4).

Table 3. Coefficient of phenotypic correlations between treatments for traits of 72 extra-early maturing maize accessions evaluated between 2017 and 2019 in Nigeria.

Trait	NS vs. DS	NS vs. HS	NS vs. DSHS	HS vs. DS	DS vs. DSHS	HS vs. DSHS
Grain yield	0.66***	0.75***	0.62***	0.48**	0.60***	0.37***
Anthesis days	0.90***	−0.05	0.87***	−0.04	0.82***	−0.13
Silking	0.80***	−0.07	0.84***	−0.08	0.73***	−0.08
Anthesis-Silking Interval	0.23*	−0.09	0.42***	−0.13	0.13	−0.28*
Plant height	0.48***	0.30*	0.64***	0.18	0.29*	0.23*
Ear height	0.60***	0.21	0.77***	0.24*	0.14	0.55***
Ears per plant	0.37**	0.11	0.06	0.03	0.22	−0.07
Ear aspect	0.54***	0.32**	0.33**	0.19	0.52***	0.05
Plant aspect	0.46***	0.14	0.63***	0.18	0.46***	0.04
Stay green	-	-	-	0.08	0.38***	0.03
Leaf firing	-	-	-	-	-	−0.20
Tassel blast	-	-	-	-	-	−0.10

*, **, *** Significant at 0.05, 0.01 and 0.001 probability levels, respectively.

Table 4. Genetic correlation between phenology (days to anthesis and silking) and other secondary traits of 72 extra-early maize accessions evaluated under contrasting environments between 2017 and 2019 in Nigeria.

Trait	NS	DS	HS	DSHS
	Days to anthesis			
Grain yield	0.34	−0.71	−0.42	−0.23
Silking days	0.95	0.99	0.98	0.97
Anthesis-silking interval	0.42	0.12	0.51	0.20
Husk cover	−0.48	−0.71	−0.92	−0.63
Plant height	0.58	0.80	0.91	0.79
Ear height	0.65	0.87	0.98	0.87
Plant aspect	−0.31	−0.76	−0.87	−0.62
Ear aspect	−0.32	−0.83	−0.41	−0.12
Ears per plant	0.06	−0.89	0.19	−0.34
Stay green	-	−0.49	−0.88	−0.49
Leaf firing	-	-	−0.79	−0.26
Tassel blast	-	-	−0.98	−0.12
	Days to Silking			
Grain yield	0.14	-0.65	-0.31	-0.14
Anthesis-silking interval	0.68	0.25	0.34	0.01
Husk cover	−0.24	−0.64	−0.82	−0.60
Plant height	0.35	0.80	0.87	0.81
Ear height	0.40	0.86	0.95	0.85
Plant aspect	−0.06	−0.71	−0.81	−0.59
Ear aspect	−0.09	−0.78	−0.26	−0.04
Ears per plant	−0.33	−0.77	0.08	−0.53
Stay green	−0.28	-	−0.76	−0.45
Leaf firing	-	-	−0.79	−0.20
Tassel blast	-	-	−0.98	-

NS: Non stress; DS: Drought stress; HS: Heat stress; DSHS: Combined drought and heat stress.

Under DS, stepwise multiple regression analysis identified EASP, PASP, and SL as the first order traits with significant contributions to GY (explaining 96 % of the total variation in GY) of the maize accessions (Figure 3). Of these first order traits, PASP had the highest negative direct effect on GY (−0.87) while SL contributed the least (−0.42) to GY. Traits, which contributed indirectly to GY through one or two of the first order traits, included EPP, EHT, RL, HC, SG, and AD. These traits were thus, classified into the second order. The traits classified into the third order were ASI, PLHT, SD, and EAROT, each contributing to variation in GY of the maize accessions through one or more of the second order traits.

Across the HS treatments, EASP, EPP, SG, SL, and LF were identified as first order traits responsible for 83% of the total variation in GY (Figure 4). Of these first order traits, EASP had the highest negative direct effect on GY (−0.54), while LF recorded the least direct negative effect on GY (−0.12). Only EPP had direct positive contribution to GY (0.25). Six traits, namely PASP, SD, EAROT, ASI, HC and TB contributed indirectly to GY through one or more of the first order traits and were thus categorized into the second order. Among the second order traits, ASI and TB contributed indirectly to GY through SG, and LF and SL, respectively. Plant aspect had the highest negative (−0.43) indirect contribution to GY through EPP, while the highest positive indirect contribution to GY was also observed for PASP through EASP (0.85). Four traits (EHT, AD, RL and PLHT) contributed indirectly to GY through one or two of the second order traits and were categorized into third order traits.

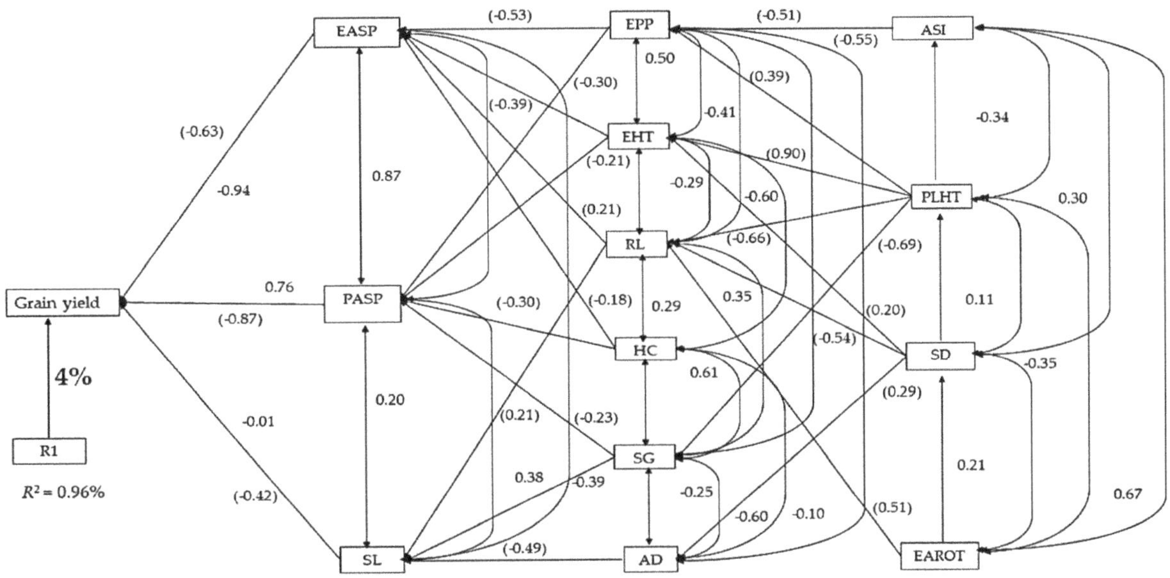

Figure 3. Path analysis diagram depicting the causal relationship of measured traits of the 72 maize accessions under drought-stressed conditions. Note: Value written in bold is the error effect; the direct path coefficients are values in parenthesis and other values are correlation coefficients. R1 is error effects, R^2 = coefficient of determination. AD: Days to 50% anthesis; SD: Days to 50% silking; ASI: Anthesis-silking interval; PLHT: Plant height; EHT: Ear height; HC: Husk cover; EPP: Ears per plant; PASP: Plant aspect; EASP: Ear aspect; SG: Stay green characteristic; RL: Root lodging; SL: Stalk lodging; TB: Tassel blast; LF: Leaf firing; EAROT: Ear rot.

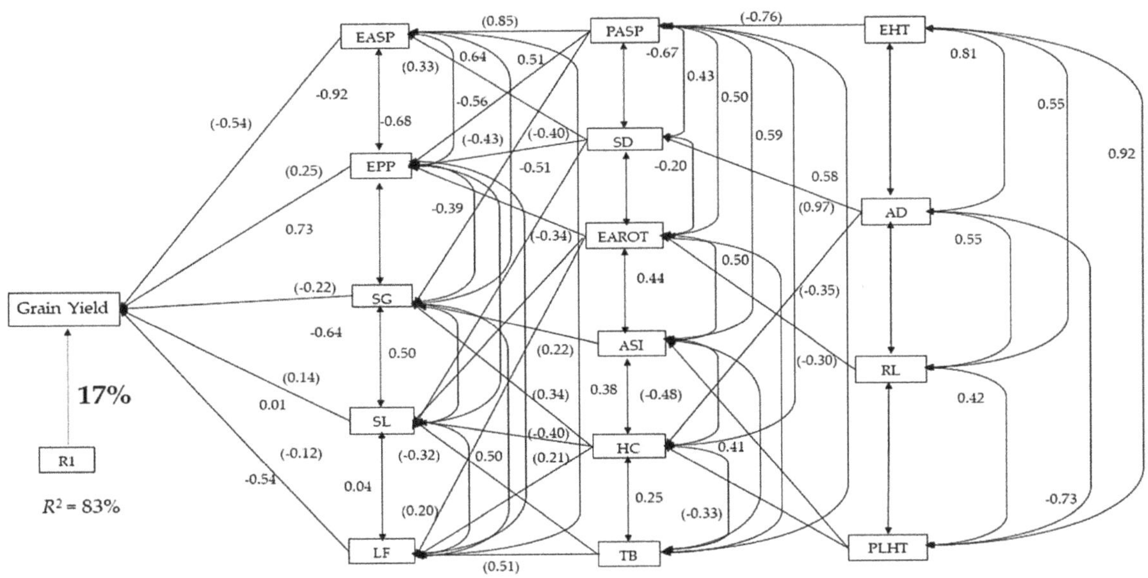

Figure 4. Path analysis diagram depicting the causal relationship of measured traits of the 72 maize accessions under heat stress conditions. Note: Values in bold are the error effect; the direct path coefficients are values in parenthesis and other values are correlation coefficients. R1 is error effects, R^2 = coefficient of determination. AD: Days to 50% anthesis; SD: Days to 50% silking; ASI: Anthesis-silking interval; PLHT: Plant height; EHT: Ear height; HC: Husk cover; EPP: Ears per plant; PASP: Plant aspect; EASP: Ear aspect; SG: Stay green characteristic; RL: Root lodging; SL: Stalk lodging; TB: Tassel blast; LF: Leaf firing; EAROT: Ear rot.

Under DSHS, traits classified in the first order (EASP, SG, PASP and ASI) explained 88% of the total variation in GY of the maize accessions (Figure 5). Each of these traits had direct negative contribution to GY with EASP being the highest contributor (−0.66). Seven traits, namely EPP, LF, AD, RL, HC, EHT and EAROT were categorized into the second order traits, each contributing to GY through one or two of the first order traits. Of the second order traits, EPP was found to have the highest negative indirect effect on GY through EASP (−0.69). Traits contributing to variation in GY through two or more of the second order traits included TB, SL, SD, and PLHT. Of these, TB contributed negatively to variation in GY through EPP (−0.35), AD (−0.26) while PLHT contributed positively to GY through EHT (0.62).

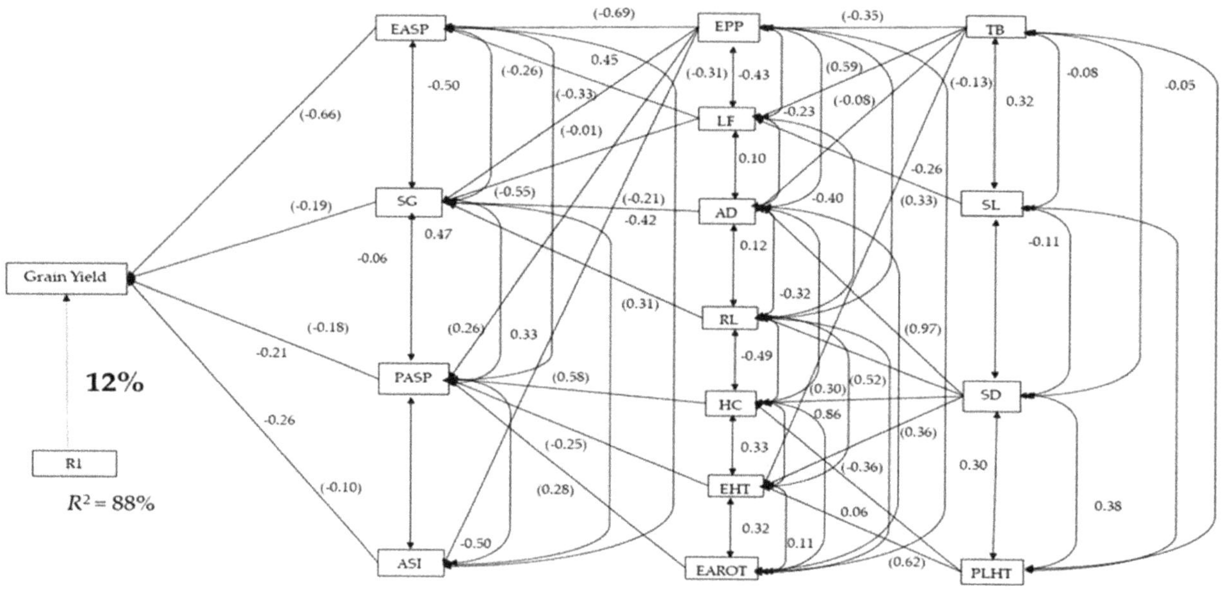

Figure 5. Path analysis diagram depicting the causal relationship of measured traits of the 72 maize accessions under combined drought and heat stressed conditions. Note: Values in bold is the error effects; the direct path coefficients are values in parenthesis while other values are correlation coefficients. R1 is error effects, and R^2 = coefficient of determination. AD: Days to 50% anthesis; SD: Days to 50% silking; ASI: Anthesis-silking interval; PLHT: Plant height; EHT: Ear height; HC: Husk cover; EPP: Ears per plant; PASP: Plant aspect; EASP: Ear aspect; SG: Stay green characteristic; RL: Root lodging; SL: Stalk lodging; TB: Tassel blast; LF: Leaf firing; EAROT: Ear rot.

3.3. Performance of Accessions under the Contrasting Environment

The distribution of the accessions in terms of GY performance under the abiotic stresses is presented in Figure 6. Under DS, DSHS and HS, about 38%, 29% and 7% of the accessions yielded below 1000 kg/ha, respectively while about 47%, 42% and 31% of the accessions produced yields between 1001 to 2000 kg/ha under DSHS, DS, and HS, respectively. Whereas under HS, 42% of the maize accessions yielded between 2001–3000 kg/ha, only a small proportion of them produced yields between 2001–3000 kg/ha (9% under DS and 18% under DSHS). None of the landrace accessions yielded >4000kg/ha under any of the stresses.

Given that selection solely for GY potential under abiotic stress condition is considered inefficient for accelerating genetic gain [42], a base index that integrated GY with other important secondary traits (ASI, EPP, PASP, EASP, and SG) was used as criterion to select accessions tolerant to each of the stresses (accessions with positive BI values) as well as across the contrasting environments. Summary of the top accessions (best check, and 15 landraces) and worse five landraces identified by the base index under each research condition is presented in Table 5. Under DS, the BI values ranged from −10.6 for TZm−1510 (with GY of ~543 kg/ha) to 13.4 for the check 3-TZEE-W HDT C3 STR C5 (with GY of ~3863 kg/ha). Of the top 15 landrace accessions based on the BI values, four (TZm-1440, TZm-1163, TZm-1162, TZm-1500) yielded between 3000 and 3487 kg/ha, six (TZm-1486, TZm-1160, TZm-1508, TZm-1449, TZm-1472 and TZm-1159) yielded between 2000 kg/ha and 3000 kg/ha while the remaining yielded below 2000 kg/ha. All accessions with negative BI values yielded below 1000 kg/ha (Table 5).

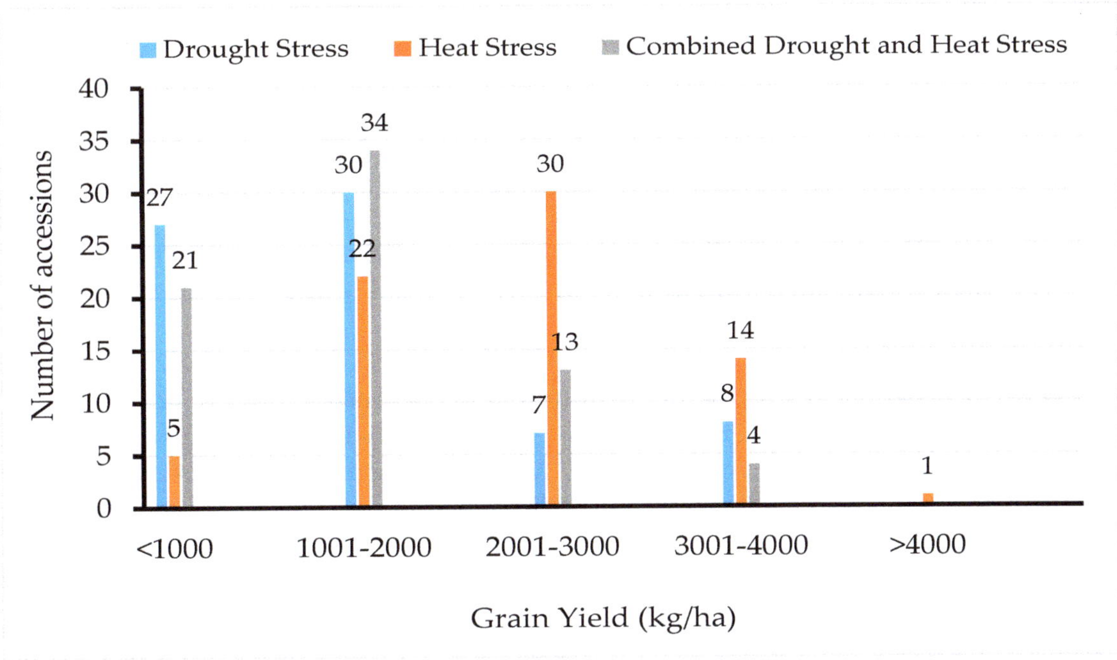

Figure 6. Distribution of grain yield of 72 extra-early maize accessions evaluated under drought, heat and combined drought and heat stress conditions during 2017 and 2019 at Ikenne and Kadawa, Nigeria.

Table 5. Grain yield and base index values of the best check and the top 15, and worse five landrace accessions evaluated under drought, heat, and combined drought and heat stress environments at Ikenne and Kadawa, Nigeria between 2017 and 2019.

Drought Stress (DS)			**Heat Stress (HS)**			**Combined Drought and Heat Stress (DSHS)**		
Accession	**GY (kg/ha)**	**BI**	**Accession**	**GY (kg/ha)**	**BI**	**Accession**	**GY (kg/ha)**	**BI**
Check 3	3863	13.4	Check 5	4723	14.5	Check 4	3899	10.9
TZm-1440	3287	11.4	TZm-1167	3895	12.3	TZm-1486	3167	8.2
TZm-1163	3487	11.3	TZm-1157	3614	9.1	TZm-1162	3174	7.7
TZm-1500	3086	11.1	TZm-1178	3896	8.0	TZm-1472	2334	4.8
TZm-1162	3256	10.5	TZm-1472	3238	7.8	TZm-1171	2070	4.7
TZm-1486	2815	9.2	TZm-1163	3673	6.4	TZm-1440	2373	4.5
TZm-1160	2215	7.4	TZm-1158	3247	6.3	TZm-1470	2391	4.2
TZm-1174	1886	6.5	TZm-1352	2524	6.1	TZm-1160	2187	3.9
TZm-1349	1932	5.1	TZm-1162	3137	4.7	TZm-1481	2039	3.8
TZm-1496	1685	4.9	TZm-1179	3956	4.5	TZm-1508	1797	3.1
TZm-1449	2404	4.9	TZm-1508	3350	4.2	TZm-1483	2042	2.6
TZm-1508	2521	4.5	TZm-1329	3415	3.9	TZm-1485	1735	2.5
TZm-1472	2015	4.3	TZm-1443	2984	3.9	TZm-1167	1700	2.5
TZm-1159	2026	4.3	TZm-1561	3277	3.9	TZm-1496	1802	2.4
TZm-1511	1861	4.1	TZm-1511	2793	3.9	TZm-1506	2210	2.2
TZm-1167	1926	3.7	TZm-1454	3194	3.4	TZm-1448	1853	2.0
TZm-1169	530	−8.8	TZm-1497	907	−9.36	TZm-1510	336	−6.8
TZm-1493	571	−9.0	TZm-1493	907	−9.37	TZm-1509	281	−7.3
GH-4863	430	−9.6	TZm-1177	936	−9.41	TZm-1176	366	−7.3
TZm-1165	501	−10.4	TZm-1170	932	−12.08	TZm-1480	467	−8.6
TZm-1510	543	−10.6	TZm-1498	501	−12.29	TZm-1173	152	−11.3

GY = Grain yield; BI = Base index; Check 3 = TZEE-W HDT C3 STR C5; Check 4 = TZEE-Y HDT C3 STR C5; Check 5 = 2014 TZEE-Y DTH STR.

Under HS, the BI values ranged from −12.29 for TZm-1498 (GY = 501 kg/ha) to 14.5 for the check5 -2014 TZEE-Y DTH STR (GY = 4723 kg/ha). All the top 15 landrace accessions based on the BI values yielded above 2500 kg/ha with TZm-1167, TZm-1157, TZm-1178, TZm-1163, and TZm-1179 yielding above 3500 kg/ha. As with DS, all accessions with negative BI values under HS yielded below 1000 kg/ha (Table 5).

Similarly, under DSHS, the BI values ranged from −11.3 for TZm-1173 (GY = 152 kg/ha) to 10.9 for check 4-TZEE-Y HDT C3 STR C5 (GY = 3899 kg/ha). Two of the top-yielding 15 landrace accessions identified by the BI (TZm-1486 and TZm-1162) yielded approximately 3000 kg/ha while TZm-1472, TZm-1440 and TZm-1470 yielded above 2300 kg/ha. The worse five landrace accessions under DSHS yielded below 500 kg/ha (Table 5).

Based on the BI values, seven landrace accessions (TZm-1159, TZm-1162, TZm-1163, TZm-1167, TZm-1472, TZm-1500 and TZm-1508) were tolerant to both DS and HS, eight (TZm-1160, TZm-1162, TZm-1167, TZm-1440, TZm-1472, TZm-1486, TZm-1496 and TZm-1508) were tolerant to both DS and DSHS. Only five landrace accessions (TZm-1167, TZm-1162, TZm-1472, TZm-1508 and TZm-1506) were tolerant to both HS and DSHS, while four (TZm-1162, TZm-1167, TZm-1472 and TZm-1508) showed good performance across all the individual and combined stresses.

3.4. Principal Component Biplot and Cluster Analysis

The biplot of principal components 1 and 2 under the different treatment conditions are presented in Figure 7. Under DS, the PCA biplot explained ~68% of the total variability among the genotypes (Figure 7a), 63% under HS (Figure 7b) and 55% under DSHS (Figure 7c). Under each stress condition, tolerant to very tolerant accessions were largely located in the lower left of the plot and were mainly associated with increased GY and EPP whiles their susceptible to very susceptible counterparts, which were characterized by increased ASI, LF, TB, SG, RL, and SL, poor HC, EASP, and PASP, and to some extent increased AD, and SD were found sparingly distributed in the upper and lower right sides of the biplot. The strong positive correlation between GY and EPP was evident by the acute angle between their respective vectors of similar length. Similarly, the negative correlation between GY and ASI, LF, TB, SG, RL, SL, HC, EASP, and PASP can be seen from the opposite direction of the variable vectors.

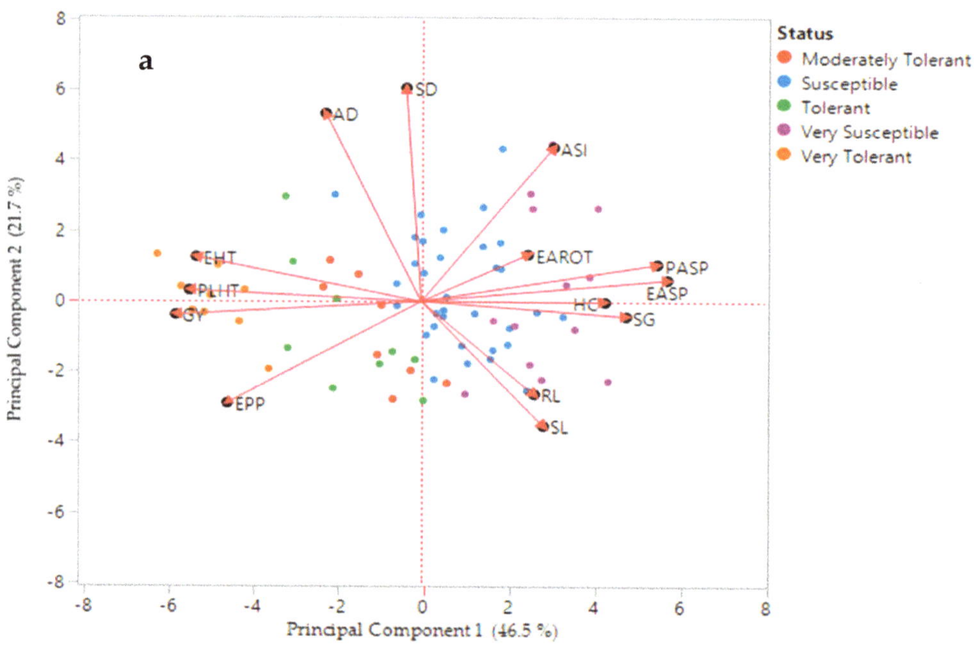

Figure 7. *Cont*

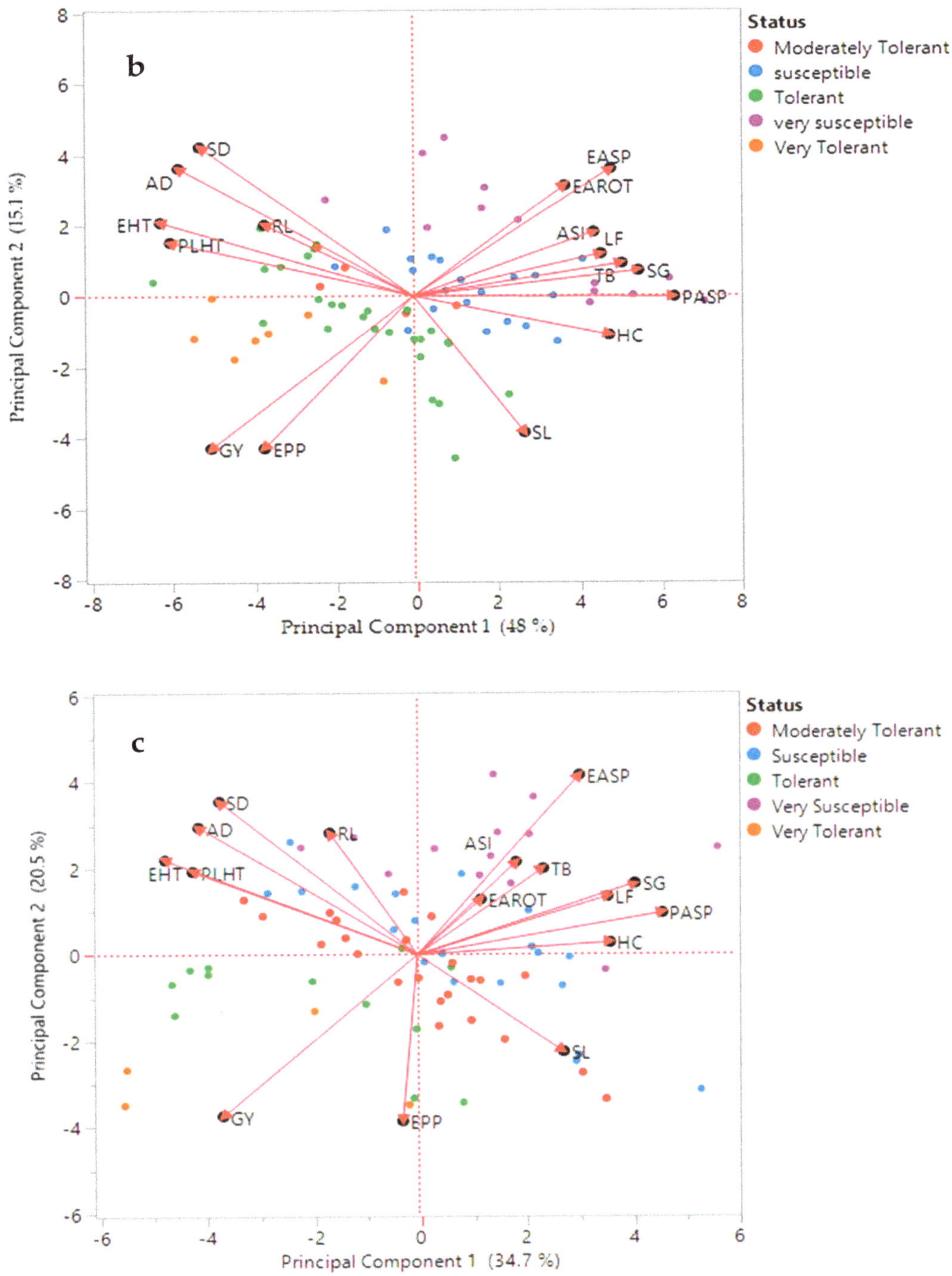

Figure 7. Biplot of 72 extra-early maize accessions and variables under drought (**a**), heat (**b**), and combined drought and heat stress (**c**) conditions. NB: The length of each variable vector is proportional to its contribution to total variation of the accessions, and the direction of the vector indicates its relative contribution to the principal components. AD: Days to 50% anthesis; SD: Days to 50% silking; ASI: Anthesis-silking interval; PLHT: Plant height; EHT: Ear height; HC: Husk cover; EPP: Ears per plant; PASP: Plant aspect; EASP: Ear aspect; SG: Stay green characteristic; RL: Root lodging; SL: Stalk lodging; TB: Tassel blast; LF: Leaf firing; EAROT: Ear rot.

Results of the regression of GY on TB showed that, largely accessions with high TB (%) had low grain yields (Figure 8). TB accounted for 15% and 28% of the yield reduction of the extra-early maize accessions under DSHS and HS, respectively.

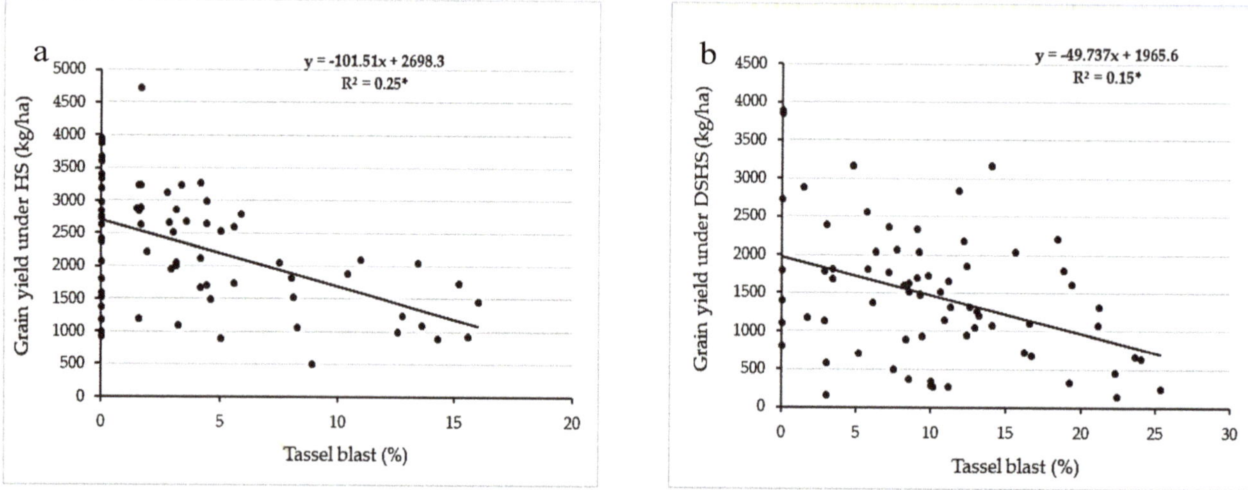

Figure 8. Regression of grain yield (kg/ ha) of 72 extra-early maize landraces including six drought and/or heat-tolerant populations/varieties, on tassel blast under heat stress (**a**) and combined drought and heat stress (**b**).

Under DS, the phylogenetic constellation plot generated using GY and the secondary traits included in the index selection classified the 72 extra-early maize accessions into five major groups, each further divided into subgroups (Figure 9). The number of accessions in the major clusters ranged from six in cluster IV to 31 in cluster III. Accessions in cluster I and II were characterized by high GY (>2000 kg/ha), increased EPP (≥0.80), delayed senescence (average rating of 3), desirable PASP and EASP (4 to 5). Consequently, the average BI values of these groups were high (averaging, 11.72 for cluster I and 5.0 for cluster II) (Supplementary Table S3).

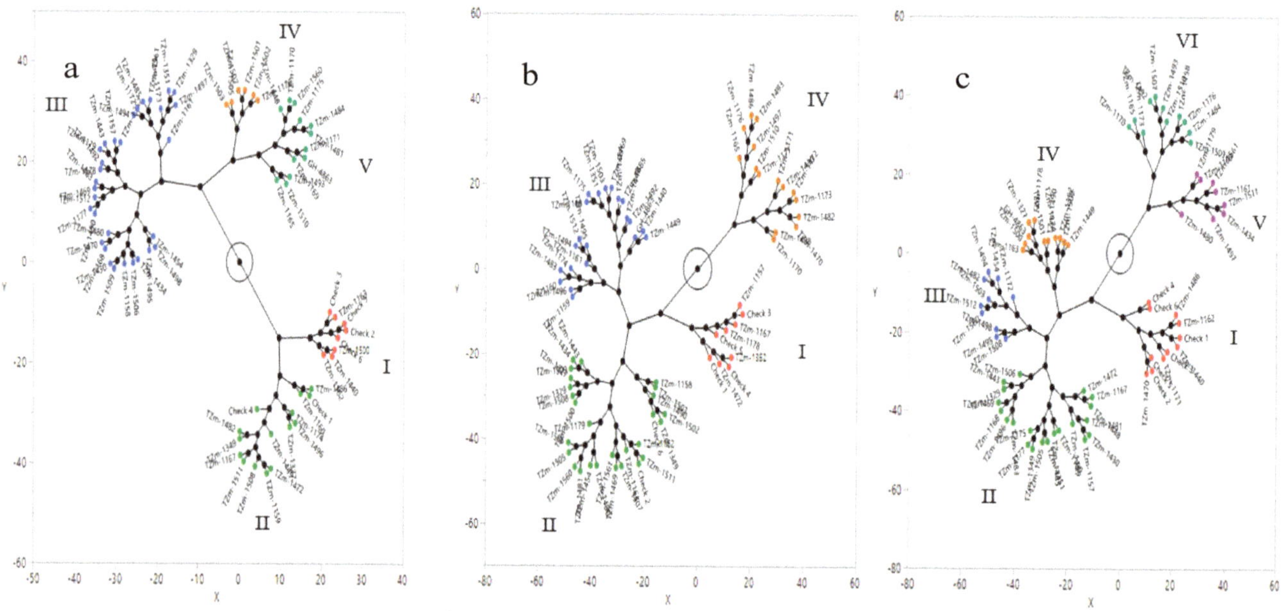

Figure 9. Phylogenetic constellation plots displaying the relationships between 66 extra-early maize landraces and six improved populations/varieties evaluated under managed drought stress (**a**), heat stress (**b**) and combined drought and heat stress (**c**). For each treatment, clusters I, and II were represented largely by tolerant accessions, while the remaining clusters consisted of susceptible accessions.

Similarly, under HS, the phylogenetic constellation plot separated the 72 accessions into five major clusters, with each further separated into sub-clusters that ranged from two in cluster I to four in cluster II (Figure 9). The first major cluster contained nine genotypes while the second consisted of 27 individuals. Individuals of the two clusters (I and II) were characterized by short ASI (2 to 3 days), good plant and ear aspect scores (4 to 5), and increased GY (>2800 kg/ha), hence the positive average BI values (9.73 for cluster I and 2.72 for II). Of the nine accessions in cluster I, five were landraces (TZm-1157, TZm-1167, TZm-1178, TZm-1352, and TZm-1472) and the other four were improved population checks. Accessions of the remaining clusters had very poor plant and ear aspect scores, reduced EPP, low GY and hence, negative BI values (Supplementary Table S3).

Under DSHS, six major groups of accessions were revealed by the phylogenetic constellation plot (Figure 9). The number of individuals in the clusters varied from eight in cluster IV to 21 in cluster II. The first major group consisted of nine accessions, including four checks, and five landrace accessions (TZm-1486, TZm-1162, TZm-1440, TZm-1171, and TZm-1470). The accessions of this group displayed desirable plant and ear aspect scores, and had good ears per plant and GY, which was reflected in their high positive index values (Supplementary Table S3). The second major group comprised accessions with good plant and ear aspect scores and had relatively good ears per plant, moderate grain yield as well as positive index values. The remaining four clusters contained accessions that recorded poor plant and ear aspect scores, reduced ears per plant, low GY, and negative BI values (Supplementary Table S3).

4. Discussion

Frequent occurrence of extreme weather conditions owing to global climate change has heightened the need for genetic improvement of major staple crops for tolerance to abiotic stresses. However, as selection pressure increases for a specific trait, genetic variability inevitably decreases [43], leading to reduced breeding gains. This is particularly true for tropical maize in which little additional gain in maximum drought-tolerance has been achieved in the last decade [29]. Addressing this issue and hence, ensuring progress in genetic improvement of maize under abiotic stress conditions requires identification of donor lines with beneficial traits that will bring new genetic variation [29,44]. Landrace gene pools of maize from areas that frequently experience DS at elevated temperatures may provide a useful source of novel alleles for abiotic stress tolerance [27,45]. In the present study, we evaluated 66 extra-early landrace accessions of maize representing gene pools from Burkina Faso, Ghana, and Togo, together with six abiotic stress-tolerant populations from the IITA-MIP under field conditions of DS, HS and DSHS imposed at the reproductive stages of crop growth and development for two years.

As shown in Figure 1, the HS trials were performed under high temperatures, while the DSHS trials were exposed to prolonged DS at elevated temperatures. In particular, temperatures during flowering and early grain-filling stages substantially exceeded the optimal threshold for lowland tropical maize (34 °C during the day and 23 °C at night; Figure 2) with no incidence of rainfall, indicating that the sites selected for this study were appropriate for screening the maize accessions for tolerance to HS and DSHS. These sites were used for screening maize for high levels of tolerance to DS and/or HS in earlier studies [13,18,19].

The presence of significant genotypic differences for all measured traits of the maize accessions under DS, HS, and DSHS conditions suggested that superior genotypes and traits conferring tolerance to the stresses could be identified and selected. Even though highly significant statistical differences were detected among genotypes, genotype × environment interaction was not significant for most traits indicating that the environments were similar in stress severity. This observation could be attributed to the coincidence of the imposed stresses at stages most critical for growth and development of the maize accessions. Moreover, the high repeatability estimates observed for majority of the measured traits including GY indicated that most of the variances observed in the present study can be attributed to differences among the studied accessions. These observations largely provided credibility in the performance of the accessions for breeding purposes. Low repeatability estimates across sites for some measured traits were probably the result of inconsistent expressions as indicated by high genotype × environment variance. Similar findings were reported in maize under multiple stresses [19,29].

In agreement with previous studies [19], the wide range of variation in GY losses observed in this study indicated that indeed the applied stresses were severe and that the yield levels observed could be attributed to stress tolerance. Compared with the NS environment, HS, DS, and DSHS, on the averaged reduced GY by ~23%, 49%, and 58%, respectively (Tables 1 and 2), suggesting that the effect of the combined stresses on GY of the extra-early maize accessions was higher than the individual effects but lower than their sum (hypo-additive effect). These results corroborated the findings of earlier workers who reported higher yield losses from the combined effects of DS and HS than DS and HS applied alone in cereals including maize [17,19,29,46]. The high GY loss under DSHS compared to HS and DS could be attributed to the interaction effects of HS and DS on stomatal movements [46]. Stress-induced changes in morpho-physiological properties of the maize accessions might have caused osmotic imbalances under DSHS, resulting in the high yield losses. Under DS at elevated temperatures, plants either close their stomata to prevent water loss or keep stomata opened to cool the leaves through transpiration [46]. The long delay in anthesis and silking (by ≥8 days) under HS and DSHS was most likely the result of severe cold stress due to harmattan at the time of planting, that might have delayed seed emergence and extended pre-flowering developmental stages. Reduction in plant and ear heights were on the average, higher under DS (26%) compared to HS (6%) and DSHS (7%). These results suggested the occurrence of DS during the early stages of growth and development of the maize accessions and that the plants were only affected by HS and DSHS towards the end of the vegetative phase. Indeed, in the present study, DS was imposed at the early growth stages (25 DAP) compared to the HS and DSHS, which had the treatments imposed at 32 DAP. Under DS and DSHS, the number of ears per plant were strongly reduced (26%) relative to HS conditions, and this might have contributed to the high yield losses observed under DS (49%) and DSHS (58%) compared to HS (23%). This was further evident from the strong association of EPP and GY with the high yielding accessions as revealed by the PCA biplot analysis (Figure 7).

Grain yield measured under NS conditions was to some extent predictive of the performance under the applied stresses as indicated by the correlations across the treatments, suggesting that similar physiological mechanisms may be conditioning yield potential under NS and DS, HS and DSHS conditions. These observations, in part agreed with the results of previous studies [18,29]. The correlation between GY under NS and DS (0.66) observed in this study is similar to that (0.63) reported by Cairns et al. [29] but slightly lower than that (0.75) recorded by Meseka et al. [18]. In the present study, the moderately strong positive correlation between GY under DS and DSHS ($r = 0.60$; $p < 0.0001$) relative to DS and HS ($r = 0.48$; $p < 0.001$) as well as HS and DSHS ($r = 0.37$; $p < 0.001$) suggested that while yield performance under DS was predictive of attainable yield under DSHS, tolerance to HS was independent of tolerance to DS and DSHS in the extra-early maize accessions. Cairns et al. [29] suggested that tolerance to combined DS and HS in maize was genetically distinct from tolerance to the individual stresses, and tolerance to either stress alone did not confer tolerance to DSHS conditions. Similarly, Meseka et al. [18] found that tolerance to DS was independent of tolerance to DSHS. The discrepancy between our results and those of Cairns et al. [29] and Meseka et al. [18] particularly for the mechanism of DS and DSHS tolerance could be attributed to differences in maturity classes of the genetic materials investigated. Cairns et al. [29] evaluated early, intermediate and late maturing inbred lines while Meseka et al. [18] assessed drought-tolerant three-way cross maize hybrids, and a local variety that were intermediate to late maturing as compared to the present study where extra-early maturing maize accessions were studied. The strong correlations observed between the measured traits under the different treatments implied the presence of common genetic elements regulating the expression of these traits under the research conditions. The negative genetic correlations observed between phenology and GY, PASP, EASP and HC under DS, HS and DSHS indicated that early flowering was associated with higher GY and desirable PASP, EASP, and HC, which might have contributed resilience to the stresses.

It is widely acknowledged that selection for increased GY together with highly heritable secondary traits can lead to remarkable progress in genetic gains under abiotic stress conditions [42]. For example, significant genetic gains were reported under low nitrogen stress and DS by complementing selection for GY potential with key secondary traits [47]. In the present study, sequential multiple regression analysis identified EASP, PASP, SL, and SG, and to some extent, EPP, and LF as the principal determinants of GY, explaining more than 80% of the differences in GY levels observed under the different stresses (Figures 3–5). This together with the moderate to high repeatability estimates of these traits indicated their potential to improve selection efficiency for GY under the abiotic stresses. Regression of TB on GY revealed that TB accounted for 15% and 25% of the yield reduction of the maize accessions under DSHS and HS conditions, respectively (Figure 8). Reduction in GY by 28% was attributed to TB under DSHS [18].

To allow efficient identification and selection of accessions tolerant to the stresses, we used a base index that integrated superior grain yield, EPP, anthesis-silking interval, plant and ear aspects, and the stay-green characteristic under each applied stress [38]. Promising accessions were identified, most of them tolerant to the individual stresses (Table 4). The tolerant landraces might contain novel resistance genes or combinations of resistance gene that would be valuable for 'climate smart' maize breeding efforts. In particular, eight landrace accessions were tolerant to DS and DSHS. Only five landrace accessions performed well across HS and DSHS, while four more accessions namely, TZm-1162, TZm-1167, TZm-1472, and TZm-1508 had outstanding performance across the three stresses. Therefore, development of lines from accessions that performed well across the stresses might be a successful strategy for abiotic stress tolerance maize breeding since hybrids resulting from the combination of the parents can have high performance for both DS and HS. The fact that only four accessions were tolerant across all the stresses was most likely the result of different physiological and morphological mechanisms conditioning tolerance to the three stresses applied in this study. Therefore, it is important to evaluate maize germplasm under each abiotic stress separately.

The genetic relationships/relatedness among germplasm under abiotic stress conditions are extremely important in determining specific groups of accessions that have good levels of tolerance to a specific stress. Herein, GY and stress adaptive secondary traits included in the base index selection were used to examine these relationships. The aggregation of the genotypes into four or more clusters under each stress further highlighted the potential of the landrace accessions to provide new genetic variation for abiotic stress tolerance in maize. In agreement with previous studies [18,19,48], cluster analyses showed a clear distinction between tolerant accessions and their susceptible counterparts. Under each stress condition, the highly tolerant landrace accessions clustered together with the most tolerant drought and/or heat-tolerant checks. This result suggested that the outstanding landraces and the resistant check cultivars were genetically similar. Majority of the outstanding landraces originated from Burkina Faso (a Sahel country). Thus, natural and artificial selection under the drier and hotter climatic conditions of the Sahel might have resulted in the excellent levels of abiotic stress resistance in these landraces. This observation provided further evidence that the superior landrace accessions identified in this study might contain novel resistance genes or combinations of resistance gene that could be valuable for expanding genetic base and thus, enhancing genetic gains in maize under the abiotic stress conditions.

5. Conclusions

This study uncovered striking levels of abiotic stress tolerance diversity among the extra-early landrace maize accessions, and identified traits potentially associated with tolerance to the stresses. Drought and heat stress, either individually or combined had significant negative effect on grain yield and other morpho-physiological traits of the maize accessions. Plant and ear aspects, stay green, lodging, leaf firing, and ears per plant were key to increased yield potential under the applied stresses. The performance pattern of the superior landrace accessions was similar to those of the best drought and/or heat-tolerant populations under DS, HS, and DSHS conditions. Therefore, they may be interesting for the development of germplasm tolerant to the stresses in SSA. In particular, accessions TZm-1162, TZm-1167, TZm-1472, and TZm-1508 showed good adaptation to the three stress. These landraces should be prioritised for further improvement of key adaptive traits and their introgression into maize breeding programs in SSA can play a considerable role in addressing the effects of drought and heat stress on maize. Extensive screening for abiotic stress tolerance in extra-early landrace accessions of maize should be undertaken to unearth further sources of tolerance.

Supplementary Materials
Table S1: List of the 72 extra-early maize accessions evaluated for tolerance to drought, heat, and combined drought and heat stress between 2017 and 2019 at Ikenne and Kadawa, Nigeria. Table S2: Genotypic and residual variance, and broad-sense heritability estimates of grain yield (kg/ha) of individual trials. Table S3: Cluster means (base index values and other secondary traits) of 72 extra-early maize accessions evaluated under non stress, drought, heat stress, and combined drought and heat stress conditions between 2017 and 2019, in Nigeria. Table S4: List of abbreviations and their explanations. Figure S1: Susceptible genotype (**left**) showing symptoms of leaf firing and tassel blasting under drought stress at elevated temperature at Kadawa, Nigeria. Figure S2: Ear aspect rating (plot basis) of extra-early maize accessions evaluated under heat stress conditions between 2017 and 2019 at Kadawa, Nigeria.

Author Contributions: Conceptualization: A.S.-P.N. and A.Y.T.; Methodology: C.N., B.B.-A. and A.L.G.-O.; Supervision: B.B.-A., A.S.-P.N., A.Y.T. and A.L.G.-O.; Data analysis and manuscript draft: C.N.; Manuscript review and editing, B.B.-A., A.Y.T., A.S.-P.N. and A.L.G.-O. All authors have read and agreed to the published version of the manuscript.

Acknowledgments: The first author is grateful to WASCAL for the PhD fellowship. We are grateful to the Genetic Resource Centre at IITA, Ibadan, Nigeria and the Plant Genetics Resources Institute at Bunso, Ghana for providing the maize accessions used in this study.

References

1. McCann, J.C. *Maize and Grace: Africa's Encounter with a New World Crop, 1500–2000*; Harvard University Press: Cambridge, MA, USA, 2005.
2. Warburton, M.L.; Reif, J.C.; Frisch, M.; Bohn, M.; Bedoya, C.; Xia, X.C.; Crossa, J.; Franco, J.; Hoisington, D.; Pixley, K.; et al. Genetic diversity in CIMMYT non-temperate maize germplasm: Landraces, open pollinated varieties, and inbred lines. *Crop Sci.* **2008**, *48*, 617–624. [CrossRef]
3. Abate, T.; Fisher, M.; Abdoulaye, T.; Kassie, G.T.; Luduka, R.; Marenya, P.; Asnake, W. Characteristics of maize cultivars in Africa: How modern are they and how many do smallholder farmers grow? *Agric. Food Secur.* **2017**, *6*, 30. [CrossRef]
4. Edmeades, G.O.; Trevisan, W.; Prasanna, B.M.; Campos, H. Tropical maize (*Zea mays* L.). In *Genetic Improvement of Tropical Crops*; Campos, H., Caligari, P.D.S., Eds.; Springer: Cham, Switzerland, 2017; pp. 57–109.
5. Wu, X.; Wang, A.; Guo, X.; Liu, P.; Zhu, Y.; Li, X.; Chen, Z. Genetic characterization of maize germplasm derived from Suwan population and temperate resources. *Hereditas* **2019**, *156*, 2. [CrossRef]
6. Singh, S.; Vikram, P.; Sehgal, D.; Burgueño, J.; Sharma, A.; Singh, S.K.; Sansaloni, C.P.; Joynson, R.; Brabbs, T.; Ortiz, C.; et al. Harnessing genetic potential of wheat germplasm banks through impact-oriented-prebreeding for future food and nutritional security. *Sci. Rep.* **2018**, *8*, 12527. [CrossRef]
7. Masuka, B.P.; van Biljon, A.; Cairns, J.E.; Das, B.; Labuschagne, M.; MacRobert, J.; Olsen, M. Genetic diversity among selected elite CIMMYT maize hybrids in East and Southern Africa. *Crop Sci.* **2017**, *57*, 2395–2404. [CrossRef]
8. Edenhofer, O. (Ed.) *Climate Change 2014: Mitigation of Climate Change*; Cambridge University Press: Cambridge, UK, 2015; p. 3.
9. Muller, C.; Cramer, W.; Hare, W.L.; Lotze-Campen, H. Climate change risks for African agriculture. *Proc. Natl. Acad. Sci. USA* **2011**, *108*, 4313–4315. [CrossRef]
10. Kurukulasuriya, P.; Mendelsohn, R.; Hassan, R.; Benhin, J.; Deressa, T.; Diop, M.; Eid, H.M.; Fosu, K.Y.; Gbetibouo, G.; Jain, S.; et al. Will African agriculture survive climate change? *World Bank Econ. Rev.* **2006**, *20*, 367–388. [CrossRef]
11. Cairns, J.E.; Hellin, J.; Sonder, K.; Araus, J.L.; MacRobert, J.F.; Thierfelder, C.; Prasanna, B.M. Adapting maize production to climate change in Sub-Saharan Africa. *Food Secur.* **2012**, *5*, 345–360. [CrossRef]
12. Knox, J.; Hess, T.; Daccache, A.; Wheeler, T. Climate change impacts on crop productivity in Africa and South Asia. *Environ. Res.* **2012**, *7*, 034032. [CrossRef]

13. Badu-Apraku, B.; Fakorede, M.A.B. Improvement of Early and Extra-Early Maize for Combined Tolerance to Drought and Heat Stress in Sub-Saharan Africa. In *Advances in Genetic Enhancement of Early and Extra-Early Maize for Sub-Saharan Africa*; Springer: Cham, Switzerland, 2017; pp. 311–358.
14. Cicchino, M.; Rattalino-Edreria, J.I.; Uribelarrea, M.; Otegui, M.E. Heat stress in field-grown maize: Response of physiological determinants of grain yield. *Crop Sci.* **2011**, *50*, 1438–1448. [CrossRef]
15. Zaidi, P.H.; Zaman-Allah, M.; Trachsel, S.; Seetharam, K.; Cairns, J.E.; Vinayan, M.T. *Phenotyping for Abiotic Stress Tolerance in Maize Heat Stress: A Field Manual*; CIMMYT: Mexico City, Mexico, 2016.
16. Araus, J.L.; Serret, M.D.; Edmeades, G.O. Phenotyping maize for adaptation to drought. *Front. Physiol.* **2012**, *3*, 305. [CrossRef]
17. NeSmith, D.S.; Ritchie, J.T. Effects of soil water-deficits during tassel emergence on development and yield components of maize (*Zea mays* L.). *Field Crop Res.* **1992**, *28*, 251–256. [CrossRef]
18. Meseka, S.; Menkir, A.; Bossey, B.; Mengesha, W. Performance Assessment of Drought Tolerant Maize Hybrids under Combined Drought and Heat Stress. *Agronomy* **2018**, *8*, 274. [CrossRef]
19. Nelimor, C.; Badu-Apraku, B.; Tetteh, A.Y.; N'guetta, A.S.P. Assessment of Genetic Diversity for Drought, Heat and Combined Drought and Heat Stress Tolerance in Early Maturing Maize Landraces. *Plants* **2019**, *8*, 518. [CrossRef]
20. Bassu, S.; Brisson, N.; Durand, J.L.; Boote, K.; Lizaso, J.; Jones, J.W.; Basso, B. How do various maize crop models vary in their responses to climate change factors? *Glob. Chang. Biol.* **2014**, *20*, 2301–2320. [CrossRef]
21. Wardlaw, I.F. Interaction between drought and chronic high temperature during kernel filling in wheat in a controlled environment. *Ann. Bot.* **2002**, *90*, 469–476. [CrossRef]
22. Shah, N.H.; Paulsen, G.M. Interaction of drought and high temperature on photosynthesis and grain-filling of wheat. *Plant Soil* **2003**, *257*, 219–226. [CrossRef]
23. Prasad, P.V.V.; Pisipati, S.R.; Momicilovic, I.; Ristic, Z. Independent and combined effects of high temperature and drought stress during grain filling on plant yield and chloroplast EF-Tu expression in spring wheat. *J. Agron. Crop Sci.* **2011**, *197*, 430–441. [CrossRef]
24. Pradhan, G.P.; Prasad, P.V.V.; Fritz, A.K.; Kirkham, M.B.; Gil, B.S. Effects of drought and high temperature stress on synthetic hexaploid wheat. *Funct. Plant Biol.* **2012**, *39*, 190–198. [CrossRef]
25. Heiniger, R.W. *The Impact of Early Drought on Corn Yield*; North Carolina State University: Raleigh, NC, USA, 2001.
26. Dwivedi, S.L.; Ceccarelli, S.; Blair, M.W.; Upadhyaya, H.D.; Are, A.K.; Ortiz, R. Landrace germplasm for improving yield and abiotic stress adaptation. *Trends Plant Sci.* **2016**, *21*, 31–42. [CrossRef]
27. Castro-Nava, S.; Ramos-Ortiz, V.H.; Reyes-Méndez, C.A.; Briones-Encinia, C.A.F.; López-Santillán, J.A. Preliminary field screening of maize landrace germplasm from northeastern Mexico under high temperatures. *Maydica* **2011**, *56*, 77–82.
28. Gedil, M.; Menkir, A. An integrated molecular and conventional breeding scheme for enhancing genetic gain in maize in Africa. *Front. Plant Sci.* **2019**, *10*. [CrossRef] [PubMed]
29. Cairns, J.E.; Crossa, J.; Zaidi, P.H.; Grudloyma, P.; Sanchez, C.; Araus, J.L.; Menkir, A. Identification of drought, heat, and combined drought and heat tolerant donors in maize. *Crop Sci.* **2013**, *53*, 1335–1346. [CrossRef]
30. Alam, M.S.; Seetharam, K.; Zaidi, P.H.; Dinesh, A.; Vinayan, M.T.; Nath, U.K. Dissecting heat stress tolerance in tropical maize (*Zea mays* L.). *Field Crop. Res.* **2017**, *204*, 110–119. [CrossRef]
31. Nelimor, C.; Badu-Apraku, B.; Nguetta, S.P.A.; Tetteh, A.Y.; Garcia-Oliveira, A.L. Phenotypic characterization of maize landraces from Sahel and Coastal West Africa reveals marked diversity and potential for genetic improvement. *J. Crop Improv.* **2019**. [CrossRef]
32. SAS Institute Inc. *SAS User's Guide: Statistics*; Version 9.4; SAS Institute Inc.: Cary, NC, USA, 2017.
33. Falconer, D.S.; Mackay, T.F.C. *Introduction to Quantitative Genetics*, 4th ed.; Longman: New York, NY, USA, 1996.
34. Peterson, B.G.; Carl, P.; Boudt, K.; Bennett, R.; Ulrich, J.; Zivot, E.; Lestel, M.; Balkissoon, K.; Wuertz, D.; Peterson, M.B.G. Performance Analytics: Econometric Tools for Performance and Risk Analysis. R Package Version 1. Available online: https://cran.r-project.org/web/packages/PerformanceAnalytics/index.html (accessed on 15 August 2014).
35. Cooper, M.; DeLacy, I.H.; Basford, K.E. Relationship among analytical methods used to analyse genotypic adaptation in multi-environment trials. In *Plant Adaptation and Crop Improvement*; Cooper, M., Hammer, G.L., Eds.; Cambridge University Press: Cambridge, UK, 1996; pp. 193–224.

36. Vargas, M.; Combs, E.; Alvarado, G.; Atlin, G.; Mathews, K.; Crossa, J. META: A suit of SAS programs to analyse multi-environment breeding trials. *Agron. J.* **2013**, *105*, 11–19. [CrossRef]
37. Mohammadi, S.A.; Prasanna, B.M.; Singh, N.N. Sequential path model for determining interrelationships among grain yield and related characters in maize. *Crop Sci.* **2003**, *43*, 1690–1697. [CrossRef]
38. Badu-Apraku, B.; Akinwale, R.O.; Oyekunle, M. Efficiency of secondary traits in selecting for improved grain yield in extra-early maize under Striga-infested and Striga-free environments. *Plant Breed.* **2014**, *133*, 373–380. [CrossRef]
39. Talabi, A.O.; Badu-Apraku, B.; Fakorede, M.A.B. Genetic variances and relationship among traits of an early-maturing maize population under drought-stress and low-N environments. *Crop Sci.* **2016**, *57*, 681–692. [CrossRef]
40. Badu-Apraku, B.; Fakorede, M.A.B.; Talabi, A.O.; Oyekunle, M.; Akaogu, I.C.; Akinwale, R.O.; Annor, B.; Melaku, G.; Fasanmade, Y.; Aderounmu, M. Gene action and heterotic groups of early white quality protein maize inbreds under multiple stress environments. *Crop Sci.* **2015**, *56*, 183–199. [CrossRef]
41. Statistical Package for Social Sciences (SPSS) Inc. *SPSS Base 17.0 for Windows User's Guide*; SPSS Inc.: Chicago, IL, USA, 2007.
42. Bolaños, J.; Edmeades, G.O. Eight cycles of selection for drought tolerance in lowland tropical maize. I. Responses in grain yield, biomass, and radiation utilization. *Field Crop. Res.* **1993**, *31*, 233–252. [CrossRef]
43. Araus, J.L.; Kefauver, S.C.; Zaman-Allah, M.; Olsen, M.S.; Cairns, J.E. Translating high-throughput phenotyping into genetic gain. *Trends Plant Sci.* **2018**, *23*, 451–466. [CrossRef] [PubMed]
44. Das, B.; Atlin, G.N.; Olsen, M.; Burgueño, J.; Tarekegne, A.; Babu, R.; Ndou, E.N.; Mashingaidze, K.; Moremoholo, L.; Ligeyo, D.; et al. Identification of donors for low-nitrogen stress with maize lethal necrosis (MLN) tolerance for maize breeding in sub-Saharan Africa. *Euphytica* **2019**, *215*, 80. [CrossRef]
45. Gouesnard, B.; Zanetto, B.; Welcker, C. Identification of adaptation traits to drought in collections of maize landraces from southern Europe and temperate regions. *Euphytica* **2015**, *209*, 565–584. [CrossRef]
46. Mahrookashani, A.; Siebert, S.; Hüging, H.; Ewert, F. Independent and combined effects of high temperature and drought stress around anthesis on wheat. *J. Agron. Crop Sci.* **2017**, *203*, 453–463. [CrossRef]
47. Edmeades, G.O.; Bolaños, J.; Chapman, S.C.; Lafiite, H.R.; Bänziger, M. Selection improves drought tolerance in tropical maize populations. I. Gains in biomass, grain yield and harvest index. *Crop Sci.* **1999**, *39*, 1306–1315. [CrossRef]
48. Tandzi, L.N.; Bradley, G.; Mutengwa, C. Morphological responses of maize to drought, heat and combined drought and heat stresses. *J. Biol. Sci.* **2019**, *19*, 7–16. [CrossRef]

5

Transcriptomic Analysis Reveals the Temporal and Spatial Changes in Physiological Process and Gene Expression in Common Buckwheat (*Fagopyrum esculentum* Moench) Grown under Drought Stress

Zehao Hou [1,†], Junliang Yin [1,†], Yifei Lu [1], Jinghan Song [1], Shuping Wang [1], Shudong Wei [2], Zhixiong Liu [3], Yingxin Zhang [1] and Zhengwu Fang [1,*]

1 College of Agriculture, Yangtze University, Jingzhou 434000, Hubei, China; 201771374@yangtzeu.edu.cn (Z.H.); yinjunliang@yangtzeu.edu.cn (J.Y.); 201872410@yangtzeu.edu.cn (Y.L.); 201875095@yangtzeu.edu.cn (J.S.); wangshuping@yangtzeu.edu.cn (S.W.); 518009@yangtzeu.edu.cn (Y.Z.)
2 College of Life Science, Yangtze University, Jingzhou 434000, Hubei, China; 500896@yangtzeu.edu.cn
3 College of Horticulture and Gardening, Yangtze University, Jingzhou 434000, Hubei, China; zxliu77@yahoo.com
* Correspondence: fangzhengwu88@yangtzeu.edu.cn
† These authors contributed equally to this work.

Abstract: Common buckwheat is a traditional alternative crop that originated from the northwest of China and is widely cultivated worldwide. However, common buckwheat is highly sensitive to drought stress, especially at the seedling stage, and the molecular mechanisms underlying the response to drought stress still remain elusive. In this study, we analyzed the stress phenotypes of buckwheat seedlings under drought condition. The results showed the wrinkled cotyledon due to the decrease of relative water content (RWC) in response to the increased activity of antioxidant enzymes. Transcriptomic analysis was further performed to analyze the regulation patterns of stress-responding genes in common buckwheat cotyledons and roots under drought stress conditions. Characterizations of the differentially expressed genes (DEGs) revealed differential regulation of genes involved in the photosynthesis and oxidoreductase activity in cotyledon, and that they were highly related to the post-transcriptional modification and metabolic process in root. There were 180 drought-inducible transcription factors identified in both cotyledons and roots of the common buckwheat. Our analysis not only identified the drought responsive DEGs and indicated their possible roles in stress adaption, but also primarily studied the molecular mechanisms regulating the drought stress response in common buckwheat.

Keywords: common buckwheat; cotyledon; root; drought stress; transcriptome analysis

---

1. Introduction

Among the forms of environmental stress, drought stress has been considered as one of the major constraints in plant growth, survival, and production [1,2]. A lack of water not only disturbs photosynthesis, limits metabolic reactions, and inhibits CO_2 exchange, but also results in stress-related damage to chloroplasts [3–5]. In order to adapt to the extreme environments, plants have evolved several mechanisms (e.g., drought escape, avoidance, and tolerance) to ensure high survival rates under drought stress [5,6]. Specifically, plants recruit a variety of responding mechanisms to deal with drought stress [7–9], such as stomatal closure, leaf rolling, and alteration in biosynthetic and antioxidant pathways, which are highly regulated by complex transcriptional networks [10].

Drought stress affects several physiological and biochemical pathways in plants [11]. Previous research has shown that the water deficits not only affect the chlorophyll biosynthesis, but also the level of malondialdehyde (MDA) and the relative water contents (RWC) of the plant, brining detrimental effects to the lipid peroxidation, and membrane constitution [12,13]. Also, the abiotic stresses can further induce the oxidative stress through generating reactive oxygen species (ROS), a prevalently recognized destroyer in cellular metabolism [14–16]. ROS generate the oxidation of photosynthetic pigments, initiate lipid peroxidation, and degrade proteins in plants, and thereby cause damage to cell structures and metabolism, particularly those associated with photosynthesis [17,18]. To counteract the effects of oxidative stress, plants have developed an efficient detoxification defense system consisting of non-enzymatic scavengers and enzymatic components to scavenge free ROS [19,20]. In terms of the enzymatic scavenging, a series of antioxidative enzymes, including peroxidase (POD), catalase (CAT), superoxide dismutase (SOD), and ascorbate peroxidase (APX), have been reported to play a vital role in reducing the damage effects (i.e., water deficiency) caused by drought stress [21]. There is evidence that keeping a high antioxidative enzyme activity level to reduce the damaging effects caused by water deficit stress may be associated with the osmotic stress tolerance of plants [22], which is also found to be positively related to plant drought tolerance [23].

Presently, there are several drought-inducible genes, including stress responses and resistance, which have been identified through transcriptome analyses [1]. These genes can be divided into two groups according their functions. The first group is composed of function proteins that include the late embryogenesis abundant (LEA) proteins, ROS detoxification enzymes, molecular chaperones, heat shock proteins (HSP), and lipid-transfer proteins [24,25]. The second group is involved in regulatory proteins or transcriptional factors (TFs), which correlate with the signal transduction and stress-responsive gene expressions, for example, the phospholipases and dehydration-responsive elements [26,27]. In order to elucidate the biological functions of these genes, several transgenic plants overexpressing various drought-resistant genes have been generated, which have both shown enhanced drought tolerance and growth retardation [28–31], demonstrating that plants may adapt to the drought environment at the expense of normal growth [1].

Common buckwheat (*Fagopyrum esculentum* Moench) is an important dual-purpose alternativecrops originated from Yunnan Province of China [32] and is widely cultivated around the world, especially in China, Japan, and Russia [33]. Because of its abundant nutrients in seeds, common buckwheat is considered as one of the sources of flour, groats, and whole grain foods. However, common buckwheat is highly sensitive to drought, especially at the seedling stage [34,35], and short-term drought occurs frequently in China, posing a threat to domestic food safety [36]. Thus, it would be important and necessary to study the physiological and molecular bases of osmotic stress tolerance in common buckwheat. We have previously identified *FeDREB1L* (GenBank: JN600617.1), a CBF/DREB homologous gene, from common buckwheat, and overexpression of the *FeDREB1L* gene was found to significantly increase the water deficit resistance of transgenic *Arabidopsis* [31]. In order to further understand the drought-resistant mechanism and identify novel water-deficit-related genes in common buckwheat, a transcriptomics analysis was carried out to investigate the variations in common buckwheat growth under short-term drought treatment, and the phenotypes and biochemical traits of seedlings were also analyzed. Our results may provide more information with regard to the transcriptional control of common buckwheat under the abiotic stresses, and help to identify the novel genes that are potentially valuable for future common buckwheat breeding.

2. Materials and Methods

2.1. Plant Material and Drought Treatments

Common buckwheats (cv. Xi'nong 9976) were germinated in Petri dishes in an incubator (plant growth incubator JY412L, Shanghai, China) in darkness (25 °C) and relative humidity of approximately 60%. After germination for 36 h, when the root length of the seedlings grew to approximately 2 cm, the seedlings were transplanted for hydroponics in an incubator with 12 h photoperiods (25 °C/20 °C, day/night temperature) and relative humidity of approximately 60%. The 7-day old buckwheat seedlings were treated with 15% polyethylene glycol 6000 (PEG 6000) solution for 1 d, 3 d, and 5 d. After treatment, the cotyledons and roots were collected and quickly frozen in liquid nitrogen and stored at −80 °C until used. The seedlings before drought treatment were served as the control. Each treatment was carried out in three biological replicates.

2.2. *Physiological Measurement*

Relative water content (RWC) was determined according to the formula described by Pan et al. [36]. The chlorophyll content and chlorophyll a/b ratio was calculated using to the method described by Harper et al. [37]. The changes of malondialdehyde (MDA) concentration were determined using the thiobarbituric acid (TBA) reaction [38], and the activities of POD and CAT were detected according to the description of Harper et al. [37]. The Rubisco activities were assayed with Rubisco assay kits (Beijing Solarbio Science and Technology Co., Ltd., Beijing, China) according to the manufacturer's instructions.

2.3. *RNA Isolation and Transcriptome Sequencing*

Total RNA was isolated from the non-treated control and drought-stressed cotyledon and root samples using EASYspin Plus Plant RNA Kit (Aidlab, Wuhan, China). The RNA quality was checked by Agilent bioanalyzer 2100 (Agilent Technologies, Santa Clara, CA, USA), and Nanodrop 2000 r spectrophotometer (Nanodrop Technologies, Wilmington, DE, USA) was used for RNA quantification.

Transcriptome sequencing was performed at Beijing Allwegene Technology Co. Ltd. (Beijing, China), following manufacturer protocols. Briefly, mRNA was enriched from total RNA using Oligo (dT) magnetic beads, and the mRNA was fragmented into small pieces using a fragmentation buffer. Then, these fragments were used as reverse transcription to synthesized the first- and second-strand cDNA, and the second-stand cDNA were purified with AMPure XP Beads Kit, repaired, poly (A) added, and ligated to paired-end adapters. Finally, the cDNA libraries were sequenced on Illumina HiSeqTM 2500 platform. Each sample had three biological replicates. The raw reads were submitted to the National Center for Biotechnology Information (NCBI) Sequence Read Archive with a Bioproject ID: PRJNA555746.

The raw reads in FASTQ format were processed using in-house Perl scripts, and the high-quality clean data were obtained by removing the low-quality data, which included the reads that contained the adapter, and more than 10% of N nucleotides, and the low-quality reads that contained more than 50% of low quality bases (Q-value ≤ 20). In addition, we calculated the Q30, GC content, and sequence duplication levels for the clean data. Cleaned and qualified reads were aligned against the *F. esculentum* reference genome [39] using Tophat2 software [40]. Then, these sequences were subjected to functional annotation and coding sequence (CDS) prediction [41], and the resulting sequences were called genes. Finally, fragments per kilobase of transcript permillionmapped reads (FPKM) method was used to calculate the gene expression unit.

2.4. *Identification and Functional Annotation of Different Expressed Genes (DEGs)*

The differential gene expression analysis was carried out using DESeq software, and DEGs were determined by combining a q value cutoff of 0.05 and adjusting to |log2 (fold change)| ≥ 1. For DEG functional annotation, Gene Ontology (GO) enrichment analysis was carried out by GOseq software, and Kyoto Encyclopedia of Genes and Genomes (KEGG) was used to perform pathway enrichment analysis of DEGs. In addition, the gene expression profiles at the pathway were display by MapMan software (version 3.6.0) [33].

2.5. *Quantitative Real-Time PCR (qRT-PCR) Analysis*

Total RNA from cotyledons and roots of both samples were extracted using EASYspin Plus Plant RNA Kit (Aidlab, Wuhan, China) following the manufacturer's protocols, and the first-strand cDNA for qPCR analysis was synthesized from 500 ng of total RNA using PrimeScript RT Reagent Kit with gDNA Eraser (Takara, Dalian, China) following the manufacturer's instructions, and cDNA was diluted 10-fold and used as the template for qRT-PCR. The primers were designed using Primer Premier 5.0 and beta-actin was used as a reference gene, with the primer information listed in Table S1. qRT-PCR was performed on a CFX 96 real-time PCR system (BioRad, Hercules, CA, USA) using TB Green (TaKaRa), according to the manufacturer's protocols. PCR amplification was conducted in a volume of 20 μL, containing ~100 ng of cDNA template, 0.6 μL of each primer (10 μmol), and 10 μL PCR-mix (2×). The conditions for all reactions were as follows: 30 s at 95 °C, followed by 40 cycles of 10 s at 95 °C, and 30 s at 55 °C, and the melting curve was generated to confirm the PCR specificity. The non-treated control treatment was chosen as the control to standardizing all samples, using the $2^{-\Delta\Delta Ct}$ method to calculate the relative expression levels [42].

2.6. Statistical Analysis

A one-way ANOVA was carried out by SPSS Statistics 19.0 software (IBM Corp, Armonk, NY, USA), and means were compared using the Duncan test to determine significant differences ($p < 0.05$). The results were presented as mean ± SD (standard deviation).

3. Results

3.1. Changes in Phenotype of Common Buckwheat Seedlings at Drought Stress

To investigate the dynamic phenotypic changes of common buckwheat seedlings in response to drought stress treatments, plant height, root length, and relative water content (RWC) were measured under 15% PEG 6000 solution treatments across four time-points (0, 1, 3, and 5 days). There was no significant change in plant height, root length, and RWC of control samples after 1, 3, and 5 days (Table S2). Under drought stress, the buckwheat seedlings showed stress phenotypes of wrinkled cotyledon (Figure 1a), but the plant height and root length did not significantly change during the treatment (Figure 1b,c). The RCW is generally used as an important indicator of plant water status under osmotic conditions, and in this study, the RCW values were clearly decreased in the 3 and 5 day-treated (DPT3d and DPT5d) seedlings, but were not significantly different among the control (CK) and the 1 day-treated (DPT1d) seedlings (Figure 1d).

3.2. Changes in Physiology of Common Buckwheat Seedlings under Drought Conditions

To investigate the physiological changes under different levels of water deficit conditions, the MDA content and the activities of POD and CAT of cotyledons were measured after drought treatment. Compared with the drought treatment, these physiological traits were not significantly changed during the whole treatment under control conditions, in which the content of chlorophyll a and chlorophyll b were increased in the 1, 3, and 5 day control plants; however, the chlorophyll a/b ratios were not significantly changed during the whole treatment under the control condition (Table S3). Under water deficit condition, the MDA content was greatly increased from 0 to 1 days, and then slightly increased until 5 days (Figure 2a). Meanwhile, the activities of POD and CAT were significantly increased under PEG treatment (Figure 2b,c). The chlorophyll a content was elevated in the 3 and 5 dy-treated plants (Figure 2d), while the content of chlorophyll b content was increased in the 1, 3, and 5 day-treated plants (Figure 2e). In addition, there was no difference between the control and 1 and 3 day-treated seedlings in chlorophyll a/b ratios, but marked decreases were observed in the 5 day-treated plants (Figure 2f). These results indicated that there were significant changes in the physiology of the common buckwheat seedlings in response to osmotic stress.

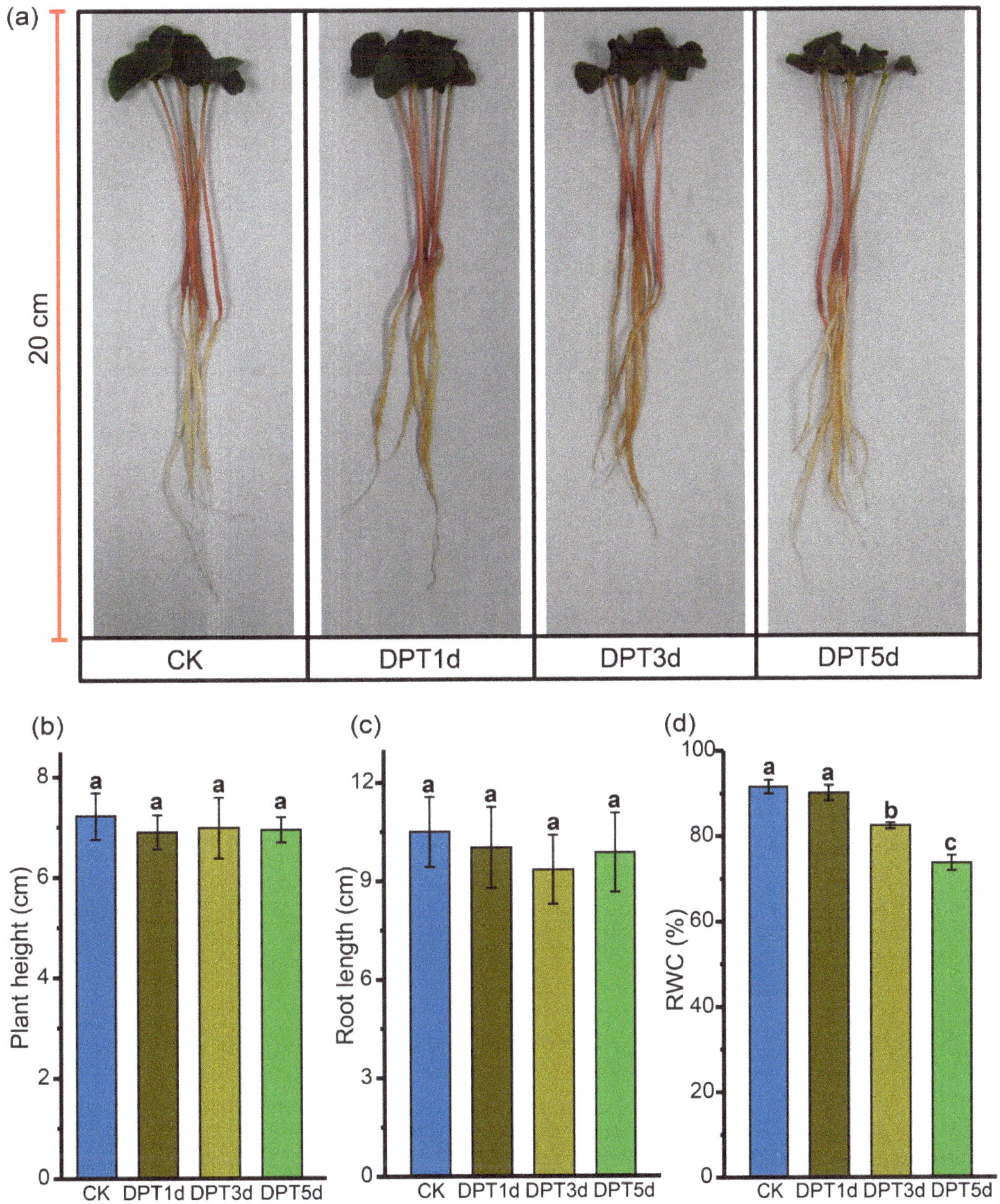

Figure 1. Changes in phenotype of common buckwheat seedlings under drought stress. (**a**) A photograph of common buckwheat seedlings after PEG treatments; CK, non-stressed control; DPT1d, DPT3d, and DPT5d, drought treatment with PEG solution for 1, 3, and 5 days, respectively; bar = 20 cm. The change of (**b**) plant height, (**c**) root length, and (**d**) relative water content of cotyledon during drought stress treatment. Bars represent means of three replicates ± SD (standard deviation). Different letters indicate means that are significantly different at the $p < 0.05$ level among different drought conditions.

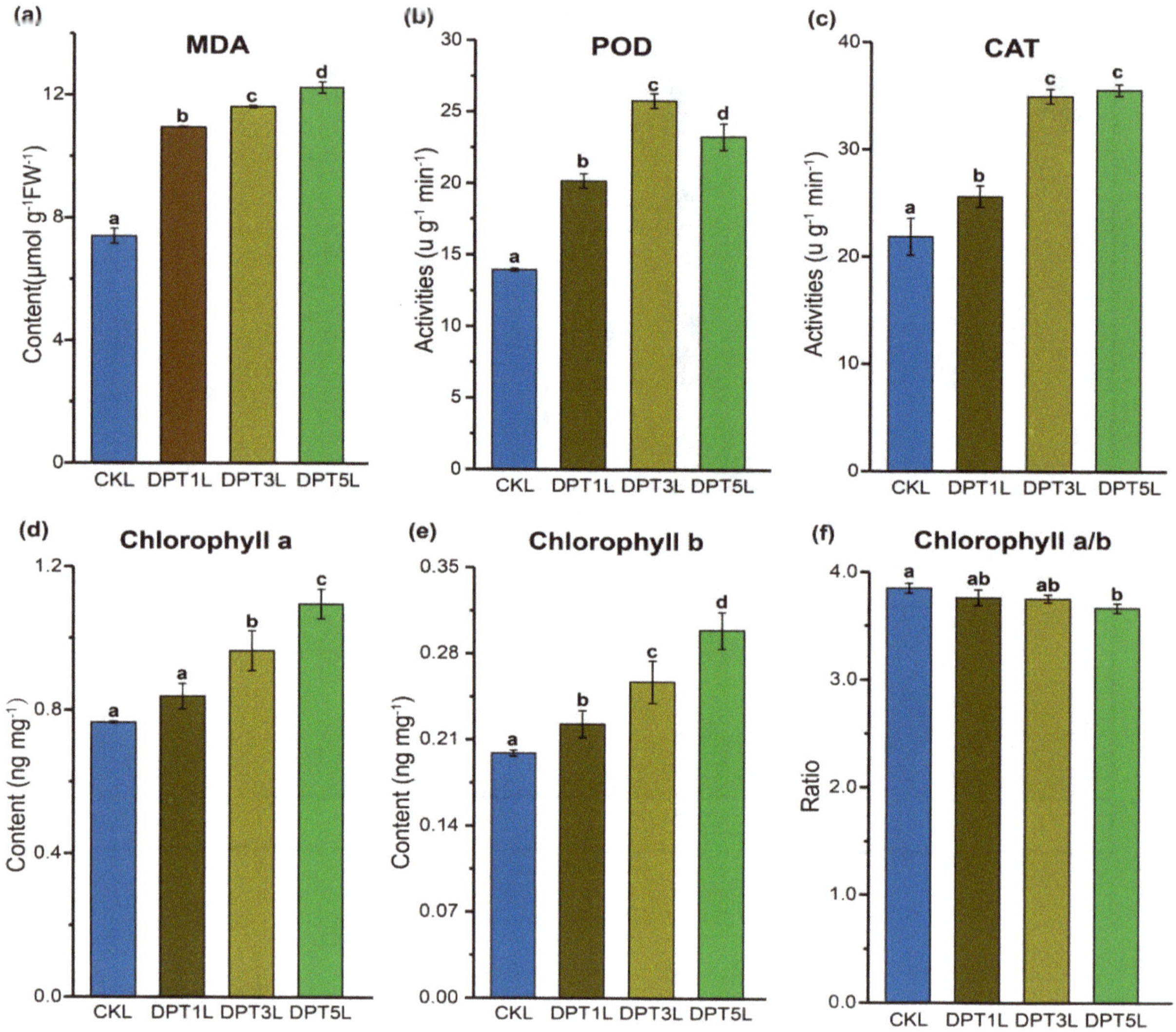

Figure 2. Changes in physiology in cotyledons of common buckwheat seedlings under different drought stress conditions. (**a**) Changes in the malondialdehyde (MDA) content of cotyledons, (**b**) changes in the peroxidase (POD) activities of cotyledons, (**c**) changes in the catalase (CAT) activities of cotyledons, (**d**) chlorophyll a content in the cotyledons, (**e**) chlorophyll b content in the cotyledons, and (**f**) ratios of chlorophyll a/b.

3.3. Overview of the Common Buckwheat Transcriptome and Identification of DEGs

To reveal the expression changes in common buckwheat cotyledons and roots at 0, 1, 3, and 5 days after PEG treatments, 24 common buckwheat samples (including 12 cotyledon samples and 12 root samples) were used for RNA-Seq analysis to further investigate the changes at the transcriptional level. A full-scale sequencing analysis from 24 cDNA samples is shown in Table S4, 88.76 gigabytes (Gb) of clean reads were abstained, and the percentages of the Q30 base of these 24 common buckwheat samples were greater than or equal to 95.70%. Furthermore, there was a highly mapped efficiency between the samples and reference genome (79.46%–87.91%), which met the requirements for information analysis. There were 877,111 CDS (coding sequence)-encoded proteins, and the length of CDS is shown in Figure 3a. Of these CDSs, only a minority (3729 CDSs, 0.43%) were more than 1500 nt, and 90.57% of CDSs appeared with a length ranging from 0 to 500 nt. In addition, the principal component analysis (PCA) was performed and the results demonstrated that the control treatment was clearly separated from the drought stress treatment in the cotyledons or roots (Figure 3b), suggesting that the gene expression pattern of common buckwheat was greatly changed under drought condition.

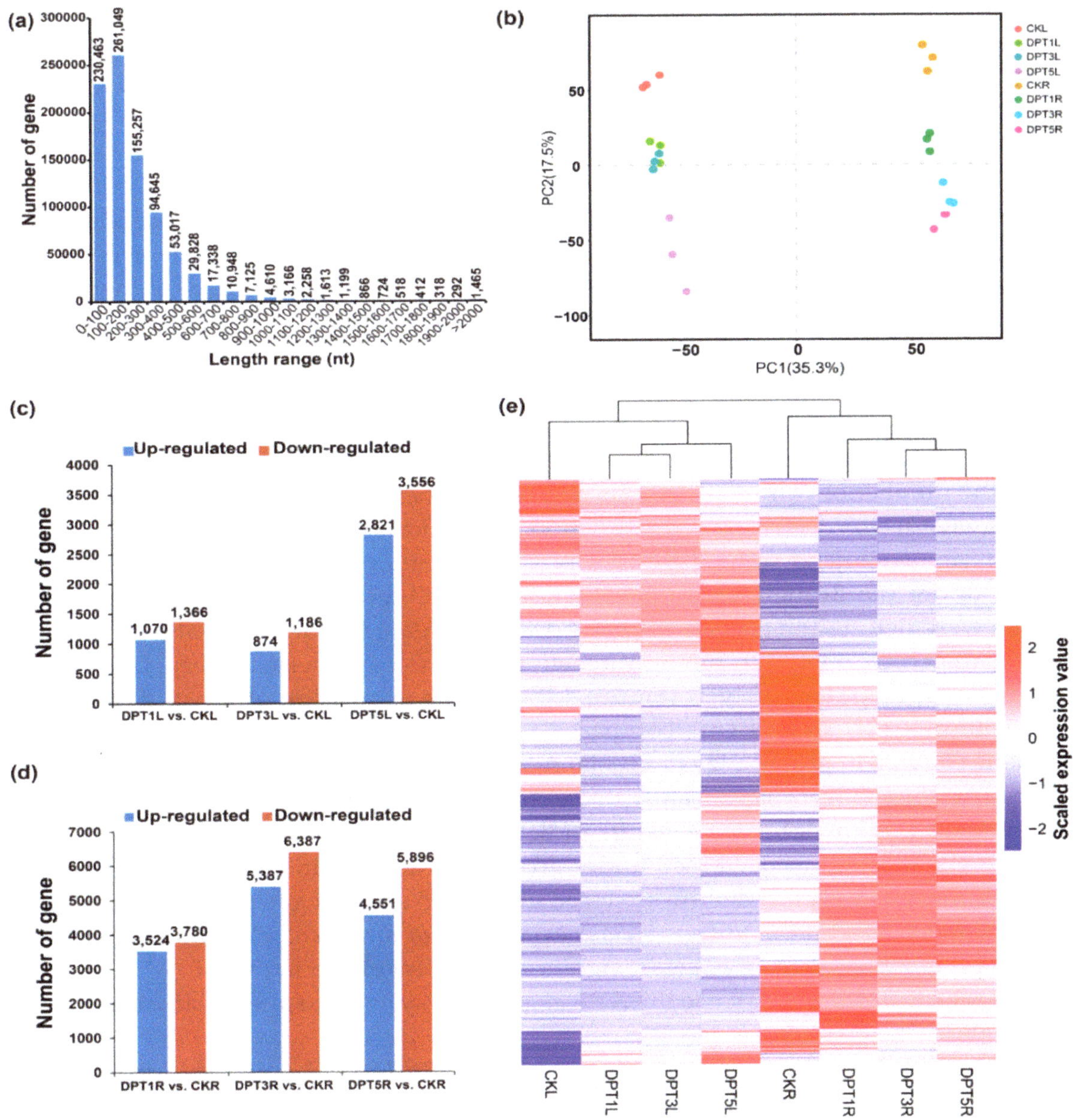

Figure 3. Overview of the transcriptomic results and changes in gene expression profiles in cotyledons and roots after drought stress treatment. (**a**) Predicted length distribution map of coding sequence (CDS)-encoded protein nucleotide (nt). (**b**) Principal component (PC) analysis of gene expression at different drought stress conditions. (**c**) Numbers of differently expressed genes (DEGs) in cotyledons of common buckwheat seedlings at different drought stress conditions in pairwise comparisons. (**d**) DEGs in roots of common buckwheat seedlings at different drought stress conditions in pairwise comparisons. (**e**) Heat-map graphics exhibiting the gene expression levels of total DEGs. CKL and CKR are the cotyledon samples and root samples of the non-stressed control, respectively. DPT1L, DPT3L, and DPT5L are the cotyledon samples of drought treatment for 1, 3, and 5 days, respectively. DPT1R, DPT3R, and DPT5R are the root samples of drought treatment for 1, 3, and 5 days, respectively. DPT1L vs. CKL, DPT3L vs. CKL, and DPT5L vs. CKL are the cotyledon samples of drought treatment for 1, 3, and 5 days compared to the non-stressed control, respectively. DPT1R vs. CKR, DPT3R vs. CKR, and DPT5R vs. CKR are the root samples of drought treatment for 1 day compared to the non-stressed control, respectively. Up-regulated means that genes were up-regulated in drought stress conditions compared to the non-stressed control and down-regulated means that genes were down-regulated in the drought stress condition compared to the non-stressed control.

The gene expression levels were calculated as FPKM values via HTSeq software analysis, and the differential gene expression analysis was carried out using DESeq software. There were 2436, 2060, and 6377 DEGs identified in the cotyledons after drought treated for 1, 3, and 5 days, respectively (Figure 3c). In the roots, compared with the control treatment, 7304, 11,774, and 10,447 DEGs were identified after drought treatment for 1, 3, and 5 days, respectively (Figure 3d). Upon drought stress exposure, more DEGs were identified in the roots than in cotyledons, suggesting that there were different drought stress response mechanisms between roots and leaves in common buckwheat. Furthermore, in order to provide a comprehensive understanding of the change in gene expression of common buckwheat under drought conditions, a heat map was developed, as shown in Figure 3e, to exhibit the overall changes of the gene expression under water deficit conditions.

3.4. Comprehensive Sets of DEGs in the Cotyledons of Drought-Treated Common Buckwheat Seedlings

Venn diagrams showed that the number of genes commonly up-regulated under the DPT1L and DPT5L were greater than the number of genes commonly up-regulated under the DPT3L and DPT5L, and the number of commonly down-regulated genes showed the same trend (Figure 4a,b). Gene Ontology (GO) was used to find the functional significance of the identified DEGs (Figure 4c), and the GO terms related to signaling and DNA modification were detected in the set of genes up-regulated under the drought conditions, while the GO terms related to light harvesting and light reaction were detected in the sets of genes down-regulated under the water deficit conditions. Furthermore, the DEGs related to the light reactions of photosynthesis and the Calvin cycle were visualized through MapMan analysis (Figure 4d), and the expression of most of these DEGs was decreased in the DPT3L and DPT5L, compared with control treatment. RubisCO, as the major photosynthetic enzyme in plants, plays a crucial role in photosynthesis of green plants. In this study, the activities of RubisCO were significantly decreased at 3 days and greatly declined at 5 days under drought treatment (Figure 4e). These results indicate that photosynthesis in the cotyledons of the common buckwheat seedlings decreased under drought stress conditions. The representative genes related to ABA (abscisic acid) metabolism are listed in Table S5 according to their functional description, and most of these (including 6 *NCED* (*9-cis-epoxycarotenoid dioxygenase*), 3 B3 domain-containing protein, and 3 protein phosphatase 2C) were significantly up-regulated in DPT5L, which indicated that common buckwheat seedlings may use the ABA regulatory systems to affect leaf wilting and defense against the water-deficit stress.

To confirm and investigate the transcriptomic data, qRT-PCR was performed to check the expression levels of several genes, and the expression of *LCHb*, a gene encode chlorophyll a/b binding protein, was markedly decreased under DPT3L and DPT5L conditions (Figure 4f). The expression of *NCED*, a gene involved in ABA biosynthesis, was observably increased under drought stress conditions (Figure 4i). Furthermore, the *DREB1L* gene, which encodes a stress tolerance-related protein, was markedly enhanced in its expression level under drought stress treatment, compared with the non-stressed control plants (Figure 4m). The correlation coefficient (R^2) between RNA-Seq data and qPCR results for the 24 total plots was 0.8743 (Figure S1). These analyses of gene expression confirmed that the transcriptomic datasets were efficacious (Figure 4f–n).

3.5. Comprehensive Sets of DEGs in the Roots of Drought Treated Common Buckwheat Seedlings

Three comparison groups were constructed to further understanding the universal response in root of common buckwheat to drought stress. As shown in Figure 5a, 1731 genes were both up-regulated in DPT1R, DPT3R, and DPT5R, compared with CKR. Furthermore, 2857 genes were collectively down-regulated in DPT1R, DPT3R, and DPT5R (Figure 5b). All non-overlapped DEGs in the three comparison groups were subjected to GO enrichment analysis, and 430, 606, and 621 GO accessions classified into three categories comprising "molecular function", "biological process", and "cellular component" were identified in DPT1R vs. CKR, DPT3R vs. CKR, and DPT5R vs. CKR, respectively (Table S6). The drought-induced DEGs were mainly involved in the nucleotide binding,

ATP binding, macromolecule modification, protein phosphorylation, protein modification, protein metabolic process, and cellular protein metabolic process (Figure 5c). According to Mapman software analysis, there were 92 protein modification and phosphorylation-related DEGs that were up-regulated after drought treatment (Table S7), suggesting that they were candidate genes for protein modification in roots of drought-treated seedlings.

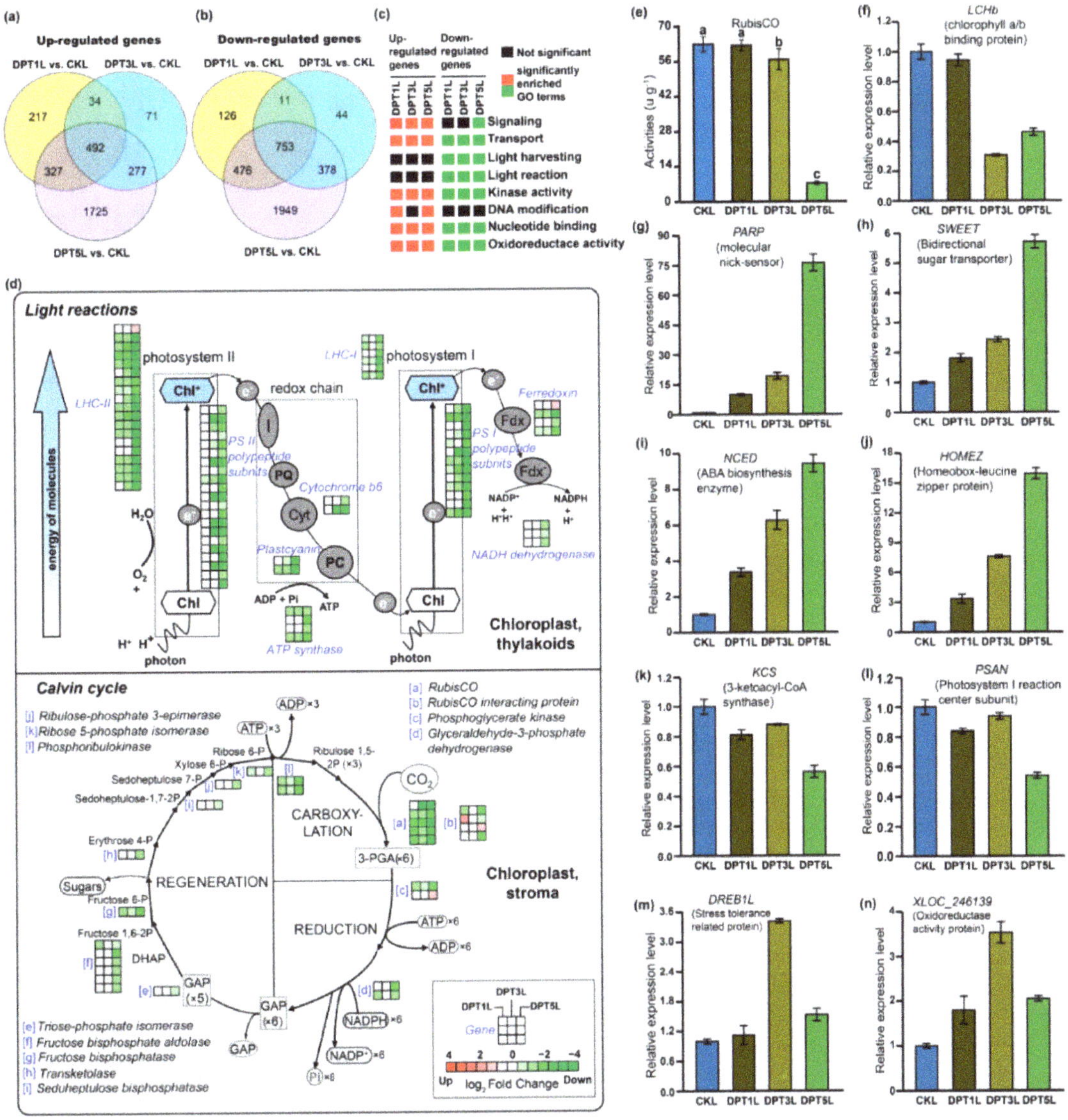

Figure 4. Comprehensive expression patterns of DEGs in the cotyledons of common buckwheat seedlings under drought stress conditions. (**a**) Venn diagrams of the numbers of up-regulated [|log2 (Fold Change)| > 1 and q-value < 0.005] genes acquired through the transcriptome analysis. (**b**) Venn diagrams of the numbers of down-regulated genes acquired through the transcriptome analysis. (**c**) Over-represented Gene Ontology (GO) terms estimated using GOseq software. (**d**) Changes in the expression of photosynthesis-related genes. Pathway diagram of light and dark reactions of photosynthesis with superimposed color-coded squares showing DEGs, drawn using MapMan. (**e**) Changes in the RubisCO activities of cotyledons. (**f–n**) Expression profiles of the selected DEGs, *LCHb* (**f**), *PARP* (**g**), *SWEET* (**h**), *NCED* (**i**), *HOMEZ* (**j**), *KCS* (**k**), *PSAN* (**l**), *DREB1L* (**m**), and *XLOC_246139* (**n**) determined using qRT-PCR analyses.

The expression patterns of several genes related to stress tolerance in the roots were analyzed via qRT-PCR (Figure 5d–k). The expression of *AAO*, a gene encoding L-ascorbate oxidase that plays a crucial role in plant cell growth, was markedly decreased under drought stress conditions (Figure 5h), and the expression of some transcription factors, such as *HOMEZ* and *DREB1L*, were significantly induced by water deficit (Figure 5e,j). In addition, the R^2 between the two experiments was 0.9385 (Figure S2). These results confirmed the effectiveness of the transcriptomic datasets and indicated that drought stress strongly affected the expression level of the genes that are related to stress tolerance in the roots.

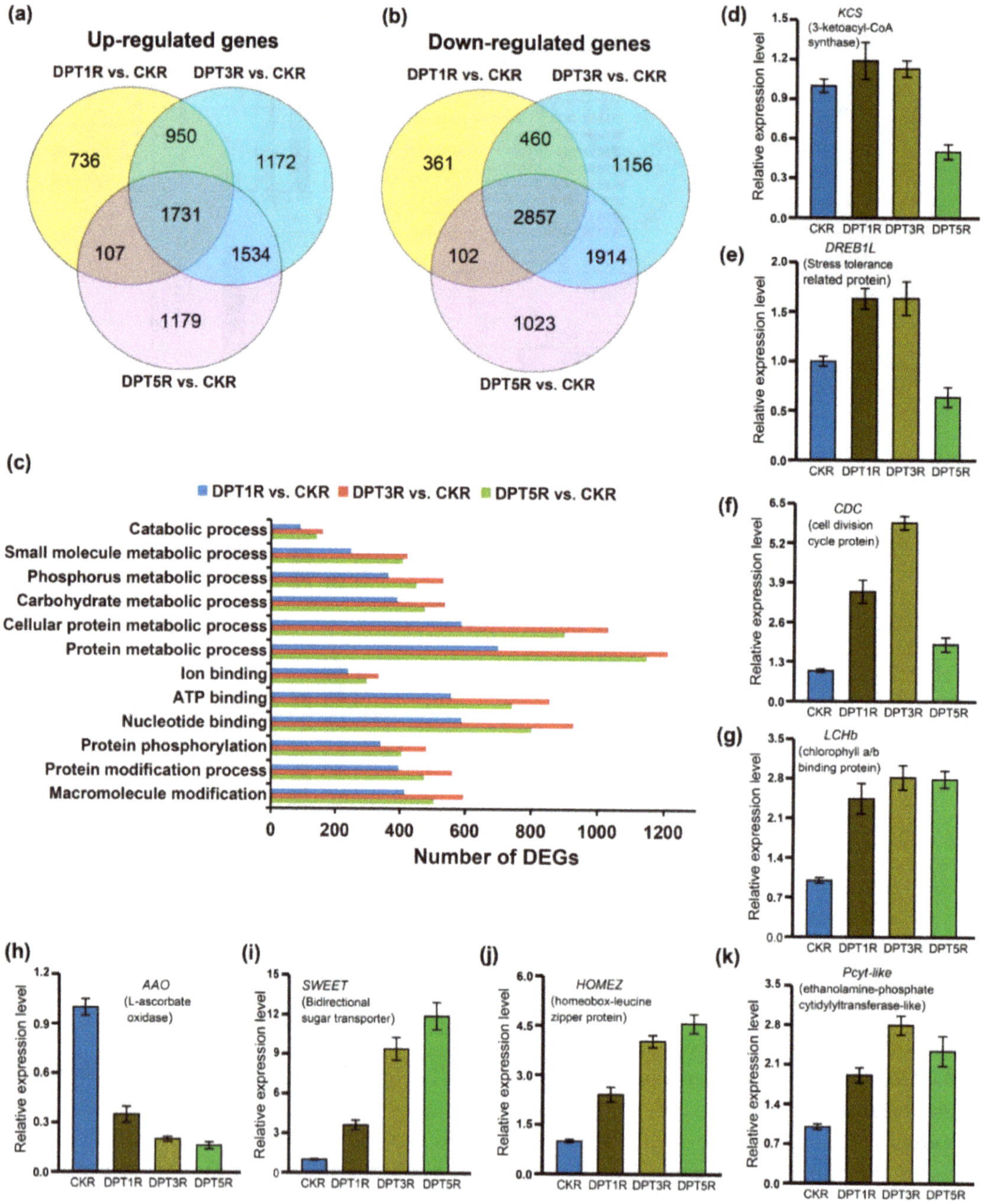

Figure 5. Comprehensive expression patterns of DEGs in the roots of common buckwheat seedlings under drought stress conditions. (**a**) Venn diagrams of the numbers of up-regulated [|log2 (Fold Change)| > 1 and qvalue < 0.005] genes acquired through the transcriptome analysis. (**b**) Venn diagrams of the numbers of down-regulated genes acquired through the transcriptome analysis. (**c**) Classification of DEGs based on metabolism, binding and modification categories. (**d**–**k**) Expression profiles of the selected DEGs, *KCS* (**d**), *DREB1L* (**e**), *CDC* (**f**), *LCHb* (**g**), *AAO* (**h**), *SWEET* (**i**), *HOMEZ* (**j**), *Pcyt-like* (**k**), determined by qRT-PCR.

3.6. Change in the Expression of Transcription Factors (TFs) Associated with Drought-Stress Response in Common Buckwheat Seedlings

Transcription factors (TFs) are important for regulating plant response to abiotic and biotic stresses. In the roots of the drought treated seedlings, large numbers of TFs were identified as DEGs, compared to the cotyledons (Figure 6a), and there were 180 TFs that were commonly identified in response to drought stress in both cotyledons and roots. Among them, the most differentially expressed TF families were the C2C2 family, follow by MYB, bZIP, HB and AP2/ERF (Figure 6b, Table S8). According to the GO enrichment analysis, 30.0%, 12.2%, and 9.4% of TFs were classified into "biological regulation", "intracellular", and "nucleic acid binding", respectively (Table S8). In addition, in order to reflect the major trends and patterns, 180 TFs were assigned to six clusters on the basis of their expression patterns. Those in cluster 1 and 2 were up-regulated by drought conditions, but the expression levels of the cluster 1 genes were high at cotyledons, while the cluster 2 genes were highly expressed at roots. Meanwhile, the cluster 3 and 4 genes showed the lowest expression level at cotyledons and roots. The cluster 5 genes were down-regulated by the drought condition. In contrast, there were 25 TFs in cluster 6, and these genes were up-regulated by water deficit, and the genes' expression was most high at cotyledons and roots. These results indicate that the expression of TFs was greatly affected by water deficit in the cotyledons and roots, and had different patterns between cotyledon and root tissues.

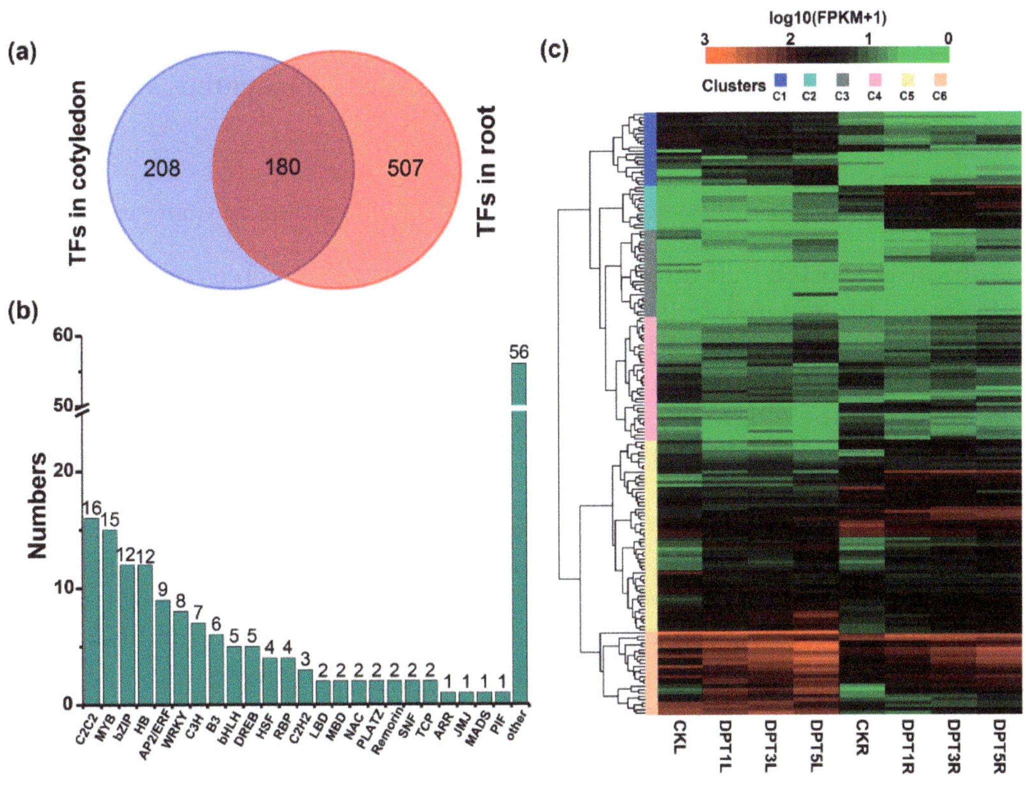

Figure 6. The differentially expressed TFs in cotyledons and roots responsive to drought stress. (**a**) Venn diagrams of TFs between cotyledons and roots. (**b**) Classification of TFs that were commonly identified in both cotyledon and root transcriptome libraries. (**c**) Expression pattern of TFs that were commonly identified in both cotyledons and roots in response to drought stress.

4. Discussion

Drought stress is one of the most detrimental environmental factors disturbing crop growth and production, and therefore understanding the drought-tolerance mechanism is pivotal for crop breeding [4,43]. Currently, RNA sequencing has been widely used to identify the drought-responsive pathway and genes that are activated during the seedling stage when exposed to abiotic stresses [5].

In this study, the phenotypic and physiological alterations of common buckwheat seedlings during drought stress were analyzed and characterized with the transcriptome analysis. Our results indicate that the common buckwheat seedlings relied on complex biological process to tackle the drought stress.

4.1. Morphological and Physiological Characteristics Related to Drought Stress in Common Buckwheat

Under drought stress, plant seedlings exhibit certain physiological and morphological variations [2,8], shown by leaf rolling and wilting [44,45], as well as the decreased RCW and wrinkled leaves (Figure 1a,d). Previous studies have demonstrated that drought stress also inhibits the photosynthesis of plants by affecting chlorophyll biosynthesis and facilitating stomatal closure [46,47], leading to the accumulation of MDA and ROS, which is harmful to the chloroplast photosystem II (PSII) [17,48]. As a result, plants have evolved antioxidant enzyme systems, such as superoxide dismutase (SOD), guaiacol peroxidase (GPX), CAT, and POD, to counteract the damage caused by drought stress [49]. MDA content has been considered important to reflect the drought tolerance ability of plants [12]. An active antioxidant capability in scavenging the cytotoxic ROS is preferred by plants in drought stress [50]. In common buckwheat seedlings, the ratio of chlorophyll a/b and RubisCO activities were significantly decreased compared to the control treatment in 5 days of treatment (Figure 2f, Figure 4e), and the expression levels of DEGs involved in the photosynthesis were correspondingly decreased DPT3L and DPT5L (Figure 4d), which may be due to the decrease in photosynthesis of the common buckwheat seedlings. By contrast, the CAT and POD activities were enhanced (Figure 2b,c). These observations suggest that the drought stress induced evident perturbations in the photosynthesis and ROS scavenging enzyme activities.

4.2. Multiple Biological Processes Are Involved in Drought Stress Responses in Common Buckwheat

Previous studies reported that multiple biological processes, such as oxidoreductase activities, and carbohydrate and protein metabolic processes, could be influenced when the ROS accumulation increased [51,52]. Our results indicated that the oxidoreductase activity, kinase activity, and DNA modification were upregulated by the drought treatment in common buckwheat cotyledons, whereas the genes were classified into "light harvesting" and "light reaction" (Figure 4c), and the expression of *LCHb* (encode chlorophyll a/b binding protein) and *PASN* (encode photosystem I reaction center subunit N protein) were markedly downregulated under drought conditions (Figure 4f,l). Repression of photosynthesis under drought stress also occurred in order to help the plants survive the water deficiency [53]. Cellular water deficit in plants caused by drought stress results in weakened carbon fixation, which may be physiologically ascribed to the stomatal closure and the biochemical inhibition of photosynthetic activities, further impacting the carbohydrate metabolism [54]. In addition, phosphorylation, as one of the reversible post-translational protein modification mechanisms, plays an important role in signaling the plant adaptation to osmotic stress [55,56], and a fine-tuned control of protein activity and function [56,57] has also been found to be altered in the common buckwheat root under drought stress. There were 92 DEGs involved in the protein modification and phosphorylation that were up-regulated (Figure 5c, Table S7), which may provide insights in the future study of the protein phosphorylation and modification events in plant drought stress.

4.3. Genes and Functional Proteins Responsive to Drought Stress

Previous studies have descried cellular changes that occur upon exposure to drought stress in plants, and the gene responses to drought stress have been studied spaciously in various species [1,9,11,53,58]. Moreover, the plant hormone ABA, as a signal-sensing molecule, can control the expression levels of stress-responsive genes, leading to cellular and physiological changes in response to water deficiency [59–61]. Moreover, previous studies have reported two regulatory systems: ABA-dependent and ABA-independent pathways, that play a major role for plant adaptation to drought stress [62]. Genes upstream and within the ABA pathway can be increased under drought conditions, and the *NCED* gene encoded 9-*cis* epoxycarotenoid dioxygenase has been shown to be induced under dehydration stress [61]. Furthermore, the changes of several metabolite levels under the water deficit condition were associated with the changes of biosynthetic gene expression, many of which were regulated by the changes of ABA accumulation levels [1,63]. In this study, a large number of DEGs were involved in ABA signaling and regulation (Table S5), and the detection of the up-regulation of the *NCED* genes in this study set identified the candidate genes for further studies of drought resistance in common buckwheat seedlings.

Recently, many drought-inducible genes involved in stress tolerance and stress responses have already been identified in several plant species [1], revealing that the transcription factors (TFs) play a central role in the biotic/abiotic stress responses [64–66]. In our previous study, we isolated and identified a *FeDREB1L* gene encoding a DREB-like transcription factor, which was simultaneously involved in the cold stress, drought stress, and ABA-mediated regulations [31]. The increased expression level of *FeDREB1L* during the earlier stage of drought stress displayed in this study (Figure 4m, Figure 5e) demonstrated that *FeDREB1L* could be a positive factor underpinning the drought stress resistance. Other TFs, including AP2/ERF, MYB, and bZIP families, were also identified to be differentially expressed in this study, for example, 180 DEGs that encoded TFs were identified in response to drought stress in both cotyledons and roots of the common buckwheat seedling, as well as the members of C2H2, MYB, bZIP, and WRKY families (Figure 6a, Table S8). Further studies are thereby required to elucidate the functions and gene-regulatory mechanisms of these TFs in response to plant drought stress.

5. Conclusions

To summarize, a comprehensive transcriptome profile of common buckwheat seedlings under drought stress was obtained using RNA-Seq technology. Phenotypic and physiological changes were determined, and the differentially expressed genes were analyzed to understand the regulatory mechanism of common buckwheat seedlings in response drought stress. The photosynthesis of the common buckwheat seedlings decreased, and the activities of antioxidant enzymes such as CAT and POD were increased under drought conditions. DEGs derived from important regulatory metabolisms were characterized. The results reflected in this study may provide useful information to better understand the molecular mechanism underlying the drought resistance in common buckwheat.

Supplementary Materials
Table S1: List of RT-qPCR primers; Table S2: Phenotypic changes of plant height, root length, and RWC under control conditions across four time points (0, 1, 3, and 5 days); Table S3: Physiological investigation of MDA content, POD and CAT activity, and chlorophyll content under control conditions across four time points (0, 1, 3, and 5 days); Table S4: Summary of the sequencing data of common buckwheat transcriptome; Table S5: Genes related to ABA metabolism in response to drought stress in common buckwheat cotyledons; Table S6: GO enrichment of differentially expressed genes in root transcriptome of common buckwheat seedlings under drought stress; Table S7: Genes related to protein modification and phosphorylation in response to drought stress in common buckwheat roots; Table S8: Classification of TFs that were commonly identified in both cotyledon and root transcriptome libraries; Figure S1: Confirmation of transcriptome data in cotyledons by qPCR analysis; Figure S2: Confirmation of transcriptome data in roots by qPCR analysis.

Author Contributions: Z.F., Z.H., and J.Y. designed the study and wrote the manuscript. Z.F., Y.L., J.S., S.W. (Shuping Wang), Z.L., S.W. (Shudong Wei), Y.Z., Z.H., and J.Y. participated in experiments. Z.H. submitted the raw data to Sequence Read Archive (SRA). Z.F., J.S., Z.L., Z.H., and J.Y. discussed the results and revised the manuscript. All authors have read and approved the final manuscript.

Acknowledgments: Authors acknowledge Xiaoyu Xu for critical reading of the manuscript.

References

1. Todaka, D.; Zhao, Y.; Yoshida, T.; Kudo, M.; Kidokoro, S.; Mizoi, J.; Kodaira, K.S.; Takebayashi, Y.; Kojima, M.; Sakakibara, H.; et al. Temporal and spatial changes in gene expression, metabolite accumulation and phytohormone content in rice seedlings grown under drought stress conditions. *Plant J.* **2017**, *90*, 61–78. [CrossRef] [PubMed]
2. Pan, J.; Li, Z.; Wang, Q.; Garrell, A.K.; Liu, M.; Guan, Y.; Zhou, W.; Liu, W. Comparative proteomic investigation of drought responses in foxtail millet. *BMC Plant Biol.* **2018**, *18*, 315. [CrossRef] [PubMed]
3. Xu, K.; Chen, S.; Li, T.; Ma, X.; Liang, X.; Ding, X.; Liu, H.; Luo, L. *OsGRAS23*, a rice GRAS transcription factor gene, is involved in drought stress response through regulating expression of stress-responsive genes. *BMC Plant Biol.* **2015**, *15*, 141. [CrossRef] [PubMed]
4. Chen, G.; Wang, Y.; Wang, X.; Yang, Q.; Quan, X.; Zeng, J.; Dai, F.; Zeng, F.; Wu, F.; Zhang, G.; et al. Leaf epidermis transcriptome reveals drought-Induced hormonal signaling for stomatal regulation in wild barley. *Plant Growth Regul.* **2018**, *87*, 39–54. [CrossRef]
5. Zhang, X.; Lei, L.; Lai, J.; Zhao, H.; Song, W. Effects of drought stress and water recovery on physiological responses and gene expression in maize seedlings. *BMC Plant Biol.* **2018**, *18*, 68. [CrossRef] [PubMed]
6. Turner, N.C.; Stern, W.R.; Evans, P. Water relations and osmotic adjustment of leaves and roots of lupins in response to water deficits. *Crop Sci.* **1987**, *27*, 977. [CrossRef]
7. Beck, E.H.; Fettig, S.; Knake, C.; Hartig, K.; Bhattarai, T. Specific and unspecific responses of plants to cold and drought. *J. Biosci.* **2007**, *32*, 501–510. [CrossRef]
8. Xiang, D.B.; Peng, L.X.; Zhao, J.L.; Zou, L.; Zhao, G.; Song, C. Effect of drought stress on yield, chlorophyll contents and photosynthesis in tartary buckwheat (*Fagopyrum tataricum*). *J. Food Agric. Environ.* **2013**, *11*, 1358–1363. [CrossRef]
9. Eom, S.H.; Baek, S.A.; Kim, J.K.; Hyun, T.K. Transcriptome analysis in Chinese Cabbage (*Brassica rapa* ssp. *pekinensis*) provides the role of glucosinolate metabolism in response to drought stress. *Molecules* **2018**, *23*, 1186. [CrossRef]
10. Joshi, R.; Wani, S.H.; Singh, B.; Bohra, A.; Dar, Z.A.; Lone, A.A.; Pareek, A.; Singla-Pareek, S.L. Transcription factors and plants response to drought stress: Current understanding and future directions. *Front. Plant Sci.* **2016**, *7*, 1029. [CrossRef]
11. Kumar, J.; Gunapati, S.; Kianian, S.F.; Singh, S.P. Comparative analysis of transcriptome in two wheat genotypes with contrasting levels of drought tolerance. *Protoplasma* **2018**, *255*, 1487–1504. [CrossRef] [PubMed]
12. Yin, J.; Jia, J.; Lian, Z.; Hu, Y.; Guo, J.; Huo, H.; Zhu, Y.; Gong, H. Silicon enhances the salt tolerance of cucumber through increasing polyamine accumulation and decreasing oxidative damage. *Ecotoxicol. Environ. Saf.* **2019**, *169*, 8–17. [CrossRef] [PubMed]
13. Luo, W.; Song, F.; Xie, Y. Trade-off between tolerance to drought and tolerance to flooding in three wetland plants. *Weltlands* **2008**, *28*, 866–873. [CrossRef]
14. Zhang, J.; Kirkham, M.B. Antioxidant responses to drought in sunflower and sorghum seedlings. *New Phytol.* **1996**, *132*, 361–373. [CrossRef] [PubMed]
15. Pagter, M.; Bragato, C.; Brix, H. Tolerance and physiological responses of *Phragmites australis* to water deficit. *Aquat. Bot.* **2005**, *81*, 285–299. [CrossRef]
16. Moller, I.M.; Jensen, P.E.; Hansson, A. Oxidative modifications to cellular components in plants. *Annu. Rev. Plant Biol.* **2007**, *58*, 459–481. [CrossRef]
17. Smirnoff, N. The role of active oxygen in the response of plants to water deficit and desiccation. *New Phytol.* **1993**, *125*, 27–58. [CrossRef]
18. Chakhchar, A.; Lamaoui, M.; Aissam, S.; Ferradous, A.; Wahbi, S.; El Mousadik, A.; Ibnsouda-Koraichi, S.; Filali-Maltouf, A.; El Modafar, C. Differential physiological and antioxidative responses to drought stress and recovery among four contrasting *Argania spinosa* ecotypes. *J. Plant Interact.* **2016**, *11*, 30–40. [CrossRef]
19. Prochazkova, D.; Sairam, R.K.; Srivastava, G.C.; Singh, D.V. Oxidative stress and antioxidant activity as the basis of senescence in maize leaves. *Plant Sci.* **2001**, *161*, 765–771. [CrossRef]
20. Kholová, J.; Hash, C.T.; Kočová, M.; Vadez, V. Does a terminal drought tolerance QTL contribute to differences in ROS scavenging enzymes and photosynthetic pigments in pearl millet exposed to drought. *Environ. Exp. Bot.* **2011**, *71*, 99–106. [CrossRef]

21. Apel, K.; Hirt, H. Reactive oxygen species: Metabolism, oxidative stress, and signal transduction. *Annu. Rev. Plant Biol.* **2004**, *55*, 373–399. [CrossRef] [PubMed]
22. Sharma, P.; Dubey, R.S. Drought induces oxidative stress and enhances the activities of antioxidant enzymes in growing rice seedlings. *Plant Growth Regul.* **2005**, *46*, 209–221. [CrossRef]
23. Ajithkumar, I.P.; Panneerselvam, R. ROS scavenging system, osmotic maintenance, pigment and growth status of Panicum sumatrense roth. Under drought stress. *Cell Biochem. Biophys.* **2014**, *68*, 587–595. [CrossRef] [PubMed]
24. Shinozaki, K.; Yamaguchi-Shinozaki, K. Gene Expression and signal transduction in water-stress response. *Plant Physiol.* **1997**, *115*, 327–334. [CrossRef] [PubMed]
25. Shinozaki, K.; Yamaguchi-Shinozaki, K. Gene networks involved in drought stress response and tolerance. *J. Exp. Bot.* **2007**, *58*, 221–227. [CrossRef] [PubMed]
26. Liu, X.; Zhang, R.; Ou, H.; Gui, Y.; Wei, J.; Zhou, H.; Tan, H.; Li, Y. Comprehensive transcriptome analysis reveals genes in response to water deficit in the leaves of *Saccharum narenga* (Nees ex Steud.) hack. *BMC Plant Biol.* **2018**, *18*, 250. [CrossRef]
27. Ji, T.; Li, S.; Li, L.; Huang, M.; Wang, X.; Wei, M.; Shi, Q.; Li, Y.; Gong, B.; Yang, F. Cucumber *Phospholipase D* alpha gene overexpression in tobacco enhanced drought stress tolerance by regulating stomatal closure and lipid peroxidation. *BMC Plant Biol.* **2018**, *18*, 355. [CrossRef]
28. Zhu, Y.; Yin, J.; Liang, Y.; Liu, J.; Jia, J.; Huo, H.; Wu, Z.; Yang, R.; Gong, H. Transcriptomic dynamics provide an insight into the mechanism for silicon-mediated alleviation of salt stress in cucumber plants. *Ecotoxicol. Environ. Saf.* **2019**, *174*, 245–254. [CrossRef]
29. Ito, Y.; Katsura, K.; Maruyama, K.; Taji, T.; Kobayashi, M.; Seki, M.; Shinozaki, K.; Yamaguchi-Shinozaki, K. Functional analysis of rice DREB1/CBF-type transcription factors involved in cold-responsive gene expression in transgenic rice. *Plant Cell Physiol.* **2006**, *47*, 141–153. [CrossRef]
30. Sakuma, Y.; Maruyama, K.; Osakabe, Y.; Qin, F.; Seki, M.; Shinozaki, K.; Yamaguchi-Shinozaki, K. Functional analysis of an *Arabidopsis* transcription factor, *DREB2A*, involved in drought-responsive gene expression. *Plant Cell* **2006**, *18*, 1292–1309. [CrossRef]
31. Fang, Z.; Zhang, X.; Gao, J.; Wang, P.; Xu, X.; Liu, Z.; Shen, S.; Feng, B. A buckwheat (*Fagopyrum esculentum*) DRE-Binding transcription factor gene, *FeDREB1*, enhances freezing and drought tolerance of transgenic *Arabidopsis*. *Plant Mol. Biol. Rep.* **2015**, *33*, 1510–1525. [CrossRef]
32. Germ, M.; Gaberščik, A. Chapter twenty one—The effect of environmental factors on buckwheat. *Mol. Breed. Nutr. Asp. Buckwheat* **2016**, 273–281. [CrossRef]
33. Fang, Z.; Hou, Z.; Wang, S.; Liu, Z.; Wei, S.; Zhang, Y.; Song, J.; Yin, J. Transcriptome analysis reveals the accumulation mechanism of anthocyanins in buckwheat (*Fagopyrum esculentum* Moench) cotyledons and flowers. *Int. J. Mol. Sci.* **2019**, *20*, 1493. [CrossRef] [PubMed]
34. Delpérée, C.; Kinet, J.M.; Lutts, S. Low irradiance modifies the effect of water stress on survival and growth-related parameters during the early developmental stages of buckwheat (*Fagopyrum esculentum*). *Physiol. Plant.* **2003**, *119*, 211–220. [CrossRef]
35. Cawoy, V.; Lutts, S.; Kinet, J.M. Osmotic stress at seedling stage impairs reproductive development in buckwheat (*Fagopyrum esculentum*). *Physiol. Plant.* **2006**, *128*, 689–700. [CrossRef]
36. Pan, L.; Meng, C.; Wang, J.; Ma, X.; Fan, X.; Yang, Z.; Zhou, M.; Zhang, X. Integrated omics data of two annual ryegrass (*Lolium multiflorum* L.) genotypes reveals core metabolic processes under drought stress. *BMC Plant Biol.* **2018**, *18*, 26. [CrossRef]
37. Harper, A.L.; von Gesjen, S.E.; Linford, A.S.; Peterson, M.P.; Faircloth, R.S.; Thissen, M.M.; Brusslan, J.A. Chlorophyllide a oxygenase mrna and protein levels correlate with the chlorophyll a/b ratio in *Arabidopsis Thaliana*. *Photosynth. Res.* **2004**, *79*, 149–159. [CrossRef]
38. Spanic, V.; Viljevac Vuletic, M.; Abicic, I.; Marcek, T. Early response of wheat antioxidant system with special reference to *Fusarium* head blight stress. *Plant Physiol. Biochem.* **2017**, *115*, 34–43. [CrossRef]
39. Yasui, Y.; Hirakawa, H.; Ueno, M.; Matsui, K.; Katsube-Tanaka, T.; Yang, S.J.; Aii, J.; Sato, S.; Mori, M. Assembly of the draft genome of buckwheat and its applications in identifying agronomically useful genes. *DNA Res.* **2016**, *23*, 215–224. [CrossRef]
40. Kim, D.; Pertea, G.; Trapnell, C.; Pimentel, H.; Kelley, R.; Salzberg, S.L. TopHat2: Accurate alignment of transcriptomes in the presence of insertions, deletions and gene fusions. *Genome Biol.* **2013**, *14*, R36. [CrossRef]

41. Lu, Q.-H.; Wang, Y.-Q.; Song, J.-N.; Yang, H.-B. Transcriptomic identification of salt-related genes and de novo assembly in common buckwheat (*F. esculentum*). *Plant Physiol. Biochem.* **2018**, *127*, 299–309. [CrossRef]
42. Yin, J.; Liu, M.; Ma, D.; Wu, J.; Li, S.; Zhu, Y.; Han, B. Identification of circular RNAs and their targets during tomato fruit ripening. *Postharvest Biol. Technol.* **2018**, *136*, 90–98. [CrossRef]
43. Ye, G.; Ma, Y.; Feng, Z.; Zhang, X. Transcriptomic analysis of drought stress responses of sea buckthorn (*Hippophae rhamnoides* subsp. sinensis) by RNA-Seq. *PLoS ONE* **2018**, *13*, e0202213. [CrossRef]
44. Lafitte, H.R.; Price, A.H.; Courtois, B. Yield response to water deficit in an upland rice mapping population: Associations among traits and genetic markers. *Theor. Appl. Genet.* **2004**, *109*, 1237–1246. [CrossRef]
45. Kadioglu, A.; Terzi, R.; Saruhan, N.; Saglam, A. Current advances in the investigation of leaf rolling caused by biotic and abiotic stress factors. *Plant Sci.* **2012**, *182*, 42–48. [CrossRef]
46. Flexas, J.; Bota, J.; Loreto, F.; Cornic, G.; Sharkey, T.D. Diffusive and metabolic limitations to photosynthesis under drought and salinity in C3 plants. *Plant Biol.* **2004**, *6*, 269–279. [CrossRef]
47. Iturbe-Ormaetxe, I.; Escuredo, P.R.; Arrese-Igor, C.; Becana, M. Oxidative damage in pea plants exposed to water deficit or paraquat. *Plant Physiol.* **1998**, *116*, 173–181. [CrossRef]
48. Demmig-Adams, B.; Adams, W.W. Photoprotection and other responses of plants to high light stress. *Annu. Rev. Plant Physiol. Plant Mol. Biol.* **1992**, *43*, 599–626. [CrossRef]
49. Sharma, P.; Jha, A.B.; Dubey, R.S.; Pessarakli, M. Reactive oxygen species, oxidative damage, and antioxidative defense mechanism in plants under stressful conditions. *J. Bot.* **2012**, *2012*, 1–26. [CrossRef]
50. Zaefyzadeh, M.; Quliyev, R.A.; Babayeva, S.M.; Abbasov, M.A. The effect of the interaction between genotypes and drought stress on the superoxide dismutase and chlorophyll content in durum wheat landraces. *Turk. J. Biol.* **2009**, *33*, 1–7. [CrossRef]
51. Hayano-Kanashiro, C.; Calderon-Vazquez, C.; Ibarra-Laclette, E.; Herrera-Estrella, L.; Simpson, J. Analysis of gene expression and physiological responses in three Mexican maize landraces under drought stress and recovery irrigation. *PLoS ONE* **2009**, *4*, e7531. [CrossRef]
52. Min, H.; Chen, C.; Wei, S.; Shang, X.; Sun, M.; Xia, R.; Liu, X.; Hao, D.; Chen, H.; Xie, Q. Identification of drought tolerant mechanisms in maize seedlings based on transcriptome analysis of recombination inbred lines. *Front. Plant Sci.* **2016**, *7*, 1080. [CrossRef]
53. Zhu, Y.; Gong, H.; Yin, J. Role of Silicon in mediating salt tolerance in plants: A Review. *Plants* **2019**, *8*, 147. [CrossRef]
54. Xue, G.P.; McIntyre, C.L.; Glassop, D.; Shorter, R. Use of expression analysis to dissect alterations in carbohydrate metabolism in wheat leaves during drought stress. *Plant Mol. Biol.* **2008**, *67*, 197–214. [CrossRef]
55. Hoyos, M.E.; Zhang, S. Calcium-Independent activation of salicylic acid-induced protein kinase and a 40-kilodalton protein kinase by hyperosmotic stress. *Plant Physiol.* **2000**, *122*, 1355–1363. [CrossRef]
56. Ren, J.; Mao, J.; Zuo, C.; Calderon-Urrea, A.; Dawuda, M.M.; Zhao, X.; Li, X.; Chen, B. Significant and unique changes in phosphorylation levels of four phosphoproteins in two apple rootstock genotypes under drought stress. *Mol. Genet. Genom.* **2017**, *292*, 1307–1322. [CrossRef]
57. Schulze, W.X. Proteomics approaches to understand protein phosphorylation in pathway modulation. *Curr. Opin. Plant Biol.* **2010**, *13*, 279–286. [CrossRef]
58. Zhu, Y.; Jia, J.; Yang, L.; Xia, Y.; Zhang, H.; Jia, J.; Zhou, R.; Nie, P.; Yin, J.; Ma, D.; et al. Identification of cucumber circular RNAs responsive to salt stress. *BMC Plant Biol.* **2019**, *19*, 164. [CrossRef]
59. Bartels, D.; Sunkar, R. Drought and Salt Tolerance in Plants. *Crit. Rev. Plant. Sci.* **2005**, *24*, 23–58. [CrossRef]
60. Schachtman, D.P.; Goodger, J.Q. Chemical root to shoot signaling under drought. *Trends Plant Sci.* **2008**, *13*, 281–287. [CrossRef]
61. Li, J.; Li, Y.; Yin, Z.; Jiang, J.; Zhang, M.; Guo, X.; Ye, Z.; Zhao, Y.; Xiong, H.; Zhang, Z.; et al. *OsASR5* enhances drought tolerance through a stomatal closure pathway associated with ABA and H_2O_2 signalling in rice. *Plant Biotechnol. J.* **2017**, *15*, 183–196. [CrossRef]
62. Yamaguchi-Shinozaki, K.; Shinozaki, K. Transcriptional regulatory networks in cellular responses and tolerance to dehydration and cold stresses. *Annu. Rev. Plant Biol.* **2006**, *57*, 781–803. [CrossRef]
63. Urano, K.; Maruyama, K.; Ogata, Y.; Morishita, Y.; Takeda, M.; Sakurai, N.; Suzuki, H.; Saito, K.; Shibata, D.; Kobayashi, M.; et al. Characterization of the ABA-regulated global responses to dehydration in *Arabidopsis* by metabolomics. *Plant J.* **2009**, *57*, 1065–1078. [CrossRef]

64. Shinozaki, K.; Yamaguchi-Shinozaki, K.; Seki, M. Regulatory network of gene expression in the drought and cold stress responses. *Curr. Opin. Plant Biol.* **2003**, *6*, 410–417. [CrossRef]
65. Kong, X.M.; Zhou, Q.; Luo, F.; Wei, B.D.; Wang, Y.J.; Sun, H.J.; Zhao, Y.B.; Ji, S.J. Transcriptome analysis of harvested bell peppers (*Capsicum annuum* L.) in response to cold stress. *Plant Physiol. Biochem.* **2019**, *139*, 314–324. [CrossRef]
66. Seo, P.J.; Xiang, F.; Qiao, M.; Park, J.Y.; Lee, Y.N.; Kim, S.G.; Lee, Y.H.; Park, W.J.; Park, C.M. The MYB96 transcription factor mediates abscisic acid signaling during drought stress response in *Arabidopsis*. *Plant Physiol.* **2009**, *151*, 275–289. [CrossRef]

6

Phytohormone-Mediated Stomatal Response, Escape and Quiescence Strategies in Plants under Flooding Stress

Kazi Khayrul Bashar [1,*], Md. Zablul Tareq [2], Md. Ruhul Amin [1], Ummay Honi [1], Md. Tahjib-Ul-Arif [3], Md. Abu Sadat [1] and Quazi Md. Mosaddeque Hossen [1]

1 Bangladesh Jute Research Institute, Manik Mia Avenue, Dhaka 1207, Bangladesh; shironbge03@gmail.com (M.R.A.); lethe_0000@yahoo.com (U.H.); sadat@snu.ac.kr (M.A.S.); mosaddequebjri@gmail.com (Q.M.M.H.)

2 Bangladesh Jute Research Institute, Jagir, Manikganj 1800, Bangladesh; zablulbarj@gmail.com

3 Department of Biochemistry and Molecular Biology, Bangladesh Agricultural University, Mymensingh 2202, Bangladesh; arif1002215@gmail.com

* Correspondence: kazi.khayrulbashar@gmail.com

Abstract: Generally, flooding causes waterlogging or submergence stress which is considered as one of the most important abiotic factors that severely hinders plant growth and development. Plants might not complete their life cycle even in short duration of flooding. As biologically intelligent organisms, plants always try to resist or survive under such adverse circumstances by adapting a wide array of mechanisms including hormonal homeostasis. Under this mechanism, plants try to adapt through diverse morphological, physiological and molecular changes, including the closing of stomata, elongating of petioles, hollow stems or internodes, or maintaining minimum physiological activity to store energy to combat post-flooding stress and to continue normal growth and development. Mainly, ethylene, gibberellins (GA) and abscisic acid (ABA) are directly and/or indirectly involved in hormonal homeostasis mechanisms. Responses of specific genes or transcription factors or reactive oxygen species (ROS) maintain the equilibrium between stomatal opening and closing, which is one of the fastest responses in plants when encountering flooding stress conditions. In this review paper, the sequential steps of some of the hormone-dependent survival mechanisms of plants under flooding stress conditions have been critically discussed.

Keywords: flood; plants; hormonal homeostasis; physiological activity

---

1. Introduction

Several oxygen limiting factors, such as flooding, waterlogging, and partial or full submergence are detrimental for normal growth and development of plants [1]. Sea, river belt and low land areas experience limited or reduced crop production due to the flooding stress. Plants try to adapt to these adverse conditions by applying several strategies, like the storage of energy, elongation of the petiole or internodes, maintenance of water level by regulating stomatal movements, the formation of adventitious roots, development of aerenchyma etc. [2–5]. Crop plants simultaneously activate various biochemical reactions, molecular and signaling pathways, and physiological processes to cope with this oxygen-limiting condition [1,6].

Phytohormone plays a central role in all morphological, anatomical, biochemical, molecular and signaling mechanisms for plant survival under oxygen-limiting stress conditions. Predominantly, ethylene, gibberellins (GA) and abscisic acid (ABA) play the most crucial roles during submergence stress conditions in plants [7]. Ethylene directly and/or indirectly induces GA expression that aids plants in carrying out the escape and/or quiescence strategy. Petiole/internode elongation and storage of carbohydrates results from the escape and quiescence strategy, respectively in plants under submergence stress [8]. GAs are directly involved in escaping submergence stress in both DELLA (N-terminal D-E-L-L-A amino acid sequence) dependent and independent pathways [9]. Furthermore, ethylene and ABA are directly responsible for the stomatal closure under waterlogging or post-waterlogging stress [10]. On the other hand, GAs are responsible for the stomatal opening under the same stress conditions [11]. So, balanced expression of these three hormones is highly indispensable to maintain the ratio of stomatal opening and closing in plants under excess water stress conditions.

Based on the above facts, this review paper proposes flowcharts of sequential steps for stomatal closing and escaping and quiescencing strategies to offer the best possible explanations. This might help plant scientists to consider several factors during the development of waterlogging-tolerant plant varieties.

2. Ethylene, GA and ABA Interactions in Plants under Submergence Stress

Under submergence stress, ethylene, GA and ABA play influential roles in the survivability of the submerged plants, where ABA biosynthesis is reduced and GA signaling is induced for shoot elongation, especially in rice plants (variety: C9285) [12,13].

Plants have evolved two types of strategies, i.e., escape strategy and quiescence strategy, to survive under flooding stress. In the escape strategy, the rice plant elongates its internodes under slow progressive flooding conditions. On the other hand, rice plants reserve energy under deep transient flash flooding conditions to escape the unfavorable conditions, which is termed the quiescence strategy [14,15]. It is quite interesting that two different functions of two distinct gene families under the same subgroup of a transcription factor are involved in submergence tolerance in plants. This is a complex process, but one that is interesting to study, and required to unveil this mechanism by functional genomics. Under the *AP2/ERF* (*Apelata2/Ethylene response factor*) transcription factor (TF) subgroup, the *Snorkel (SK)* gene is responsible for internode elongation, whereas *Sub1A* (*Submergence 1A*) is related to shoot elongation restriction [16].

In low land rice varieties, the *ERF* transcription factor, *Sub1*, is considered as a major player in submergence tolerance [17,18]. As a quiescence strategy in rice genotypes, ethylene directly enhances *Sub1A* expression. *Sub1A* induces the over accumulation of GA signaling repressors, *Slender Rice-1* (*SLR1*) and *SLR1 Like-1* (*SLRL1*). *Sub1A*, a group under VII *AP2/ERF* transcription factor, restricts shoot elongation by suppressing the *SLR1* and *SLRL2* for saving energy that is necessary for growth and development under desubmergence conditions. Therefore, ethylene is indirectly responsible for the induction of these GA signaling repressors and the reduction of GA responsive gene expression under submergence stress through the *Sub1A*-dependent pathway. *Sub1A* may also play an important role in limiting ethylene production during submergence stress conditions, resulting in the restriction of ethylene-induced enhancement of GA responsiveness in submergence tolerant varieties [17]. The function of *Sub1A1* is also regulated by MPK3 (Mitogen-activated protein kinase3). MPK3-dependent phosphorylated *Sub1A1* binds to the G-box of the MPK3 promoter to regulate its activity [19].

On the other hand, ethylene can directly enhance GA responsive shoot elongation in rice genotypes as an escape strategy of submergence stress, which is not only a highly energy-consuming process, but also requires the continuous production of energy. Due to the lack of *Sub1A* in submergence-susceptible genotypes, plants' survivability in the desubmergence stage is very low or limited [20]. Thus, *Sub1A* acts as a limiting factor for ethylene-promoted GA responsive shoot elongation in tolerant genotypes during submergence conditions to store or save the energy required for normal physiological and biochemical activities [21].

Moreover, in rice, *Sub1A* actively participates in maintaining chlorophyll contents and carbohydrate reserves in photosynthetic tissues [22]. *Sub1A* increases brassinosteroid (BR) levels in rice plants under submergence stress. BR induces GA catabolic *SLR1* proteins which restrict shoot elongation under oxygen-limiting conditions. In addition, BR enhances the expression of *GA2ox7* (*GA 2 oxidase7*) as an early response (within 1 d after submergence), which is responsible for catabolic GA degradation of endogenous GA_4 [23] (Figure 1). *OsAP2-39*, an *Apelata 2* (*AP2*) transcription factor, directly regulates the ABA biosynthetic gene *OsNCED1* (*9-cis-epoxycarotenoid dioxygenase 1*) and the GA repressing *EUI* (*Elongation of uppermost internode I*) gene. Over-expression of *OsAP2-39* has been shown to enhance drought resistance in rice by producing more ABA and degrading GA [24], which supports the antagonistic crosstalk between ABA and GA as a crucial mechanism to control plant growth and development under abiotic stress conditions [25]. From the above discussion, it can be said that *OsAP2-39* might have great scope to restrict GA signaling in plants under waterlogging or submergence conditions, and stomatal control through ABA signaling pathways under desubmergence conditions.

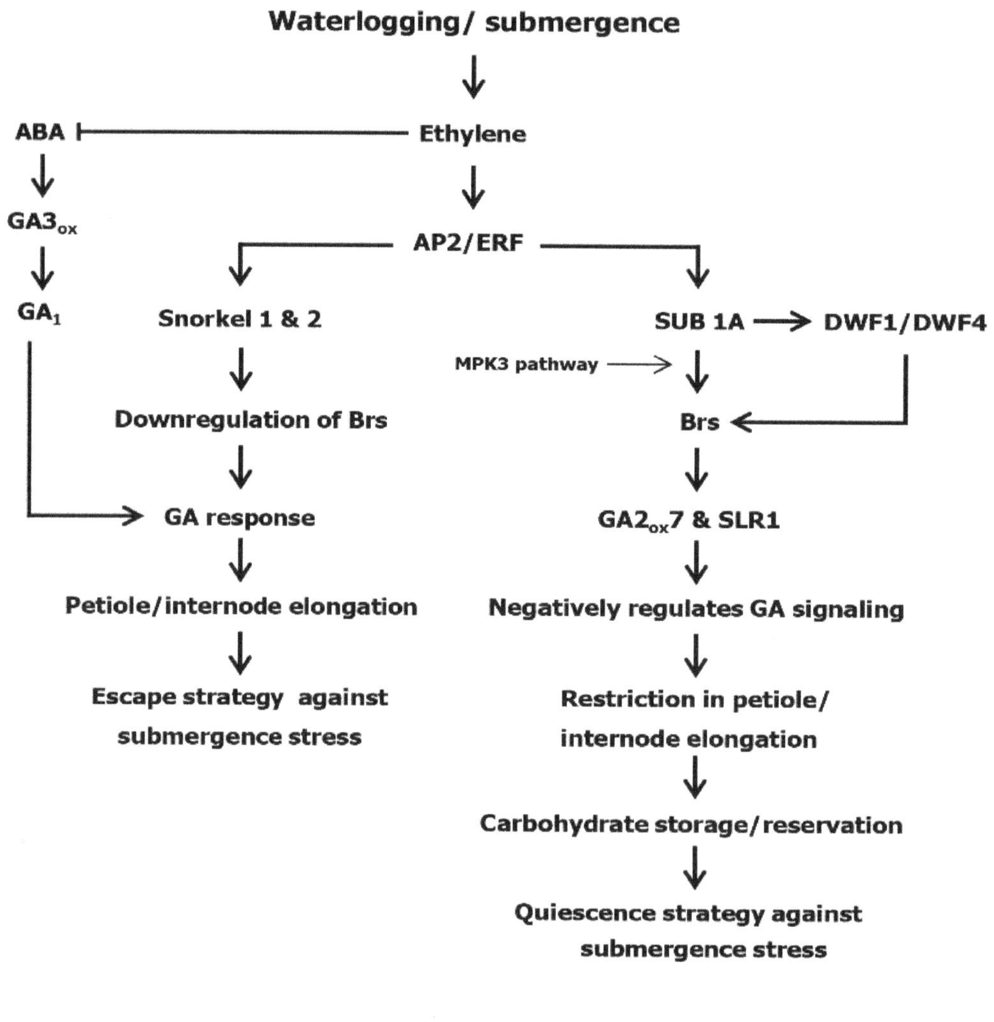

Figure 1. Ethylene-mediated escape and quiescence strategies in plants against submergence stress. Enhanced ethylene expression under submergence stress induces *Snorkel* and *Sub1A* genes for escape and quiescence strategies respectively. Ethylene suppresses ABA expression, which triggers GA_1 expression for escape strategy in plants. ABA: Abscisic acid; AP2/ERF: Apelata2/Ethylene response factor; Brs: Brassinosteroids; DWF1/4: Dwarf 1/4; GA_1: Gibberellins 1; GA2ox7: GA 2 oxidase7; GA3ox: GA 3 oxidase; Sub1A: Submergence1A; SLRL1: SLR1 Like-1.

In deep water rice, ethylene enhances the expression of *Snorkel1* (*SK1*) and *Snorkel2* (*SK2*) which are responsible for significant internode elongation via GA signaling pathways [26]. *Snorkel* genes are only present in lowland deepwater rice accessions for their internode elongation through downregulation of BR biosynthesis as an escape mechanism under submergence stress [1] (Figure 2).

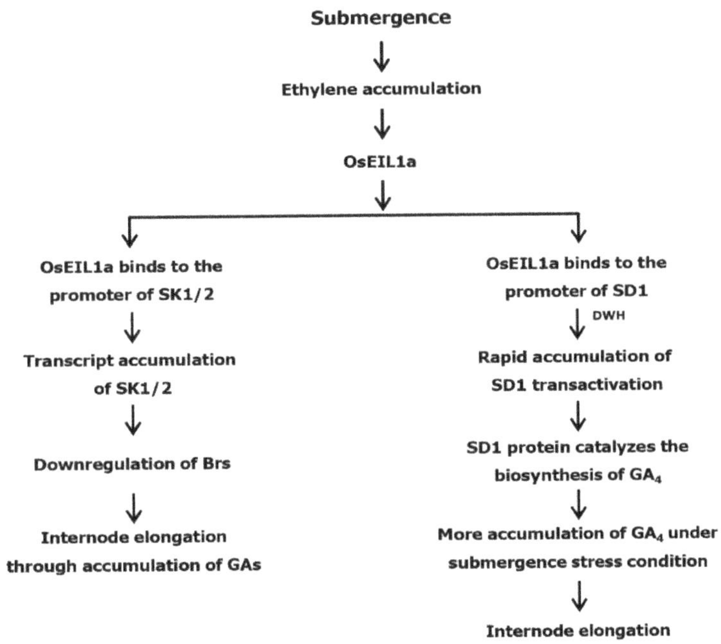

Figure 2. Proposed model for rice internode elongation under submergence stress conditions. Modified figure [13]. *EIL1a* binds with the both promoters of *SK1/2* and *SD1* in rice under submergence stress conditions. *SK1/2* downregulates Brs that induce internode elongation through the accumulation of GAs. On the other hand, *SD1* induces internode elongation through the accumulation of GA_4. Brs: brassinosteroids; EIL1a: Ethylene Insensitive 3-like 1a; GA: Gibberellins; SD1: Semidwarf1; SK1/2: Snorkel 1/2.

In rice, *Snorkel* dependent (variety: C9285) and independent (variety: T65) internode elongation for escaping the submergence stress conditions was discovered by [13]. In this escaping strategy the accumulated ethylene enhances *OsEIL1a* (*Ethylene Insensitive 3-like 1a*) in rice plants. In the *Snorkel* dependent escaping strategy, *OsEIL1a* binds to the promoter of *Snorkel1/2* to accumulate the transcript of *Snorkel1/2* [27]. This leads to the downregulation of BR to induce GA-mediated (mainly GA_1) internode elongation [22,25]. GA response enhances the expression of cyclins transcription factor, which leads to rapid cell division in lotus (*Nelumbo nucifera*) under submergence stress conditions [28]. However, as a *Snorkel* independent escaping strategy, *OsEIL1a* binds to the promoter of *SD1 (Semidwarf1)* for DWH (deepwater rice-specific haplotype) mediated rapid amplification of *SD1* transactivation. The SD1 protein catalyzes the biosynthesis of bioactive GA species, GA_4 that increases GA_4 level in addition to GA_1 after submergence. GA_4 is more capable of internode elongation than GA_1 [13]. So, *Snorkel* independent *SD1* mediated internode elongation in rice is comparatively faster than that of the *Snorkel* dependent pathway (Figure 2).

Rumex plants showed *Snorkel* independent petiole elongation. In flood-tolerant *Rumex palustris,* ethylene reduces *RpNCED* expression which inhibits ABA biosynthesis. Thus, *R. palustris* elongates its petiole by degrading ABA into phaseic acid and enhancing GA_1-mediated gene expression in an ethylene-mediated pathway under oxygen-limited condition [29]. *R. palustris* maintains gas exchange between the submerged tissues and the atmosphere by elongating shoots under long-term flooding stress condition [29,30]. In this stress condition, the accumulation of ethylene not only breaks down ABA into phaseic acid, but also downregulates ABA expression by inhibiting 9-cis-epoxycarotenoid dioxygenase expression. Elevated content of ethylene independently degrades ABA through ABA 8′ hydroxylase pathway under submergence stress conditions [31]. Inhibition of ABA stimulates GA 3-oxidase to produce bioactive GA (GA_1). The downstream function of GA is to mobilize food materials by the breakdown of starch and cell wall loosening, which ultimately elongates internode or leaf sheath to escape submergence or waterlogging stress conditions [30]. In rice, a similar type of ABA-dependent GA expression was also reported [32]. Both the quiescence strategy and escape strategy are considered as the survival mechanisms of plants under submergence stress, but escape strategy is considered as a yield limiting factor in rice plants under de-submergence stress [33].

3. DELLA-Dependent GA Expression under Submergence Stress

Gibberellin-insensitive dwarf 1 (GID1), a soluble receptor for GA signaling, is involved in GA-mediated signaling pathways in plants under stress conditions, especially during abiotic stress conditions [34,35]. GA binds to GID1 and generates a GID1-GA complex which has the ability to interact with different growth repressors like DELLA or SLENDER1 (SLR1), a DELLA ortholog in rice [33]. Five types of DELLA are present in the model plant *Arabidopsis thaliana*, namely Gibberellin-insensitive (GAI), Repressor of GA1-3 (RGA), RGA-like1 (RGL1), RGL2 and RGL3 [36]. GID1-GA complex facilitates GA-mediated interaction between GID1 and DELLA protein, which is responsible for conformational changes in DELLA proteins. The *Sleepy1* (*SLY1*) gene contain the F-box domain, which is a positive regulator of GA signaling in *Arabidopsis*. *Sleepy1* (*SLY1*) in *Arabidopsis* and GID2 in rice are capable of recognizing this change of DELLA proteins where *SCF* (*SLY1*) (Skp1, Cullin, F-box), E3 ubiquitin ligase ubiquitinates DELLA protein, targeting DELLA for degradation through proteolysis by the 26S proteasome [34,35,37]. Thus GA expression is continued in plants by suppressing the expression of DELLA protein (Figure 3A).

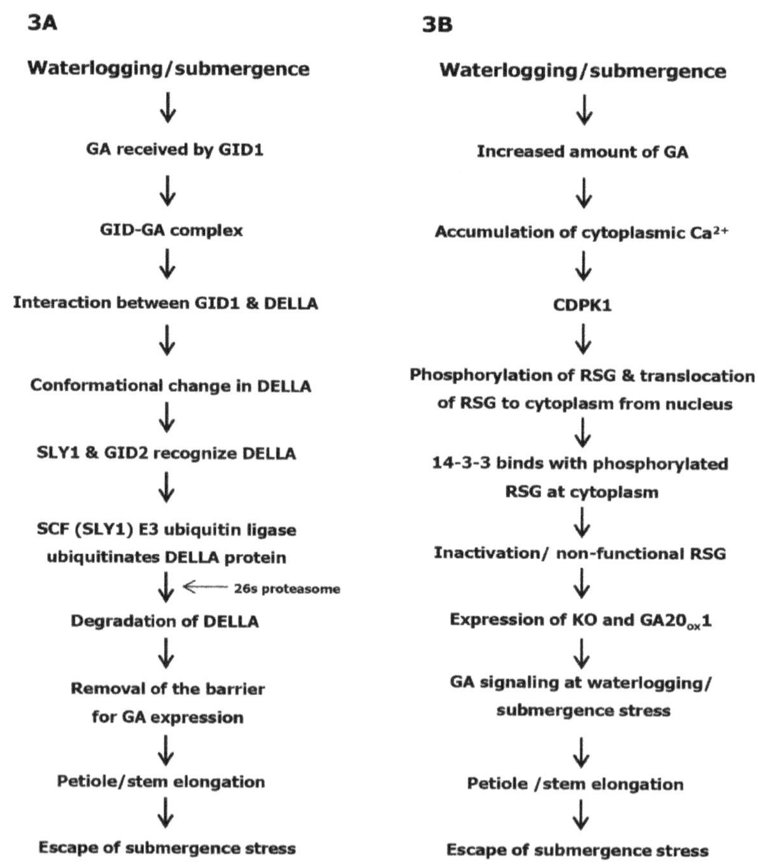

Figure 3. DELLA-dependent and -independent GA signaling in plants under submergence stress conditions. (**3A**) DELLA-dependent escape strategy in plants; GID-GA complex makes conformational change in DELLA. After that SCF ubiquitinates DELLA protein which induces GA expression for petiole/stem elongation. (**3B**) DELLA-independent escape strategy in plants; Ca^{2+} accumulation enhances CDPK1 production for the phosphorylation of RSG to translocate it into the cytoplasm from nucleus. 14-3-3 inactivates RSG as a GA-mediated submergence escaping strategy. CDPK1: Ca2+-dependent protein kinase 1; DELLA: N-terminal D-E-L-L-A amino acid sequence; GA: Gibberellins; GA20ox1: GA 20 oxidase1; GID1: Gibberellin-insensitive dwarf 1; GID2: Gibberellin-insensitive dwarf 2; RSG: Repression of shoot growth; SCF: Skp1, Cullin, F-box; SLY1: Sleepy1.

4. DELLA Independent GA Expression under Submergence Stress

GAs can increase cytoplasmic Ca^{2+} very rapidly during various cellular processes occurring inside the cell. However, the mechanism of increasing GA-mediated cytoplasmic Ca^{2+} in a DELLA-independent manner is still unknown. The GA-mediated increase of Ca^{2+} and degradation of DELLAs are completely independent processes [38]. An elevated level of cytoplasmic Ca^{2+} activates Ca^{2+}-dependent protein kinase (NtCDPK1) via a DELLA-independent GA pathway in tobacco [39]. NtCDPK1 is responsible for the translocation of Repression of shoot growth (RSG) from the nucleus to the cytoplasm [40,41]. RSG represses the expression of two important GA biosynthetic genes, *NtKO* and *NtGA20ox1*, which are considered as GA enhancing genes [42,43]. RSG binds to the promoter region of *NtGA20ox1* through GA-mediated approach while RSG binds to *NtKO* promoter in the independent manner of GA concentrations [43]. NtCDPK1 acts as a RSG kinase and phosphorylates RSG, which promotes 14-3-3 to bind with RSG at cytoplasm [40,44,45]. In rice plants, CDPK is induced under low oxygen stress for survivability under anaerobic conditions [46]. 14-3-3 proteins directly bind to the RSG in the cytoplasm and regulate RSG function negatively, making non-functional RSG [41,47]. So, inactivation of RSG by 14-3-3 proteins indirectly helps to express both *NtKO* and *NtGA20ox1* to continue GA biosynthesis for different cellular processes (Figure 3B).

5. ABA in Plants under Waterlogging Stress

Imbalanced conditions between leaf transpiration and root water uptake creates dehydration (physiological drought) stress in plants, that is noticed in plants under waterlogging stress with partially or fully damaged root tissues. The plant hormone ABA has the ability to modify root hydraulic properties [48]. For example, ABA downregulation and upregulation in tomato plants expresses lower and higher hydraulic conductance, respectively [49,50]. Depending on both the flooding duration and plant species, ABA differentially responds in leaves and in the roots of plants. *Malus sieversii* is considered as less tolerant to hypoxia than *Malus hupehensis,* showing a larger increase in ABA in both the leaf and root tissues [51]. In *Gerbera jamesonii,* ABA levels were increased in both leaf and root, where a transient increase in root follows a sharp decrease in the recovery period [52]. ABA is also increased in roots and leaves of *Triticum aestivum* L., but gradually decreases after reaching certain levels [53].

6. Stomatal Regulation at Waterlogging Stress

Stomata is a specialized epidermal pore-like structure consisting of two guard cells through which plants exchange both CO_2 and O_2 with the environment [54,55]. Stomatal conductance was insignificantly affected by flooding stress despite a significant reduction of photosynthesis in both flood-tolerant and flood-sensitive poplar genotypes [56]. A similar finding was also reported in maize plants where flooding resulted in a significant decrease in photosynthesis and ribulose-1,5-bisphosphate carboxylase activity without a noticeable reduction in the rates of stomatal conductance [57]. In GID1 mutated rice plant, increased chlorophyll content has been found, which is subsequently responsible for the increase of carbohydrate production and decrease of reactive oxygen species (ROS) accumulation under submergence stress [58]. Reduction in leaf transpiration is a common phenomenon of flooding stress, which affects the lowered stomatal aperture [59–61]. It is difficult to believe that flooding paradoxically causes leaf dehydration in plants [62–64]. Stomatal closing is operated by the turgor pressure and volume of guard cells. ABA-dependent signaling effluxes of anions, potassium ions and conversion of malate into starch trigger the reduction of turgor pressure, as well as changing the volume of guard cells close the stomata [65]. Under hypoxic stress, *hypoxia responsive universal stress protein 1* (*HRU1*) activates *AtrbohD* (*Respiratory burst oxidase homolog protein D*), following interaction between *ROP2* (*Rho of plants 2*) and *AtrbohD* in *Arabidopsis* under low oxygen stress conditions. NADPH oxidase *AtrbohD* is responsible for alcoholic fermentation and ABA-dependent stomatal closure in plants under abundant water conditions [66,67]. It has been reported that barley plants swiftly close the stomata after flooding stress imposition [68]. A similar phenomenon was also observed in pea plants under flooding conditions, where a prompt closure of stomata from older leaves was recorded. Wilting of younger leaves was protected by older pea leaves by increasing ABA-dependent stomatal closure [69]. Unwilted younger pea leaves might result from the ABA transportation from older to younger leaves or de novo biosynthesis of ABA in the younger leaves [70]. Though *Populus deltoides* is considered as a waterlogging tolerant plant species, it showed a significant reduction in stomatal conductance at both waterlogging and de-waterlogging stress, with an exception at recovery period after 90 days of waterlogging stress [71].

Hormonal Regulation in Stomatal Closing of Plants under Waterlogging Stress

Several waterlogging related experiments showed a positive correlation between ABA accumulation and increase in ROS, in soybean roots [72], barley roots and leaves [73], maize leaves [74], bread wheat roots [75]. The enzymatic mechanisms responsible for ABA-triggered ROS generation in guard cells at the molecular level are little known at present. ABA initiates H_2O_2 generation by using the plasma membrane NADPH oxidase [76]. H_2O_2 activates plasma membrane Ca^{2+} channels, resulting in an increase in Ca^{2+} level in guard cells. [77]. Inhibition of inward K^+ channels in guard cells is the result of increased Ca^{2+} level in the cytoplasm of guard cells [78,79]. This results in reduced solute accumulation following a reduced amount of water entrance in the guard cells and ultimately leading to stomata closure [80]. In *Arabidopsis*, H_2O_2 can also stimulate NO (nitric oxide) production to induce stomatal closure [81]. Ethylene and ABA activate CuAO (copper amine oxidase) in *Vicia faba* [82,83]. Oxidation of putrescine by CuAO produces H_2O_2, follows stomatal closure in *V. faba* [82] (Figure 4). ABA is also responsible for stomatal closing in plants as a survival mechanism under post-waterlogging stress [84].

Moreover, H_2O_2 is also produced by extracellular calmodulin (ExtCaM) which is activated by heterotrophic G protein [85]. In rice, accumulated ethylene induces G protein for aerenchyma formation under flooding stress [86]. G protein induces H_2O_2 production for epidermal cell death in rice under submergence stress [87]. Inactivation of CTR1 (Constitutive triple response 1) is induced by the binding of ethylene to ETR1 (Ethylene receptor 1), ERS1 (Ethylene response sensor 1) and EIN4 (Ethylene insensitive 4), resulting in the activation of G_{alpha} (G protein alpha subunit). G_{alpha} promotes H_2O_2 production in plants via NADPH oxidases. ETR1 and ERS1 translocate the signals of H_2O_2 to EIN2, EIN3 and ARR2 (2-component response regulator), which are essential for stomatal closure functioning [88]. Activated G proteins may inhibit inward K^+ channels via an elevated level of cytoplasmic Ca^{2+} in the guard cells [78,79]. From the above discussion, a proposed signaling pathway may work for stomatal closure in plants under waterlogging stress (Figure 4).

In *Arabidopsis thaliana*, flooding stress operates H_2O_2-mediated stomatal closure followed by an increase in antioxidant enzyme activities [89]. Improvement of anoxia tolerance was confirmed by applying exogenous ABA in different plants including maize, citrus, lettuce and *Arabidopsis* [90–94]. Waterlogged pea plant restricted its leaf ABA to translocate in shoot, while in non-waterlogged plants, ABA moves readily from shoots to roots [95,96]. This extra ABA in pea leaves was responsible for reducing leaf transpiration by the closing of stomata [96].

Stomata closing may occur in an ABA-independent manner. As for example, ABA was increased in citrus after 3 weeks of flooding, indicating that closing of stomata and increase of ABA are independent in citrus plants upon flooding stress, and not only that ABA was transported to younger leaves from older leaves, rather than, as expected, being transportated from roots to shoots [97].

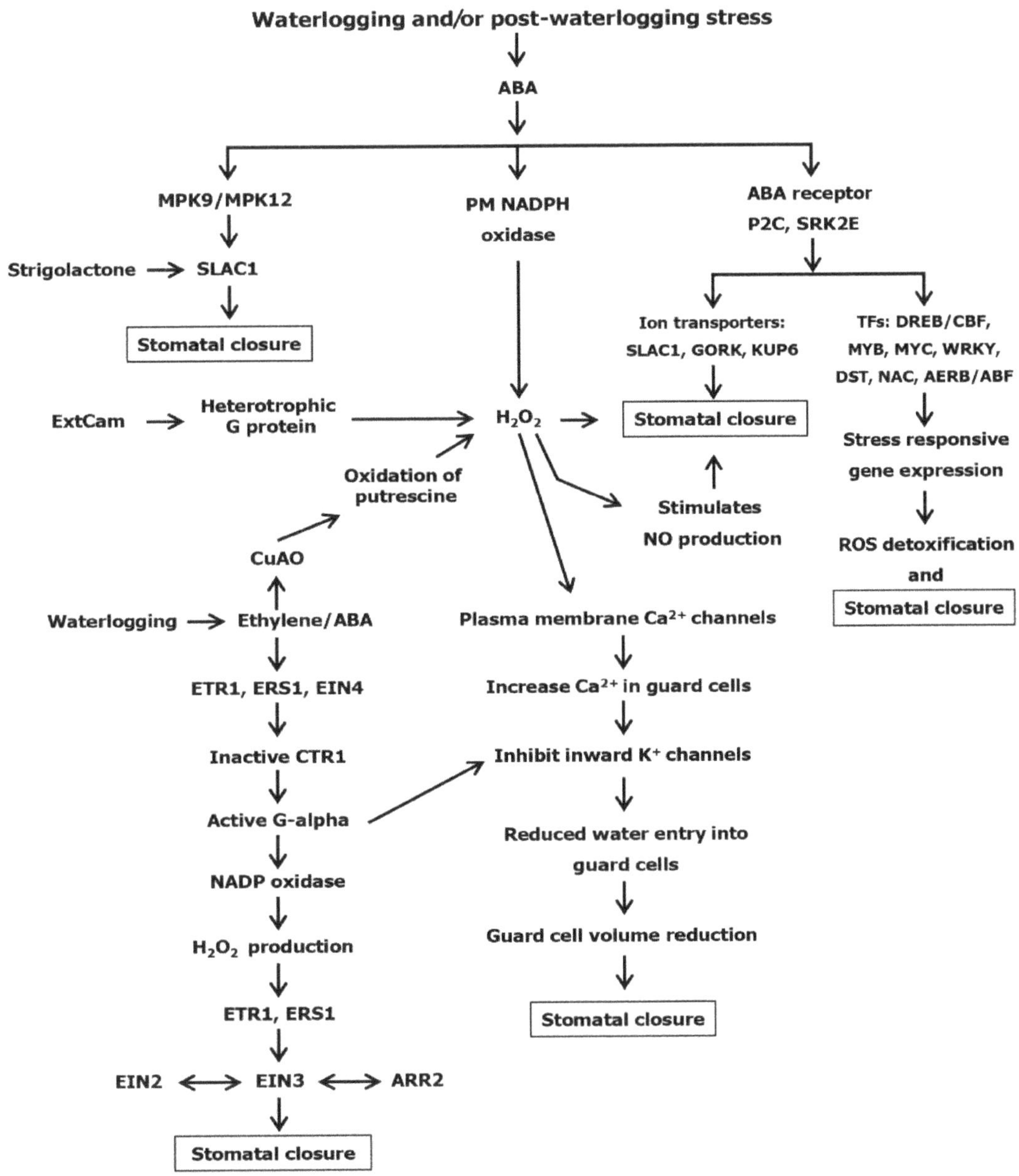

Figure 4. Proposed model for hormone mediated stomatal closure under waterlogging and/or post-waterlogging stress conditions. Modified figure [84]. Ethylene and/or ABA are directly or indirectly enhance H_2O_2 production under waterlogging and/or post-waterlogging stress conditions. H_2O_2 is directly responsible for closing of stomata or indirectly increasing Ca^{2+} in the guard cell for stomatal closure by reducing the volume of the guard cell. ABA: Abscisic acid; AREB/ABF: (ABA)-responsive element binding proteins/ABRE(ABA-responsive element)-binding factors; ARR2: 2-component response regulator; CTR1: Constitutive triple response 1; CuAO: Copper amine oxisase; DREB/CBF: Dehydration-responsive element-binding/C-repeat binding factor; DST: Drought and salt tolerance; EIN: Ethylene Insensitive; ERS1: Ethylene response sensor 1; ETR1: Ethylene receptor 1; ExtCaM: Extracellular Calmodulin; GORK: Guard cell outward-rectifying K+; KUP6: K+ uptake transporter 6; MPK3/9: Mitogen-activated protein kinase3/9; MYB: **My**elo**b**lastosis; MYC2: **My**elo**c**ytomatosis; NAC: NAM (No apical meristem), ATAF (Arabidopsis transcription activation factor) and CUC (Cup-shaped cotyledon) transcription factor; P2C: Protein phosphatase 2C; ROS: Reactive oxygen species; SLAC1: Slow anion channel-associated 1; SRK2E: SNF1 (Sucrose nonfermenting 1)-related protein kinase 2; WRKY: W-tryptophan, R-arginine, K-lysine, Y-tyrosine.

7. Conclusions

From this above discussion, it can be concluded that ethylene directly or indirectly regulates the expression of gibberellins (GAs) and abscisic acid (ABA) in plants under flooding stress conditions. However, ABA plays a major role in stomatal closing, whereas escape and quiescence strategies are controlled by the expression of GA. Finally, it can be concluded that plants maintain their internal homeostasis by balancing hormonal cross talk under excess water stress.

Author Contributions: Conceptualization and writing-original draft preparation, K.K.B.; Writing-review and editing, K.K.B.; M.Z.T, M.R.A., U.H. and M.T.-U.-A.; Supervision, M.A.S. and Q.M.M.H.

References

1. Tamang, B.G.; Fukao, T. Plant adaptation to multiple stresses during submergence and following desubmergence. *Int. J. Mol. Sci.* **2015**, *16*, 30164–30180. [CrossRef] [PubMed]
2. Bailey-Serres, J.; Fukao, T.; Ronald, P.; Ismail, A.; Heuer, S.; Mackill, D. Submergence tolerant rice: SUB1's journey from landrace to modern cultivar. *Rice* **2010**, *3*, 138–147. [CrossRef]
3. Verboven, P.; Pedersen, O.; Quang, T.H.; Nicolai, B.M.; Colmer, T.D. The mechanism of improved aeration due to gas films on leaves of submerged rice. *Plant Cell Environ.* **2014**, *37*, 2433–2452. [CrossRef] [PubMed]
4. Eysholdt-Derzsó, E.; Sauter, M. Hypoxia and the group VII ethylene response transcription factor HRE2 promote adventitious root elongation in *Arabidopsis*. *Plant biol.* **2018**, *1*, 103–108. [CrossRef] [PubMed]
5. Yamauchi, T.; Colmer, T.D.; Pedersen, O.; Nakazono, M. Regulation of root traits for internal aeration and tolerance to soil waterlogging–flooding Stress. *Plant Physiol.* **2018**, *176*, 1118–1130. [CrossRef]
6. Lin, I.; Wu, Y.; Chen, C.; Chen, G.; Hwang, S.; Jauh, G.; Tzen, J.T.C.; Yang, C. AtRBOH I confers submergence tolerance and is involved in auxin-mediated signaling pathways under hypoxic stress. *Plant Growth Regul.* **2017**, *83*, 277–285. [CrossRef]
7. Phukan, U.J.; Mishra, S.; Shukla, R.K. Waterlogging and submergence stress: Affects and acclimation. *Crit. Rev. Biotechnol.* **2015**, *36*, 956–966. [CrossRef] [PubMed]
8. Xiang, J.; Wu, H.; Zhang, Y.; Zhang, Y.; Wang, Y.; Li, Z.; Lin, H.; Chen, H.; Zhang, J.; Zhu, D. Transcriptomic analysis of gibberellin- and paclobutrazol-treated rice seedlings under submergence. *Int. J. Mol. Sci.* **2017**, *18*, 2225. [CrossRef]
9. Colebrook, E.H.; Thomas, S.G.; Phillips, A.L.; Hedden, P. The role of gibberellin signaling in plant responses to abiotic stress. *J. Exp. Biol.* **2014**, *217*, 67–75. [CrossRef] [PubMed]
10. Else, M.A.; Janowiak, F.; Atkinson, C.J.; Jackson, M.B. Root signals and stomatal closure in relation to photosynthesis, chlorophyll a fluorescence and adventitious rooting of flooded tomato plants. *Ann. Bot.* **2009**, *103*, 313–323. [CrossRef]
11. Göring, H.; Koshuchowa, S.; Deckert, C. Influence of gibberellic acid on stomatal movement. *Biochem. Physiol. Pflanzen* **1990**, *186*, 367–374. [CrossRef]
12. Bailey-Serres, J.; Voesenek, L.A.C.J. Life in the balance: A signaling network controlling survival of flooding. *Curr. Opin. Plant Biol.* **2010**, *13*, 489–494. [CrossRef] [PubMed]
13. Kuroha, T.; Nagai, K.; Gamuyao, R.; Wang, D.R.; Furuta, T.; Nakamori, M.; Kitaoka, T.; Adach, K.; Minami, A.; Mori, Y.; et al. Ethylene-gibberellin signaling underlies adaptation of rice to periodic flooding. *Science* **2018**, *361*, 181–186. [CrossRef] [PubMed]
14. Bailey-Serres, J.; Fukao, T.; Gibbs, D.J.; Holdsworth, M.J.; Lee, S.C.; Licausi, F.; Perata, P.; Voesenek, L.A.C.J.; Dongen, J.T.V. Making sense of low oxygen sensing. *Trends Plant Sci.* **2012**, *17*, 129–138. [CrossRef] [PubMed]
15. Fukao, T.; Xiong, L. Genetic mechanisms conferring adaptation to submergence and drought in rice: Simple or complex? *Curr. Opin. Plant Biol.* **2013**, *16*, 196–204. [CrossRef] [PubMed]
16. Jung, K.; Seo, Y.; Walia, H.; Cao, P.; Fukao, T.; Canlas, P.E.; Amonpant, F.; Bailey-Serres, J.; Ronald, P.C. The submergence tolerance regulator Sub1A mediates stress-responsive expression of AP2/ERF transcription factors. *Plant Physiol.* **2010**, *152*, 1674–1692. [CrossRef] [PubMed]
17. Fukao, T.; Xu, K.; Ronald, P.C.; Bailey-Serres, J. A variable cluster of ethylene response factor-like genes regulates metabolic and developmental acclimation responses to submergence in rice. *Plant Cell* **2006**, *18*, 2021–2034. [CrossRef] [PubMed]

18. Xu, K.; Xu, X.; Fukao, T.; Canlas, P.; Maghirang-Rodriguez, R.; Heuer, S.; Ismail, A.M.; Bailey-Serres, J.; Ronald, P.C.; Mackill, D.J. Sub1A is an ethylene-response-factor-like gene that confers submergence tolerance to rice. *Nature* **2006**, *442*, 705–708. [CrossRef] [PubMed]
19. Singh, P.; Sinha, A.K. A positive feedback loop governed by SUB1A1 interaction with Mitogen-activated protein kinase3 imparts submergence tolerance in rice. *Plant Cell.* **2016**, *28*, 1127–1143. [CrossRef] [PubMed]
20. Locke, A.M.; Gregory, A.B., Jr.; Sathnur, S.; Larive, C.K.; Bailey-Serres, J. Rice SUB1A constrains remodeling of the transcriptome and metabolome during submergence to facilitate post-submergence recovery. *Plant Cell Environ.* **2018**, *41*, 721–736. [CrossRef] [PubMed]
21. Fukao, T.; Bailey-Serres, J. Submergence tolerance conferred by Sub1A is mediated by SLR1 and SLRL1 restriction of gibberellin responses in rice. *PNAS* **2008**, *105*, 16814–16819. [CrossRef] [PubMed]
22. Fukao, T.; Yeung, E.; Bailey-Serres, J. The submergence tolerance gene SUB1A delays leaf senescence under prolonged darkness through hormonal regulation in rice. *Plant Physiol.* **2012**, *160*, 1795–1807. [CrossRef] [PubMed]
23. Schmitz, A.J.; Folsom, J.J.; Jikamaru, Y.; Ronald, P.; Walia, H. SUB1A-mediated submergence tolerance response in rice involves differential regulation of the brassinosteroid pathway. *New Phytol.* **2013**, *198*, 1060–1070. [CrossRef] [PubMed]
24. Yaish, M.W.; El-kereamy, A.; Zhu, T.; Beatty, P.H.; Good, A.G.; Bi, Y.; Rothstein, S.J. The APETALA-2-Like transcription factor OsAP2-39 controls key interactions between abscisic acid and gibberellin in rice. *PLoS Genet.* **2010**, *6*, e1001098. [CrossRef] [PubMed]
25. Seo, M.; Hanada, A.; Kuwahara, A.; Endo, A.; Okamoto, M.; Yamauchi, Y.; North, H.; Marion-Poll, A.; Sun, T.; Koshiba, T.; et al. Regulation of hormone metabolism in Arabidopsis seeds: Phytochrome regulation of abscisic acid metabolism and abscisic acid regulation of gibberellin metabolism. *Plant J.* **2006**, *48*, 354–366. [CrossRef] [PubMed]
26. Hattori, Y.; Nagai, K.; Furukawa, S.; Song, X.; Kawano, R.; Sakakibara, H.; Wu, J.; Matsumoto, T.; Yoshimura, A.; Kitano, H.; et al. The ethylene response factors *SNORKEL1* and *SNORKEL2* allow rice to adapt to deep water. *Nature* **2009**, *460*, 1026–1030. [CrossRef] [PubMed]
27. Voesenek, L.A.; Bailey-Serres, J. Flood adaptive traits and processes: An overview. *New Phytol.* **2015**, *206*, 57–73. [CrossRef] [PubMed]
28. Wang, B.; Jin, Q.; Zhang, X.; Mattson, N.S.; Ren, H.; Cao, J.; Wang, Y.; Yao, D.; Xu, Y. Genome-wide transcriptional analysis of submerged lotus reveals cooperative regulation and gene responses. *Sci. Rep.* **2018**, *8*, 918. [CrossRef]
29. Benschop, J.J.; Jackson, M.B.; Guhl, K.; Vreeburg, R.A.M.; Croker, S.J.; Peeters, A.J.; Voesenek, L.A.C.J. Contrasting interactions between ethylene and abscisic acid in *Rumex* species differing in submergence tolerance. *Plant J.* **2005**, *44*, 756–768. [CrossRef]
30. Benschop, J.J.; Bou, J.; Peeters, A.J.; Wagemaker, N.; Gühl, K.; Ward, D.; Hedden, P.; Moritz, T.; Voesenek, L.A.C.J. Long-term submergence-induced elongation in *Rumex palustris* requires abscisic acid-dependent biosynthesis of gibberellin. *Plant Physiol.* **2006**, *141*, 1644–1652. [CrossRef]
31. Saika, H.; Okamoto, M.; Miyoshi, K.; Kushiro, T.; Shinoda, S.; Jikumaru, Y.; Fujimoto, M.; Arikawa, T.; Takahashi, H.; Ando, M.; et al. Ethylene promotes submergence-induced expression of *OsABA8ox1*, a gene that encodes ABA 8-hydroxylase in rice. *Plant Cell Physiol.* **2007**, *48*, 287–298. [CrossRef] [PubMed]
32. Bailey-Serres, J.; Voesenek, L.A.C.J. Flooding stress: Acclimations and genetic diversity. *Ann. Rev. Plant Biol.* **2008**, *59*, 313–339. [CrossRef] [PubMed]
33. Perata, P. The rice SUB1A gene: Making adaptation to submergence and post-submergence possible. *Plant Cell Environ.* **2018**, *41*, 717–720. [CrossRef]
34. Ueguchi-Tanaka, M.; Ashikari, M.; Nakajima, M.; Itoh, H.; Katoh, E.; Kobayashi, M.; Chow, T.; Hsing, Y.C.; Kitano, H.; Yamaguchi, I.; et al. Gibberellin insensitive dwarf1 encodes a soluble receptor for gibberellin. *Nature* **2005**, *437*, 693–698. [CrossRef] [PubMed]
35. Griffiths, J.; Murase, K.; Rieu, I.; Zentella, R.; Zhang, Z.L.; Powers, S.J.; Gong, F.; Phillips, A.L.; Hedden, P.; Sun, T.; et al. Genetic characterization and functional analysis of the GID1 gibberellin receptors in Arabidopsis. *Plant Cell* **2006**, *18*, 3399–3414. [CrossRef]
36. Sun, T.P.; Gubler, F. Molecular mechanism of gibberellin signaling in plants. *Annu. Rev. Plant Biol.* **2004**, *55*, 197–223. [CrossRef]
37. Ariizumi, T.; Lawrence, P.K.; Steber, C.M. The role of two F-box proteins, SLEEPY1 and SNEEZY, in Arabidopsis gibberellin signaling. *Plant Physiol.* **2011**, *155*, 765–775. [CrossRef]

38. Okada, K.; Ito, T.; Fukazawa, J.; Takahashi, Y. Gibberellin induces an increase in cytosolic Ca^{2+} via a DELLA-independent signaling pathway. *Plant Physiol.* **2017**, *175*, 1536–1542. [CrossRef]
39. Ito, T.; Okada, K.; Fukazawa, J.; Takahashi, Y. DELLA-dependent and -independent gibberellin signaling. *Plant Signal Behav.* **2018**, *13*, e1445933. [CrossRef]
40. Ishida, S.; Yuasa, T.; Nakata, M.; Takahashi, Y. A tobacco calcium-dependent protein kinase, CDPK1, regulates the transcription factor REPRESSION OF SHOOT GROWTH in response to gibberellins. *Plant Cell* **2008**, *20*, 3273–3288. [CrossRef]
41. Ishida, S.; Fukazawa, J.; Yuasa, T.; Takahashi, Y. Involvement of 14-3-3 signaling protein binding in the functional regulation of the transcriptional activator REPRESSION OF SHOOT GROWTH by gibberellins. *Plant Cell* **2004**, *16*, 2641–2651. [CrossRef] [PubMed]
42. Fukazawa, J.; Sakai, T.; Ishida, S.; Yamaguchi, I.; Kamiya, Y.; Takahashi, Y. Repression of shoot growth, a bZIP transcriptional activator, regulates cell elongation by controlling the level of gibberellins. *Plant Cell* **2000**, *12*, 901–915. [CrossRef] [PubMed]
43. Fukazawa, J.; Nakata, M.; Ito, T.; Yamaguchi, S.; Takahashi, Y. The transcription factor RSG regulates negative feedback of *NtGA20ox1* encoding GA 20-oxidase. *Plant J.* **2010**, *62*, 1035–1045. [CrossRef] [PubMed]
44. Ito, T.; Nakata, M.; Fukazawa, J.; Ishida, S.; Takahashi, Y. Phosphorylation-independent binding of 14-3-3 to NtCDPK1 by a new mode. *Plant Signal. Behav.* **2014**, *9*, e977721. [CrossRef] [PubMed]
45. Ito, T.; Ishida, S.; Oe, S.; Fukazawa, J.; Takahashi, Y. Autophosphorylation affects substrate-binding affinity of tobacco Ca^{2+}-dependent protein kinase1. *Plant Physiol.* **2017**, *174*, 2457–2468. [CrossRef] [PubMed]
46. Morello, L.; Giani, S.; Breviario, D. The Influence of anaerobiosis on membrane-associated rice (*O. sativa* L.) protein kinase activities. *J. Plant Physiol.* **1994**, *144*, 500–504. [CrossRef]
47. Igarashi, D.; Ishida, S.; Fukazawa, J.; Takahashi, Y. 14-3-3 proteins regulate intracellular localization of the bZIP transcriptional activator RSG. *Plant Cell* **2001**, *13*, 2483–2497. [CrossRef]
48. Olaetxea, M.; Mora, V.; Bacaicoa, E.; Garnica, M.; Fuentes, M.; Casanova, E.; Zamarreño, A.M.; Iriarte, J.C.; Etayo, D.; Ederra, I.; et al. Abscisic acid regulation of root hydraulic conductivity and aquaporin gene expression is crucial to the plant shoot growth enhancement caused by rhizosphere humic acids. *Plant Physiol.* **2015**, *169*, 2587–2596. [CrossRef]
49. Nagel, O.W.; Konings, H.; Lambers, H. Growth-rate, plant development and water relations of the ABA-deficient tomato mutant sitiens. *Physiol. Plant.* **1994**, *92*, 102–108. [CrossRef]
50. Thompson, A.J.; Andrews, J.; Mullholland, B.J.; McKee, J.M.T.; Hilton, H.W.; Horridge, J.S.; Farquhar, G.D.; Smeeton, R.C.; Smillie, I.R.A.; Black, C.R.; et al. Overproduction of abscisic acid in tomato increases transpiration efficiency and root hydraulic conductivity and influences leaf expansion. *Plant Physiol.* **2007**, *143*, 1905–1917. [CrossRef]
51. Bai, T.; Yin, R.; Li, C.; Ma, F.; Yue, Z.; Shu, H. Comparative analysis of endogenous hormones in leaves and roots of two contrasting *Malus* species in response to hypoxia stress. *J. Plant Growth Regul.* **2011**, *30*, 119–127. [CrossRef]
52. Olivella, C.; Biel, C.; Vendrell, M.; Save´, R. Hormonal and physiological responses of *Gerbera jamesonii* to flooding stress. *HortScience* **2000**, *35*, 222–225.
53. Nan, R.; Carman, J.G.; Salisbury, F.B. Water stress, CO_2 and photoperiod influence hormone levels in wheat. *J. Plant Physiol.* **2002**, *159*, 307–312. [CrossRef]
54. Cowan, I.R.; Troughton, J.H. The relative role of stomata in transpiration and assimilation. *Planta* **1971**, *97*, 325–336. [CrossRef]
55. Raschke, K.; Zeevaart, J.A. Abscisic acid content, transpiration, and stomatal conductance as related to leaf age in plants of *Xanthium strumarium* L. *Plant Physiol.* **1976**, *58*, 169–174. [CrossRef]
56. Peng, Y.; Zhou, Z.; Zhang, Z.; Yu, X.; Zhang, X.; Du, K. Molecular and physiological responses in roots of two full-sib poplars uncover mechanisms that contribute to differences in partial submergence tolerance. *Sci. Rep.* **2018**, *8*, 12829. [CrossRef]
57. Yordanova, R.Y.; Popova, L.P. Flooding-induced changes in photosynthesis and oxidative status in maize plants. *Acta Physiol. Plant* **2007**, *29*, 535–541. [CrossRef]
58. Du, H.; Chang, Y.; Huang, F.; Xiong, L. GID1 modulates stomatal response and submergence tolerance involving abscisic acid and gibberellic acid signaling in rice. *J. Integr. Plant Biol.* **2015**, *57*, 954–968. [CrossRef]
59. Blanke, M.M.; Cooke, D.T. Effects of flooding and drought on stomatal activity, transpiration, photosynthesis, water potential and water channel activity in strawberry stolons and leaves. *Plant Growth Regul.* **2004**, *42*, 153–160. [CrossRef]

60. Yetisir, H.; Caliskan, M.E.; Soylu, S.; Sakar, M. Some physiological and growth responses of watermelon [*Citrullus lanatus* (Thumb.) Matsum. And Nakai] grafted onto *Langenaria siceraria* to flooding. *Environ. Exper. Bot.* **2006**, *58*, 1–8. [CrossRef]
61. Atkinson, C.J.; Harrison-Murray, R.S.; Taylor, J.M. Rapid-flood induced stomatal closure accompanies xylem sap transportation of root-derived acetaldehyde and ethanol in Forsythia. *Environ. Exp. Bot.* **2008**, *64*, 196–205. [CrossRef]
62. Ruiz-Sa´nchez, M.C.; Domingo, R.; Morales, D.; Torrecillas, A. Water relations of Fino lemon plants on two rootstocks under flooded conditions. *Plant Sci.* **1996**, *120*, 119–125. [CrossRef]
63. Domingo, R.; Pe´rez-Pastor, A.; Ruiz-Sa´nchez, C. Physiological responses of apricot plants grafted on two different rootstocks to flooding conditions. *J. Plant Physiol.* **2002**, *159*, 725–732. [CrossRef]
64. Nicolás, E.; Torrecillas, A.; Dell'Amico, J.; Alarcón, J.J. The effect of short-term flooding on the sap flow, gas exchange and hydraulic conductivity of young apricot trees. *Trees* **2005**, *19*, 51–57. [CrossRef]
65. Eisenach, C.; Angeli, A.D. Ion transport at the vacuole during stomatal movements. *Plant Physiol.* **2017**, *174*, 520–530. [CrossRef] [PubMed]
66. Gonzali, S.; Loreti, E.; Cardarelli, F.; Novi, G.; Parlanti, S.; Pucciariello, C.; Bassolino, L.; Banti, V.; Licausi, F.; Perata, P. Universal stress protein HRU1 mediates ROS homeostasis under anoxia. *Nat. Plants* **2015**, *1*, 15151. [CrossRef] [PubMed]
67. Sun, L.; Ma, L.; He, S.; Hao, F. AtrbohD functions downstream of ROP2 and positively regulates waterlogging response in *Arabidopsis*. *Plant Signal. Behav.* **2018**, *13*, e1513300. [CrossRef]
68. Yordanova, R.Y.; Uzunova, A.N.; Popova, L.P. Effects of short-term soil flooding on stomata behavior and leaf gas exchange in barley plants. *Biol. Plant.* **2005**, *49*, 317–319. [CrossRef]
69. Zang, J.; Zang, X. Can early wilting of old leaves account for much of the ABA accumulation in flooded pea plants? *J. Exp. Bot.* **1994**, *45*, 1335–1342. [CrossRef]
70. Ashraf, M.A. Waterlogging stress in plants: A review. *Afr. J Agric. Res.* **2012**, *7*, 1976–1981. [CrossRef]
71. Miao, L.F.; Yang, F.; Han, C.Y.; Pu, Y.J.; Ding, Y.; Zhang, L.J. Sex-specific responses to winter flooding, spring waterlogging and post-flooding recovery in *Populus deltoides*. *Sci. Rep.* **2017**, *7*, 2534. [CrossRef] [PubMed]
72. Vantoai, T.T.; Bolles, C.S. Postanoxic injury in soybean (*Glycine max*) seedlings. *Plant Physiol.* **1991**, *97*, 588–592. [CrossRef]
73. Kalashnikov, Y.E.; Balakhnina, T.I.; Zakrzhevsky, D.A. Effect of soil hypoxia on activation of oxygen and the system of protection from oxidative destruction in roots and leaves of *Hordeum vulgare*. *Russ. J. Plant Physiol.* **1994**, *41*, 583–588.
74. Yan, B.; Da, Q.; Liu, X.; Huang, S.; Wang, Z. Flooding induced membrane damage, lipid oxidation and activated oxygen generation in corn leaves. *Plant Soil* **1996**, *179*, 261–268. [CrossRef]
75. Biemelt, S.; Keetman, U.; Mock, H.P.; Grimm, B. Expression and activity of isoenzymes of superoxide dismutase in wheat roots in response to hypoxia and anoxia. *Plant Cell Environ.* **2000**, *23*, 135–144. [CrossRef]
76. Kwak, J.M.; Mori, I.C.; Pei, Z.M.; Leonhardt, N.; Torres, M.A.; Dangl, J.L.; Bloom, R.E.; Bodde, S.; Jones, J.D.G.; Schroeder, J.I. NADPH oxidase AtrbohD and AtrbohF genes function in ROS-dependent ABA signaling in *Arabidopsis*. *EMBO J.* **2003**, *22*, 2623–2633. [CrossRef]
77. Coelho, S.M.; Taylor, A.R.; Ryan, K.P.; Sousa-Pinto, I.; Brown, M.T.; Brownlee, C. Spatiotemporal patterning of reactive oxygen production and Ca^{2+} wave propagation in *Fucus rhizoid* cells. *Plant Cell* **2002**, *14*, 2369–2381. [CrossRef]
78. Fairley-Grenot, K.; Assmann, S.M. Evidence for G-protein regulation of inward K^+ channel current in guard-cells of fava-bean. *Plant Cell* **1991**, *3*, 1037–1044. [CrossRef]
79. Wu, W.H.; Assmann, S.M. A membrane-delimited pathway of G protein regulation of the guard-cell inward K^+ channel. *Proc. Natl. Acad. Sci. USA* **1994**, *91*, 6310–6314. [CrossRef]
80. Fischer, R.A. Stomatal opening: Role of potassium uptake by guard cells. *Science* **1968**, *160*, 784–785. [CrossRef]
81. Bright, J.; Desikan, R.; Hancock, J.T.; Weir, I.S.; Neill, S.J. ABA-induced NO generation and stomatal closure in Arabidopsis are dependent on H_2O_2 synthesis. *Plant J.* **2006**, *45*, 113–122. [CrossRef]
82. An, Z.; Jing, W.; Liu, Y.; Zhang, W. Hydrogen peroxide generated by copper amine oxidase is involved in abscisic acid-induced stomatal closure in *Vicia faba*. *J. Exp. Bot.* **2008**, *59*, 815–825. [CrossRef] [PubMed]
83. Song, X.G.; She, X.P.; Yue, M.; Liu, Y.E.; Wang, Y.X.; Zhu, X.; Huang, A.X. Involvement of Copper Amine Oxidase (CuAO)-Dependent Hydrogen Peroxide Synthesis in Ethylene-Induced Stomatal Closure in *Vicia faba*. *Russ. J. Plant Physiol.* **2014**, *61*, 390–396. [CrossRef]

84. Bashar, K.K. Hormone dependent survival mechanisms of plants during post waterlogging stress. *Plant Signal Behav.* **2018**, *13*, e1529522. [CrossRef] [PubMed]
85. Chen, Y.; Huang, R.; Xiao, Y.; Lü, P.; Chen, J.; Wang, X. Extracellular calmodulin-induced stomatal closure is mediated by heterotrimeric G protein and H_2O_2. *Plant Physiol.* **2004**, *136*, 4096–4103. [CrossRef] [PubMed]
86. Sasidharan, R.; Voesenek, L.A.C.J. Ethylene-mediated acclimations to flooding stress. *Plant Physiol.* **2015**, *169*, 3–12. [CrossRef]
87. Steffens, B.; Sauter, M. Heterotrimeric G protein signaling is required for epidermal cell death in rice. *Plant Physiol.* **2009**, *151*, 732–740. [CrossRef] [PubMed]
88. Ge, X.M.; Cai, H.L.; Lei, X.; Zhou, X.; Yue, M.; He, J.M. Heterotrimeric G protein mediates ethylene-induced stomatal closure via hydrogen peroxide synthesis in Arabidopsis. *Plant J.* **2015**, *82*, 138–150. [CrossRef]
89. Liu, P.; Sun, F.; Gao, R.; Dong, H. RAP2.6L overexpression delays waterlogging induced premature senescence by increasing stomatal closure more than antioxidant enzyme activity. *Plant Mol. Biol.* **2012**, *79*, 609–622. [CrossRef] [PubMed]
90. Hwang, S.-Y.; VanToai, T.T. Abscisic acid induces anaerobiosis tolerance in corn. *Plant Physiol.* **1991**, *97*, 593–597. [CrossRef] [PubMed]
91. Gómez-Cadenas, A.; Tadeo, F.R.; Talón, M.; Primo-Millo, E. Leaf abscission induced by ethylene in water-stressed intact seedlings of Cleopatra mandarin requires previous abscisic acid accumulation in roots. *Plant Physiol.* **1996**, *112*, 401–408. [CrossRef] [PubMed]
92. Kato-Noguchi, H. Abscisic acid and hypoxic induction of anoxia tolerance in roots of lettuce seedlings. *J. Exp. Bot.* **2000**, *51*, 1939–1944. [CrossRef] [PubMed]
93. Ellis, M.H.; Dennis, E.S.; Peacock, W.J. Arabidopsis roots and shoots have different mechanisms for hypoxic stress tolerance. *Plant Phyisol.* **1999**, *119*, 57–64. [CrossRef]
94. Dat, J.F.; Capelli, N.; Folzer, H.; Bourgeade, P.; Badot, P.M. Sensing and signaling during plant flooding. *Plant Physiol. Biochem.* **2004**, *42*, 273–282. [CrossRef] [PubMed]
95. Hocking, T.J.; Hillman, J.R.; Wilkins, M.B. Movement of abscisic acid in *phaseolus vulgaris* plants. *Nat. New Biol.* **1972**, *235*, 124–125. [CrossRef]
96. Jackson, M.B.; Hall, K.C. Early stomatal closure in waterlogged pea plants is mediated by abscisic acid in the absence of foliar water deficits. *Plant Cell Environ.* **1987**, *10*, 121–130. [CrossRef]
97. Rodriguez-Gamir, J.; Ancillo, G.; Gonzalez-Mas, M.C.; Primo-Millo, E.; Iglesias, D.J.; Forner-Giner, M.A. Root signaling and modulation of stomatal closure in flooded citrus seedlings. *Plant Physiol. and Biochem.* **2011**, *49*, 636–645. [CrossRef]

7

What has been Thought and Taught on the Lunar Influence on Plants in Agriculture? Perspective from Physics and Biology

Olga Mayoral [1,2,*], Jordi Solbes [1], José Cantó [1] and Tatiana Pina [1,*]

[1] Department of Science Education, Universitat de València (UV), Avda. Tarongers, 4, 46022 Valencia, Spain; jordi.solbes@uv.es (J.S.); jose.canto@uv.es (J.C.)
[2] Botanical Garden UV, Universitat de València, Calle Quart, 80, 46008 Valencia, Spain
* Correspondence: olga.mayoral@uv.es (O.M.); tatiana.pina@uv.es (T.P.)

Abstract: This paper reviews the beliefs which drive some agricultural sectors to consider the lunar influence as either a stress or a beneficial factor when it comes to organizing their tasks. To address the link between lunar phases and agriculture from a scientific perspective, we conducted a review of textbooks and monographs used to teach agronomy, botany, horticulture and plant physiology; we also consider the physics that address the effects of the Moon on our planet. Finally, we review the scientific literature on plant development, specifically searching for any direct or indirect reference to the influence of the Moon on plant physiology. We found that there is no reliable, science-based evidence for any relationship between lunar phases and plant physiology in any plant–science related textbooks or peer-reviewed journal articles justifying agricultural practices conditioned by the Moon. Nor does evidence from the field of physics support a causal relationship between lunar forces and plant responses. Therefore, popular agricultural practices that are tied to lunar phases have no scientific backing. We strongly encourage teachers involved in plant sciences education to objectively address pseudo-scientific ideas and promote critical thinking.

Keywords: plant growth; agriculture; traditions; pseudo-science; lunar phases; physics; biology; education

---

1. Introduction

This paper addresses the existing dichotomy between what science shows regarding agriculture protocols and past and current agricultural practices in much of Europe and Latin America. More specifically, it focuses on some pseudo-scientific questions and beliefs that impregnate a large part of agricultural traditions and agronomic practices according to which certain lunar phases encourage plant growth while others compromise their development. These beliefs share our lives with scientific and technological advances not reached ever before.

After introducing the main features of the Moon and its phases, as well as the factors that determine plant growth, this study continues with a brief historical overview about what has been thought from the agricultural sector concerning the lunar influence on plants and crops. In this overview, we have included references to both earlier ages and the most recent trends within agriculture, such as biodynamic agriculture, which bases part of its operating on the close relationship between the Moon and plant growth.

Then, we analysed monographs on botany, plant biology and physiology—considered as texts of consolidated science—, searching for any mention about the Moon being a factor influencing plant growth. At the same time, we reviewed physics handbooks, focusing on which aspects or natural processes the Moon has influence on, looking for any mention of some type of effect on living beings and, specifically, on plants.

The paper concludes with a reflection on the implications of the different existing visions lasting over time within both the field of agriculture and citizenship in general, being part of the global desirable scientific literacy.

1.1. The Moon

This section covers the basic key aspects about the Moon required to understand most of the arguments detailed in the subsequent analysis of both scientific literature and explanations provided by some agricultural sectors.

1.1.1. The Gravitational Pull

The Moon is the only natural satellite of our planet describing an elliptical orbit around it with a semi-major axis of 384,000 km, an eccentricity of 0.0549 and an angle of 5°9′ relative to the ecliptic plane. The Moon takes 29.5 days to orbit around the Earth and return to its analogous position with respect to the Sun and the Earth (lunar month or synodic month) [1]. However, it takes 24.8 h for a specific location on the Earth to rotate from one exact point beneath the Moon and back (lunar day) [2,3]. The combined action of these two cycles (lunar month and day) has different effects on the Earth such as changes in tides and in the intensity of illuminance.

So, what is the explanation for the Moon's influence on tides? Tides are due to the difference in gravitational pull (or gravity acceleration) between the part of the oceans which are nearest (A) and farthest (B) to the Moon and the relative acceleration in relation to the Earth's centre of mass (CM) in such points (Figure 1).

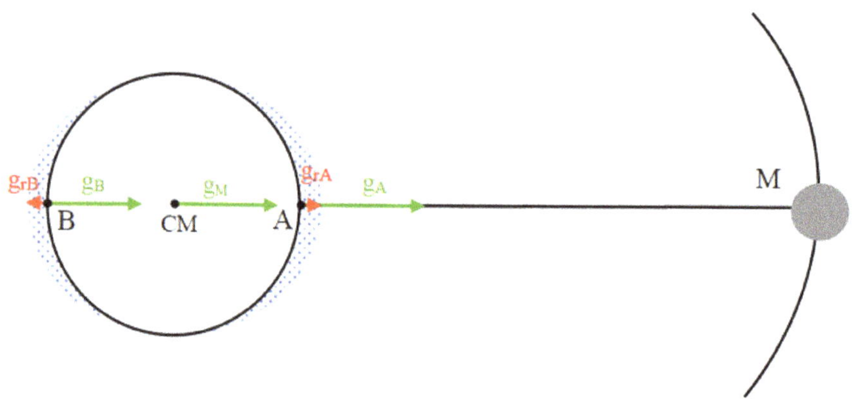

Figure 1. Representation of how tides are produced. In the drawing, g_M represents the acceleration in the Earth's centre of mass (CM) caused by lunar attraction; g_A and g_B are, respectively, the accelerations of points A and B located at both ends of the Earth's surface over the Earth–Moon line, and g_{rA} and g_{rB} are the acceleration in relation to the Earth's CM. Modified from Martínez et al. [1].

From the gravitational point of view, the effect of gravity on the Earth's CM produced by the Moon (g_M) can be calculated by means of the expression $g_M = Gm/r^2$, being G the universal gravitational constant, r the distance Earth–Moon (E–M) and m the mass of the Moon. That is to say, the value of the Moon's gravity on the Earth's surface is approximately 2,951,800 times lower than the Earth's gravity ($g_M = 3.32 \times 10^{-5}$ ms^{-2}). Therefore, the gravitational pull is negligible. Accordingly, the Sun's gravity (g_S) on the Earth is 177 times greater than the Moon's ($g_S = g/1627 = 177\ g_M = 6 \times 10^{-3}$ ms^{-2}).

We can also calculate the Moon's gravity in point A ($g_A \approx g_M(1 + 2\ R/r)$ being M the mass of the Earth and R its radius) and in point B ($g_B \approx g_M(1 - 2\ R/r)$ and their relative accelerations in A (g_{rA}) and B (g_{rB}) regarding Earth's CM ($g_{rA} = g_A - g_M \approx 2\ GmR/r^3 = 2\ Rg_M/r$ and $g_{rB} = g_B - g_M \approx -2\ GmR/r^3 = -2\ Rg_M/r$, respectively) (Figure 1). From these calculi, we observe that the relative acceleration in relation to CM depends on the distance between A and B and on the cubed distance between the Earth and the Moon, instead of squared distance, rendering identical values in both A and B points (as r = 60 R, its value is 10^{-6} ms^{-2}, 30 times lower than g_M) but in the opposite direction: the relative acceleration in point A (gr_A) is directed towards the Moon and in point B (gr_B) towards the opposite direction [1,4]. Therefore, there will be high tide in A and B, and low tide in those points located at 90° (Figure 2).

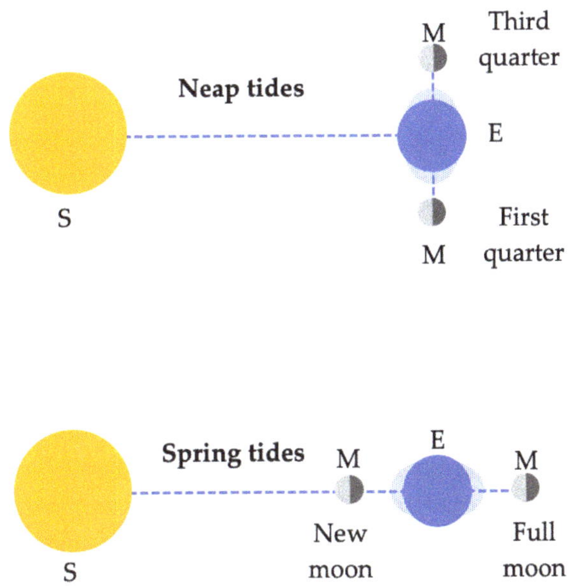

Figure 2. Diagram to illustrate the Sun (S)–Earth (E)–Moon (M) configuration regarding neap and spring tides. Source: designed by the authors.

For this reason, although the Sun's gravity on the Earth is greater than the one from the Moon, as tides are influenced by the inverse of the distance to cube ($1/r^3$), its effect is lower than the one from the Moon, as its distance is much greater. And as tides depend on the size of the object, in the Mediterranean Sea, for instance, these are negligible due to the fact that it is a semi-enclosed, shallow and small sea (with an average depth of R = 1500 m). In contrast, as all the oceans are communicated, we can consider their size being the size of the Earth and, therefore, tides are apparent. In this sense, the tidal effect of the Moon over a 2 m height living being located on the Earth is about 1000 times lower (tidal acceleration = 2 gMh/r = 3×10^{-13} ms^{-2}) than the effect produced by a mass of 1 kg at 1 m height above it (2.67×10^{-10} ms^{-2}) [5].

Thus, because of the daily rotation of the Earth, the tides rise and fall twice each lunar day in most coastal areas and estuaries at intervals of approximately 12.4 h (tidal cycles) reflecting the lunar 24.8 h day. The amplitude of successive tides is also modulated every 14.77 days, or semi-lunar cycle. So, the highest tides, or spring tides, take place when the Sun and the Moon are aligned with the Earth (i.e., full and new moon), and the lowest, or neap tides, occur when the Sun–Earth axis and the Moon–Earth axis are at right angles (90°) to each other (i.e., first and third quarter) (Figure 2) [2,6]. These tidal forces due to the Moon and the Sun are also observed in the atmosphere and the Earth's crust [7].

1.1.2. Illuminance

The moonlight we see from the Earth is the sunlight reflected on the greyish-white surface of the Moon. Since the Moon orbits the Earth and the Earth orbits the Sun, the fraction of the Moon we see changes along the lunar month giving rise to the lunar phases, being new moon, first quarter, full moon and third quarter the main ones. The illuminance (defined as the amount of luminous flux striking a surface per unit area) varies depending on the lunar phase [2,8]. In the case of the Moon, as endpoint cases we can find 0.001 lx for a new moon and 0.25 lx for a full moon to 0.01 lx for a crescent or waning moon (Table 1).

Table 1. Illuminance according to the Moon phase.

Illuminance (lx)	Description
0.001	Clear night sky, new moon
0.01	Clear night sky, crescent or waning moon
0.25	Full moon on a cloudless night
600	Sunrise or sunset on a cloudless day
32,000	Sunlight on an average day (minimum)
100,000	Sunlight on an average day (maximum)

Source: adapted from RCA Corporation [9] and Schlyter [10].

As it can be seen in Table 1, the Moon's maximum illuminance is 128,000 times lower than the minimum of sunlight on an average day or 400,000 times lower than the maximum of sunlight on an average day.

1.2. Factors Influencing Plant Growth and Development

The revision carried out considers plant growth and development from a holistic point of view. This implies all those changes in structure and function of plants and their parts, the course of genesis, assimilation, growth and development, as well as environmental, physiological and chemical modifications, maturation and decline [11–15].

Plant growth and development is regulated by both endogenous and external factors [12,14]. Regarding endogenous factors, phytohormones are in charge of the coordination of metabolic and developmental processes at the molecular and cellular level. Phytohormones can be divided into two groups depending on their functions: (i) those involved in growth-promoting activities; (ii) those in charge of responding to wounds or to biotic and abiotic stresses [16]. Synthesis or changes in the concentration of these phytohormones transduce the perception of environmental stimulus (i.e., radiation, photoperiod, temperature, gravity or stresses as cold, heat, drought or flooding). However, which plant hormones will be triggered will depend on the plant developmental state, the type of external stimulus, the part of the plant exposed, when this stimulus arrives, etc. [14]. Phytohormones, together with external factors, can activate growth and differentiation processes and allow the synchronization of plant development and seasonal changes. Furthermore, they also regulate plant growth (intensity and direction), the metabolic activity and the storage and transport of nutrients. All these endogenous factors are determined by endogenous genetic components (genome structure and gene expression, i.e., plant genotype) [17].

The growth and development of plants can also be affected by external factors such as quality, intensity, direction and duration of radiation, temperature, position with relation to Earth's gravitational field and stresses conferred by wind, water currents or snow cover, apart from other chemical influences. These external factors can initiate, complete and regulate the timing of developmental processes (inductive mode of action) but can also act quantitatively (by altering the speed and extent of growth) and formatively (by affecting morphogenesis and tropisms) [14]. These external factors are the ones which might be affected by a potential effect of the Moon—specifically, the gravitational and illuminance effects—.

Other authors propose to split the factors that determine quality and quantity into biotic and abiotic factors [18,19]. Within the biotic factors, we find arthropods, nematodes, bacteria, fungus and viruses as well as their relationships with other plants and organisms which can be competitive, mutual or parasitic types, among others [20]. In addition, within the abiotic factors, we find soil composition, salinity, pH, temperatures, pollution, humidity (water), wind and ultraviolet radiation, among others. The interaction of biotic and abiotic factors will determine plant growth, development, and productivity. Understanding their interactions is essential in agriculture when searching for the ideal growth conditions for each particular plant. In this sense, stress physiology research is very valuable, as it focuses on whether the full genetic potential of plants will be fulfilled and if plants will attain maximal growth and reproductive potential depending on different factors [21]. In particular, the study of abiotic stress originated from excess or deficit in the physical, chemical and energetic conditions to which plants are exposed provides farmers with guidelines for optimizing their harvests.

The references to the potential influence of the Moon on plants will have to be searched considering this influence as an abiotic factor. The excess or deficit of this factor should be studied taking into account that the Moon is always present, so the search should focus on those moon-derived sub-factors that can undergo substantial changes. The indirect possible effect of the Moon on the biotic components interacting with plants is a matter which falls outside the scope of the revision carried out in this paper.

2. What Has Been Thought on the Influence of the Moon on Plants

This section focuses on all those aspects related to agriculture that, according to some traditions, are determined by the Moon. To do so, we have developed a brief overview of what has been thought and written throughout history about the influence of the Moon on living beings and, in particular, plants. This analysis addresses manuals which have been used and are still used in certain agricultural sectors and information present on websites related to agriculture, gardening, agricultural machinery and so on. A special section is dedicated to biodynamic agriculture which links plant growth and lunar phases.

2.1. Brief Historical Overview

The Moon and the Sun hold a significant place in many mythologies and popular legends throughout the world. In particular, beliefs regarding the relationships between lunar phases and human and other organisms' behaviour are as ancient as human cultural heritage but have hardly ever found any solid scientific support [22].

Assertions concerning the existence of repetitive cycles in the Moon (phases), the Sun (day/night, solstices, and equinoxes/seasons), and Sirius (its heliacal rising) were extremely useful to develop lunar, lunar–solar and solar calendars, and to predict eclipses—just as it happened in the Egyptian, Babylonian, Greek or Chinese world. Such knowledge was continued in different cultures, mainly the Arab or the Mayan, the Aztec, and the Inca in America [23,24]. It is known that the Mayan carried out thorough observations of natural events, finding certain cyclical repetitions which allowed making predictions and organising when to sow or harvest [25].

Botanists and herbalists from the seventeenth century, such as Nicholas Culpeper (1616–1654), believed that plants and ailments were determined by constellations. The Sun ruled our heart, blood circulation and spine, while the Moon had influence on growth, fertility, breasts, stomach, uterus and menstrual flow. In fact, all the body fluids, as the tides, were controlled by the lunar phases. This was the prevalent belief at that time, since astrology was broadly accepted as the key to understanding the universe [26].

Surprisingly, these beliefs are still active, as shown by Phillips [26] in his *Encyclopaedia of Plants in Myth, Legend, Magic and Lore* which includes more than 200 entries linking different plant species or genera with stars or natural elements. This author states, as we have been able to check along with personal interviews made in the agricultural and rural world of the Iberian Peninsula, that garlic (*Allium sativum* L.) has been strongly associated with the Moon, and it was thought to grow stronger as the Moon waned. However, Navazio [27], in his manual dedicated to the organic seed grower, makes no mention of the Moon as an element to consider. Potato growing (*Solanum tuberosum* L.) is also supposed to be influenced by the Moon. According to different beliefs this underground crop should be planted during the black moon, that is to say, when it is waning [26]. But, once again, Navazio [27] does not mention the need to consider this aspect in the organic cultivation. Other examples would be the white clover (*Trifolium repens* L.), which has to be seeded by the darkness of the Moon or "no-Moon"—that is during the 24 h between the waning moon and the crescent moon—if you want it to grow, since if it is seeded under the moonlight, it will not sink into the ground [26]. Or corn (*Zea mays* L.), the seeds of which must be planted by moonlight in order to obtain a good performance [26].

Anglés Farrerons [28], in his work *Influence of the Moon on Agriculture and Other Topics of Main Interest for the Farmer and People from the City*, collects all the existing beliefs among elder farmers regarding the Moon. He dedicates specific chapters to the vine and the wine, the fruit growing, the cereals, the olive tree, several horticultural crops such as chard (*Beta vulgaris* var. *cicla* (L.) K.Koch), artichoke (*Cynara cardunculus* var. *scolymus* (L.) Benth.), garlic, celery (*Apium graveolens* L.), onion (*Allium cepa* L.), etc. Anglés Farrerons [28] also focuses on tree felling, forage harvesting, influence of animal manure or weather forecast. According to these traditions, he states when to sow, prune, harvest, etc., depending on the Moon phase and the crop, being true nonsense in some cases, just as the author suggests in his introduction.

Another work that requires special attention is that of Restrepo [29], a Brazilian agronomist who reflects the beliefs from Latin America and the Caribbean Area. He provides an interesting revision of the calendars of the ancient people and cultures as well as an extensive description of when to carry out all the agricultural practices (e.g., sow, layer, graft, prune, transplant) based on whether they are annual or perennial plants, vegetables, cereals and grains, tubers, bulbs and rhizomes. He also includes a description of how lunar phases and Moon illuminance affect the movement of the sap in plants (Figure 3).

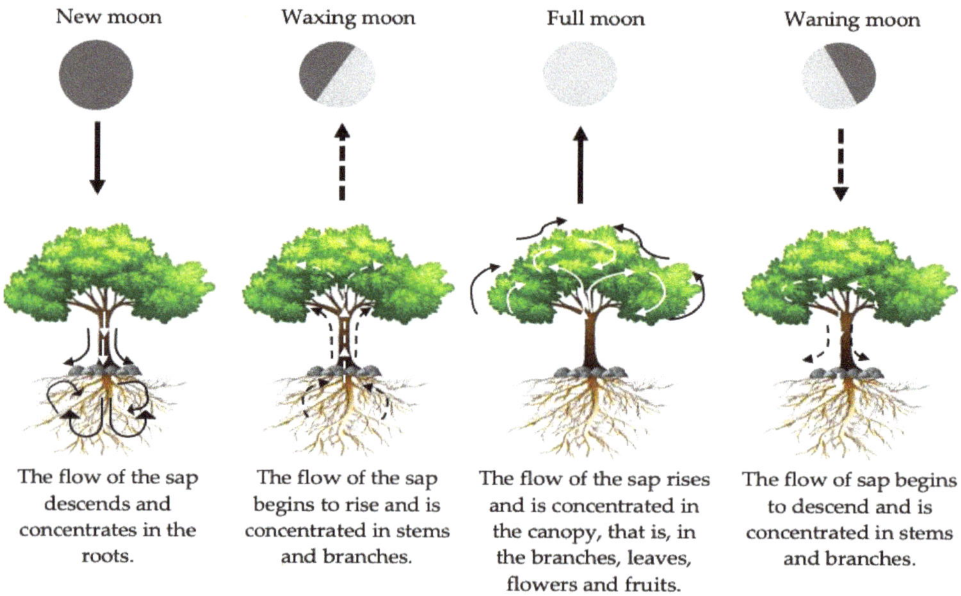

Figure 3. Explanation of how lunar phases affect sap dynamics in plants according to Restrepo [29]. Redrawn and translated from Restrepo [29].

Which is the cause–effect explanation proposed to link lunar phases and sap movement? Restrepo [29] links it with the tides:

> "Therefore, in certain positions of the Moon, the water from the oceans rises to reach a maximum height, and then goes down to a minimum level, maintaining this oscillation regularly and successively. It has also been checked that this phenomenon makes itself felt in plant sap". (Translated from Restrepo [29]).

It has already been shown that the effect of the tide of the Moon on a 2 m height living being is absolutely negligible (3×10^{-13} ms^{-2}), compared to the Earth's gravity (9.8 ms^{-2}) [5]. But considering the tides, there are two high tides and two low tides each day, so if the tide caused any effect on a plant, there should be two sap rises and falls per day and none with the lunar phases. If the latter wanted to be introduced, we have already seen that at both new moon and full moon, the tides are a bit stronger due to the fact that the Sun and the Moon are aligned (as the Sun is much farther away, its effect on the tide is much lower), but the effects of the tide are symmetrical making the water rise (Figures 1 and 2). Yet, to make matters more contradictory, Restrepo [29] assigns them different effects: the full moon takes up to the leaves the waters of the plant, and the new moon takes them to the roots. On the other hand, the illuminance is the only thing that surely varies with the Moon phases (Table 1), but it does not generate any force that can cause the movement of the sap.

Finally, it has to be pointed out that there are beliefs and practices contained in many different manuals which we did not pretend to either analyse or introduce in detail in this paper. Only as an example, we outlined best sellers, such as *The Secret Life of Plants* [30], in which certain physical, emotional and spiritual relationships between plants and our species are explained in such an appealing manner that have clearly helped to strengthen several pseudo-scientific beliefs among society and many farmers. This best seller has been explicitly refuted by texts, such as *The Not-So-Secret Life of Plants*, in which the historical and experimental myths about emotional communication between animals and plants are put to rest by researchers such as Galston and Slayman [31] or Horowitz et al. [32].

2.2. *Agricultural Astronomy Manuals and Websites*

Some authors summarise the situation such as follows [26]:

> "There has been a certain amount of interest in planting according to the phase of the Moon. The basic premise being that 'above ground crops' should be planted in the light of the Moon, i.e., on the days between the new moon and the full moon. 'Below ground crops' must be planted in the dark of the Moon, that is between the full moon and the next new moon. Refinements on this require that leaf crops are planted at the new moon and fruit crops or flowers planted at the full moon".

Apart from the agricultural traditions that could explain the use of the Moon as a calendar to organise the crops, the time of seeding, harvesting, reaping, etc., and the advice given by some authors on an individual basis, some companies have taken a step forward by developing a range of documents and even manuals appearing in the guise of scientific advice, which raise popular wisdom to the category of regulated recommendations [33–38]. Such manuals are used as reference books by many farmers at the small and large scales, and they offer a scheduled sequence of agricultural activities and, in many cases, advice regarding health care on the basis of the phases of the Moon and its ascending or descending position in the sky. But in this case, its influence is not obvious since neither gravitation nor illuminance vary.

We can also find these recommendations in the area of gardening with titles such as *Gardening by the Moon Calendar* [39]. This book provides guidance based on the following statements:

> "The best rate of germination is achieved just before a full moon, when moonlight and the Moon's gravitational pull are both at their maximum, grafting should be done on a waxing moon, because sap rises in plants during this period and this will help a graft to establish, pruning should be done on a waning moon, because the sap is now falling, and this will help cut surfaces to heal quickly and crops for storage should be harvested while the Moon is waning".

Apart from the Moon, it includes elements about astrology to guide the practice: "the planting of fruit trees and bushes should be done when the Moon is passing through a fire constellation". Besides, this author links the effectiveness of the response to the planetary influence on cultivating without chemicals, pointing out that chemical products desensitize agricultural land.

Many entries to websites link the influence of the Moon to the biodynamic agriculture and the zodiac, which has nothing to do with it since the zodiac are the constellations where the ecliptic passes by (the apparent path of the Sun among the fixed stars).

Some companies within the forestry, agriculture and gardening area also dedicate sections to providing advice on tasks related to the handling of vegetables, fruit trees, etc., on their websites (Table 2).

Table 2. Summary of the tasks recommended by a multinational company supplying forestry, agriculture, and gardening machinery according to the lunar phases.

New Moon	Waxing Moon	Full Moon	Waning Moon
Crop covering with soil	Prune diseased or fruit trees	Prune	Sow root vegetables
Fertilise	Cultivate sandy soils	Plant perennial species	Remove withered leaves
Remove weeds	Sow flowers, leafy vegetables	Transplant	Water flowering plants
Remove withered leaves	Grafting	Vegetative propagation	Fertilise
Sow grass	Avoid watering flowering plants	-	Plant longleaf trees

Source: modified and translated from [40].

Moreover, this link between lunar phases and different aspects of cultivation rank highly in major search engines on the internet and in database image repositories, which raises many of the same ideas as Restrepo [29]—that lunar gravity changes according to the phases is the only way to explain popular beliefs concerning the influence of the Moon on plant growth. However, illuminance is the only thing that varies according to the phases, as it can be seen in Table 1. The Moon's gravitational pull does not generate any force able to cause sap movement.

2.3. *Biodynamic Agriculture*

Part of the traditions regarding the Moon have been incorporated into biodynamic agriculture, an agricultural management system which is mainly based on the fact that the astronomical bodies influence crop production. As in other forms of organic farming, the use of industrial fertilizers, pesticides and herbicides is avoided. However, the difference lies in the use of plant and mineral preparations as additives to compost and soil sprays—"biodynamic preparations"—and in following a planting schedule for cultivation, sowing and harvesting based on cosmic forces and rhythms and, particularly, on Moon rhythm (Table 3) [41,42].

Table 3. Practices and products used in organic and biodynamic agriculture.

Practice or Product	Organic	Biodynamic
Crop rotation	x	x
Polyculture or intercropping	x	x
Cover cropping	x	x
Low- or no-till	x	x
Green manures	x	x
Biological, cultural, mechanical and physical means of pest control	x	x
Biodynamic preparations that involve alchemy and homeopathy		x
Lunar and astrological calendars for planting, managing and harvesting		x
Stones used for channelling cosmic energy and radiant fields		x
Burning of pests and weeds (pest ashing)		x
Sensitive testing (including biocrystallization or morphochromatography, among others)		x

Source: modified from Chalker-Scott [41].

This variant of organic agriculture, initiated in a series of lectures given by Rudolf Steiner [43] in 1924, is considered by some authors as an alternative approach to modern agriculture (see review in Brock et al. [44]), while others consider it as not being a science-based practice (see review in Chalker-Scott [41]). The latter brands it as a scam of great implantation in countries as advanced as Germany where it has its origin and from where Brussels is pressured to accept its principles. The restoration of soil quality, of the "harmony" of ecosystems, and of biodiversity can be pointed out as the main objective of biodynamic farmers. The US website for biodynamic certification marks an update of the Moon phases with a quotation from the Natural History of Pliny the Elder (23–79 CE) —the first-century Roman naturalist who wrote extensively about tides—, explaining that the Moon "replenishes the Earth; when she approaches it, she fills all bodies, while, when she recedes, she empties them" [45].

Kirchmann [46] suggests that biodynamic agriculture has a mystical origin (called spiritualistic research by Steiner, and based on mediation and clairvoyance) that drove Steiner's research to reject scientific inquiry because as he explained in the sixth lecture of the course, "We do not need any confirmation by circumstances or by external methods. Spiritualism is an extension of scientific thought broadening the prevailing one-sided scientific view, being true and correct". A good example of this can be found in what Rudolf Steiner [47] wrote in 2004 insisting on the special care that should be taken when teaching questions about "moon forces", since conventional science considers them pure superstition or mystical fantasy, being a truth that is still difficult to talk about openly. However, Kirchmann [46] maintains that "Steiner's predictions that can be scientifically tested have been found to be incorrect".

3. What Has Been Taught on the Influence of the Moon on Plants? Analyses of Handbooks and Scientific Literature

This section deals with all those aspects that are known and taught in the agronomic and biological background in relation to plant development. It also analyses the physics books that explain the influence of the Moon on our planet with the intention of clarifying whether they mention a possible relationship with plant development. Consequently, specific sections are devoted to those factors that could depend on or have an influence from the Moon, concretely in relation to the effect of gravity and the light reflected by it, both from the point of view of biology and physics. Likewise, the basic books on botany and plant physiology have been revised, paying special attention to those factors that are determining or causing stress to the development of plants.

The revision of the handbooks has been complemented with the information gathered from the analysis of different scientific articles published in data repositories, such as Web of Knowledge, Scopus or Google Scholar, using the keywords "Moon and plants" or "lunar and plants".

3.1. What Handbooks Say from the Perspective of Physics and Biology

According to traditional beliefs, the influence of the Moon on plant growth is attributed, among other factors, to the attractive forces that the satellite exerts on the Earth and more specifically on its waters. The gravitational theory of the Moon could be attributed for the first time to Kepler (1571–1630), who claimed that the ocean tides were produced by a hidden force from the Moon. Kepler believed it was due to the affinity that the Moon had for water which was one of the four basic elements [48] in [8]. Gravity was also recognised as an agent of lunar influence with the publication of "Principia" by Newton (1643–1727).

The analyses of various physics textbooks (Table 4) commonly used in science and engineering courses reveals that the term Moon appears in most of them linked to different concepts such as the distance from the Earth (as it was calculated in ancient times or as it is calculated today, with laser telemetry), the Moon's gravity, tides, etc. With regard to the origin of tides, there are many possibilities: (i) it is not approached [49]; (ii) it is introduced in a qualitative way [50,51]; (iii) the exact dependence of R/r^3 is provided where r is the distance Earth–Moon, and R is the size of the object on which the tides act—in the case of the oceans, the Earth's radius [5]; (iv) locally, high tides are shown as an effect of the resonance [52] or tidal applications to produce energy are explained [53]. In order to find correct demonstrations of the tides, we have to deal with books on Astronomy (e.g., [4]) which, due to their extraordinary specificity, are beyond our general review.

Another factor that should be considered when approaching sap movement in plants is capillary action or capillarity, described as the spontaneous ability of a liquid to flow against gravity in a narrow space such as a thin tube or pipe (in plants, vascular tissues as xylem and phloem). This rising of liquid is the outcome of two opposing forces: cohesion (the attractive forces among similar molecules or atoms) and adhesion (the attractive forces among dissimilar molecules or atoms). In our case, the contact area between the particles of the liquid and the particles forming the tube. Capillarity is high when adhesion is greater than cohesion and vice versa. There is another important factor in capillarity, which is the contact area, dependent on the diameter of the tube (i.e., vascular tissue). Capillarity interacts with other forces, as gravity, which should be included when considering possible gravitational effects of the Moon on plants. In this sense, Jurin's law is usually introduced, giving information on the height (h) reached when balancing the weight of the column of a liquid and the force $h = 2\,\gamma \cos \theta/\rho g r$, where γ is the surface tension (Nm^{-1}), θ is the contact angle, ρ is the density of the liquid (kgm^{-3}), g is the gravity acceleration (ms^{-2}) and r is the radius of the pipe (m). Therefore, the Moon's gravity would have to be subtracted from that of the Earth $g = g_E - g_M$, and since it is 288,000 times smaller, its effect on capillarity is negligible.

Table 4. Revision of some of the reference handbooks on physics in relation to possible mentions of the Moon affecting plants.

Book	Issues Regarding the Moon			
	Gravity	Tides	Capillarity	Luminosity
Feynman [50]	Law of Gravitation	Qualitative explanation of tides	No	Illuminance $I = S/r^2$
Gettys et al. [52]	Point where $g_E = g_M$	Tides in the Bay of Fundy (e.g., resonance)	No	No
Giancoli [53]	Law of Gravitation	Tidal energy	Jurin's Law and negative pressure	No
Hewitt [5]	Moon radius, distance Earth–Moon. Law of Gravitation	Compares Sun and Moon tides by distance. It makes approximations to introduce R/r^3 Distinguishes between spring and neap tides. Applied to people. Tides in the ionosphere	Qualitative capillarity from surface tension	No
Holton and Brush [51]	Law of Gravitation	Qualitative explanation of spring and neap tides	No	No
Tipler [49]	Calculation of the g_E on the Moon	No	Jurin's Law	No

Source: authors' review.

Physics books, even those studying applications of physics in biology [53], do not deal with the Moon's influence on plant growth. This may be due to the fact that the Moon's gravity is, as we have seen in the Introduction (Section 1), negligible compared to that of the Earth. Regarding illuminance, since it is a topic addressed in specialized books on optics [54], it is not usually included in physics books (only one of them does, as shown in Table 4), even less lunar illuminance.

The analysis of reference handbooks and monographs dealing with plant growth and development in the background of biology, environmental sciences, forestry, and agronomy is a key issue to understanding the extent to which this is a question that is limited to agricultural practice and/or the scientific and training field. Table 5 shows a summary of six widespread and commonly used books on botany and plant physiology, making a synoptic review of the endogenous and exogenous factors that determine and modulate plant development. In particular, the focus has been placed on those Moon-dependent factors that could be beneficial or stressful for plants, specifically in relation to Moon gravity or to the light reflected by the Moon.

As mentioned in the Introduction (Section 1) and reflected in Table 5, plant growth and development are regulated by endogenous and exogenous factors. The possible effects of the Moon should be considered as abiotic external factors, either if the effect is considered to be due to the light reflected or to gravitation. Regarding light, we searched for possible quotes of the Moon when addressing light effects on seeds, plant development, phototropism, photoperiodism, phototaxis, photonasties and quantity and quality of light, etc. Focusing on gravitational influence, the search was made on different aspects of gravitropism.

Table 5. Revision of some of the reference handbooks on botany and plant physiology in relation to possible references to the Moon's influence on plant growth.

Handbook	Phase/Process	Endogenous Factors	Exogenous or Environmental Factors		Moon Mention
			Biotic	Abiotic	
Arteca [12]	Vital cycle	Endogenous growth substances			No
	Seeds			Water, temp., aeration and light	No
	Flowering			Photoperiod and vernalization	No
	Abscission			Temperature, oxygen and nutrients	No
Evans [13]	Growth			Environment and ecosystem	No
Fosket [55]	Development	Genetics		Light versus darkness	No
	Embryogenesis, germination and development				No
	Apical meristems and development				No
	Plant development		Plant-microbe and symbiotic interactions		No
	Vital cycle			Phototropism and gravitropism	No
Raven et al. [56]	Biological rhythms	Circadian and biological clocks		Track of daylength by length of darkness	No
	Flowering			Daylength as determinant of flowering time	No
	Photoautotrophy			Sunlight	No
Strasburguer [18]	Growth and differentiation	Phytohormones		Temperature, light, gravity, hydromorphosis	No
	Biological rhythms	Circadian rhythms		Photoperiod	No
	Movement			Phototaxis, phototropisms, photonasties	No
	Daylength perception	Circadian rhythms, phytochromes		Light quality, phase setting, interaction of light	No
Thomas and Vince-Prue [56]	Flower timing			Light quantity, moonlight	Yes
	Flower development			Photoperiodism	No
	Bud dormancy				No
	Storage and propagation				No
	Germination				No
	Stem elongation				No
	Leaf growth				No

Source: authors' review.

Considering endogenous factors, we searched for possible interactions of Moon radiation and photoperiod, as well as gravity, on the transduction of the perception of those environmental stimuli as well as the possible determination by endogenous genetic components. An important internal process in plants, animals, fungi and cyanobacteria is that related to circadian rhythms that refer to any biological process that displays oscillation, driven by circadian clocks, synchronized with solar time. Plant circadian rhythms are related to seasons and determine, for example, when to flower to maximize the success of pollinator attraction. Circadian rhythms also determine leaf movement, growth, germination, gas exchange or photosynthetic activity, among others. All monographs reviewed mentioning circadian rhythms refer exclusively to synchronization with the light cycle of the surrounding environments of plants, considering the Sun as light source that can determine or influence these cycles.

This search in what is considered consolidated science and is incorporated to handbooks has revealed practically no mention of the Moon (Table 5). We have only found an anecdotic reference, in relation to the possible influence of moonlight on flowering in Thomas and Vince-Prue [57]. These authors explain the work of Salisbury [58], who had indicated that the effective red-light threshold for flowering is higher than the amount of red light produced by the Moon. In addition, it is important to consider that the shade provided by the leaves of the plant itself can reduce the radiation received to 5–10% of the direct moonlight [59]. Thomas and Vince-Prue [57] state that it seems unlikely that full moon light can influence flowering, even in the most sensitive plants, highlighting the scarcity of research on this issue. In this book the authors mention the work of Kadman-Zahavi and Peiper [60], who carried out research with *Pharbitis nil* (L.) Roth —a very sensitive short-day species—which they exposed to moonlight or shielded for different periods. They concluded that, although it is possible that moonlight is perceived, it had no effect on the experience developed with a short-day species that is particularly sensitive to radiation. The difficulty of isolating the "Moon" factor was highlighted, pointing out the possible influence of shade treatment on plants in other environmental factors that could in turn have an effect on flowering [60]. On the other hand, they indicated that the full moon was only present on very few days of the lunar cycle, so its effect should be negligible under natural conditions.

3.2. *What Research Papers Say from the Perspective of Physics and Biology*

We consider a reference and starting point for the review of scientific articles, the brief paper published in *Nature* by Cyril Beeson in 1946, entitled "The Moon and Plant Growth" [61]. In this paper, the author writes "Beliefs that phases of the Moon have a differential effect on the rate of development of plants are both ancient and world-wide" and concludes that the research carried out to that date had not been able to demonstrate a correlation between the Moon and vital processes of terrestrial plants pointing out that, if any research does, the relationship was so unclear that it has no implications for agriculture.

In the 1950s, Frank A. Brown [22,62,63] undertook different investigations in which he studied the possible lunar rhythmicity in organisms. Most of this research was carried out on marine organisms closely linked to the tides—such as algae, crustaceans, molluscs—he also studied the physiological aspects of terrestrial plants. Brown et al. [22] studied the persistent rhythms of O_2-consumption in potatoes, carrots (*Daucus carota* L.) and brown seaweed (*Fucus*) and searched for a possible influence of barometric pressure rhythms of primary lunar frequency, noting that they are of much lower amplitude than the solar ones. The study was inconclusive in relation to what external rhythmic forces are involved in the rhythms of O_2-consumption, as many of them exhibit some degree of correlation with barometric pressure. In barometric pressure $p = \rho gh$, as its expression depends on g, we would have the same case as with capillarity: the effect of g_M should be subtracted from g_E and, as we have seen, g_M is approximately 300,000 times lower than g_M, so the effect of the Moon on barometric pressure is negligible. The authors discuss the possibility that some of the responses attributed to external factors are due to endogenous rhythmic components. This connection between internal and external factors is supported by Wolfgang Schad [64], who states that "all chrono-biological rhythms are always exo-endogenous, sharing their autonomous inner clock to some degree with the periodicity of the environment, both sides being connected by the long process of evolution", remaining unanswered, the question of how the balance between endogenous and exogenous factors oscillates.

Some authors mention the influence of the lunar phases in a tangential way, without getting to clarify anything. One example explores the resistance of circadian clocks to transient fluctuations in night light levels in nature (i.e., change in cloud cover or stellar/lunar illumination) [65]. Van Norman et al. [66], when differentiating the circadian and infradian rhythms, indicate that the former are the best characterised with a period of around 24 h, while the infradians have periods of more than 24 h and can be due to the tides, lunar, seasonal, annual or longer. In other publications, the authors actively search, without finding them, for relationships between the Moon and some organisms. A paradigmatic case is the study conducted by Bitzand Sargent [67], who unsuccessfully tries to relate the growth rate of the fungus *Neurospora crassa* Shear & Dodge to the influence of a supposed lunar magnetic field (which, as we explain in detail in this article, is even more negligible than the gravitational field). Recently Mironov et al. [68] mentioned a circalunar growth rhythm in a research carried out with genus *Sphagnum*. They found an acceleration in the growth of the mosses studied near the new moon, and a slowdown in growth near the full moon.

Regarding biodynamic practices in agriculture, Hartmut Spiess carried out chronobiological investigations of crops grown under biodynamic management, developing experiments to test the effects of lunar rhythms on the growth of winter rye (*Secale cereale* L.) and little radish (*Raphanus sativus* L., cv. Parat) [69,70]. Spiess [69,70] tried to clarify some of the varying results that a number of studies conducted in the 1930s and 1940s had left unclear. This author also focused on studies made by M. and M.K. Thun [71] establishing a relationship between the position of the Moon relative to the zodiac (sidereal rhythms), planting dates and crop growth, which served as a basis for the publication of calendars. Spiess' [69,70] results pointed out that the effects of lunar rhythms were weak, and especially the effects of the sidereal rhythms described by Thun and Thun were not apparent. In contrast to these papers, Kollerstrom and Staudenmaier [72], pointed out that, although Spiess' [69,70] experiments were well designed, there was a lack of care in the data analysis. According to these authors, the results published to date of its publication suggested that lunar factors may have a practical significance for agriculture.

Without a doubt, one of the botanists who dedicated the most effort and publications to the search for relationships between the Moon and plants was Peter Barlow. Barlow [73–85] devoted part of his research to decoding the influence of the Moon on biological phenomena. Specifically those aspects that take place in plants [73], such as the movements of leaves [74–76], stem elongation [77], fluctuations in tree stem diameters [78], the growth of roots [79–81], biophoton emissions from seedlings [82–84], and chlorophyll fluorescence [85]. According to Barlow et al. [76], and other works of the same author, at least in the cases analysed, the rhythm of leaf movements seem to have been developed or entrained in synchrony with the exogenous lunisolar rhythm experienced either on the Earth or in Space. Barlow [76] believed that plant movements were related with water movements within the plant: as ocean tides are produced by lunisolar gravitational force, water movement in the pulvinus could be responsible for leaf movement, explanation that we have previously discussed.

From all external factors, the perception of light plays a significant role as it can modify biosynthesis by photostimulation and act as a trigger initiating the different stages of development (Table 6). Reversive responses of plant to changes in light conditions can allow them to adjust their leaf or flower position (photonastic and heliotropic movements, respectively) to modulate the incoming radiation. Germination is also severely affected in some plants by light exposition. In fact, some seeds only germinate when they are exposed to a particular red to far-red ratios (660/730 nm), and in a particular moment [14].

Table 6. Radiation effects on developmental processes in plants.

Process	Mode of Action [1]	Spectral Range [2]	Fluctuation [3]
Seed germination and bud break	I	R/FR, B	P
Stem elongation	Q, F	R/FR	P
Stem orientation	Q, F	B	
Leaf orientation	Q	R/FR	C
Flowering process	I	R/FR	C
Development and filling of storage organs	I	R/FR	P
Dormancy	I	R/FR	P
Enzyme synthesis	I	R/FR	
Enzyme activation	I	R/FR	
Membrane potentials	I	R/FR	

[1] I = Inductive; Q = Quantitative; F = Formative. [2] B = Blue light; R/FR = Red-to-Far-Red ratio. [3] P = Photoperiodism; C = Circadian rhythm. Source: modified from Larcher [14], Kronenberg et al. [86] and Salisbury [87].

Despite light being crucial for plant life, just a few studies have explored the effect of moonlight on plant physiology and their results are not conclusive. Kolisko [88] observed that the period and percentage of germination and subsequent plant growth was influenced by the phase of the Moon at sowing time. And according to Bünning and Moser [59], light intensities as low as 0.1 lx, which correspond approximately to moonlight intensities (see Table 1), may influence photoperiodism in plants and animals whose threshold values of photoperiodic time-measurement is on the order of 0.1 lx. They suggest that light intensity may reach 0.7 lx or even 1 lx when the altitude of the Moon is at 60° or higher altitudes in tropical and subtropical regions (respectively), clearly influencing photoperiodic reactions. However, they observed that in short-day plants such as *Perilla ocymoides* L. and *Chenopodium amaranticolor* H.J.Coste & Reyn., light intensities similar to those of the full moon favoured rather than inhibited flowering [59]. They justified the circadian leaf movements observed in *Glycine*, *Arachis* and *Trifolium* plants as an adaptive mechanism to reduce the intensity of full moon received in the upper surface of the leaf avoiding plant misinterpretations of confounding full moonlight as it would be long day [59]. However, Kadman-Zahavi and Peiper [60] rejected this hypothesis concluding "that in the natural environment moonlight may have at most only a slight delaying effect on the time of flower induction in short-day plants" (p. 621). Furthermore, Raven and Cockell [89] suggested that photosynthesis on Earth can occur in the photosynthetically active radiation (PAR) range of (10^{-8}–8×10^{-3}) mol of photons m^{-2} s^{-1}, and PAR values of moonlight at full moon goes from (0.5–5) $\times 10^{-9}$ mol of photons m^{-2} s^{-1}, suggesting that moonlight is not a significant source of energy for photosynthesis on Earth.

Recently, Breitler et al. [90] described that the photoreceptors present in *Coffea arabica* L. plants are able to perceive full moonlight and this full moonlight PAR is inadequate for photosynthetically supported growth. Plants perceive it as blue light with a very low R/FR ratio, yet this weak light has a great impact on numerous genes. In particular, it affects up to 50 genes related to photosynthesis, chlorophyll biosynthesis and chloroplast machinery at the end of the night. Moreover, full moonlight promotes the modification of the transcription of major rhythmic redox genes, many heat shock proteins and carotenoids genes suggesting that the moonlight seems to be perceived as a stress factor by the plant.

In other cases, full moonlight is correlated with a successful pollination of *Ephedra* species. Rydin and Bolinder [91] observed a correlation between pollination and the phases of the Moon on the gymnosperm *Ephedra foeminea* Forssk., specifically with the full moon of July. During that period, non-mature cones secreted enough pollination drops to apparently attract pollinators that can use the full moon to navigate and also be attracted to the glittering drops in the full moonlight. According to the authors, when insects are not used as pollinators, as it happens in other species of *Ephedra*, the adaptive value of correlating pollinating with the full moon is lost.

In the literature review carried out, some works were found that deal with two different topics that could have relationship with the Moon: polarization and magnetism. According to Semmens [92–94] during certain periods, moonlight is partially polarised, "the maximum effect being with the oblique reflexion of half-moon, or somewhat later for the waxing and earlier for the waning moon" and that polarised light can favour the diastase, which catalyses the hydrolysis, first of starch into dextrin and immediately afterwards into sugar or glucose, to favour germination, as he observed in crushed mustard seeds in the presence of this polarised light. Macht [95] studied the effect of (not lunar) polarized light on seeds of *Lupinus albus* L. and his results were consistent with previous findings of the action of diastase on starch. However, as far as we know, apart from those works no other research papers have been focused on the role of lunar polarized light. Despite, a full body of evidence supports that polarized moonlight has a biological significance in the vision and orientation of nocturnal animals [96,97]. Although we are at the very beginning of understanding the extent to which and why nocturnal animals use the lunar polarization, we do know that the land area over which it is viewable in pristine form is relentlessly shrinking due to human activity. In this sense, Kyba et al. [98] showed that urban skyglow has a great degree of linear polarization and confirmed that its presence diminishes the natural lunar polarization signal. They also observed that the misalignment between the polarization angles of the skyglow and scattered moonlight could explain the reduction of the degree of linear polarization as the Moon rises. Regarding nocturnal animal navigation systems based on perceiving polarized scattered moonlight, these authors highlighted the necessity of considering polarization pollution models in highly light-polluted areas. In any case, there is almost no doubt that the level of polarization of moonlight would be extremely small: so minimal, that its effect would be completely negligible in plants [98].

On the other hand, some studies suggest an influence of the lunar magnetic field. There is evidence that some animals, fungi, some protists and some bacteria seem to be able to react to the variation of the Earth's magnetic field [99–101]. The question that arises is whether plants are also able to respond to these fields and whether the Moon is capable of producing some magnetic field that plants can respond. There is abundant literature discussing magnetoreception in plants [102–106], but no conclusive results have been reported with direct application to agriculture.

Our planet has a magnetic field, called geomagnetic field, with an intensity of approximately $(25–65) \times 10^{-6}$ T, ridiculously small compared to a commercial magnet (about 0.01 T) or a 0.2 T neodymium magnet. Although there are studies that argue that billions of years ago the Moon generated a magnetic field probably even stronger than the current magnetic field of the Earth, the lunar dynamo ended around one billion years ago [107,108]. The intensity of the present-day magnetic field on the lunar surface is $<0.2 \times 10^{-9}$ T, indicating that the Moon currently does not have a global magnetic field [109]. A magnetic field of this numerical value is approximately 225,000 times less than the Earth's, and if divided by the distance Earth–Moon (3.84×10^{10} m), we can easily conclude that the possible effect of a hypothetical lunar magnetic field on the Earth would be much more negligible than that of the gravitational field.

Other theories claim that it is not the lunar magnetic field that affects, but the disturbance in the Earth's electromagnetic field caused by the lunar gravitational changes that take place during the full moon [4]; or also that Moon effects to the Earth's magnetosphere [110]. In both cases, the assumed effects would be (as we have seen in the calculations for the gravity case) completely insignificant.

A general analysis of the above-mentioned literature highlights the heterogeneity in the information sources regarding year of publication and discipline of the journal. On the one hand, there are very recent papers [68,90] but also literature from more than half a century ago [61,92–94]. On the other hand, there are peer-reviewed papers indexed in the Q1 of JCR in specific publications on Plant Science discipline, as *Annals of Botany* [75,79,80], *BMC Plant Biology* [90], *Frontiers in Plant Science* [104], *Journal of Plant Research* [103], *New Phytologist* [81], *Planta* [76], *Physiologia Plantarum* [68], *Plant Cell* [65,66] or *Plant Physiology* [67], with a long and consolidated trajectory in the field and with a pool of reviewers with solid expertise. Other articles are published in the Q2–Q3 of *JCR* in the same category as *Plant Biology* [77] and *Protoplasma* [78,83], or in other categories as Horticulture or Agronomy (e.g., *Biological Agriculture and Horticulture* [69,70,72]). Other papers included belong to other disciplines: *Astrobiology* [89], *Biology Letters* [91], *Icarus* [109], *Philosophical Transactions of the Royal Society B: Biological Sciences* [96], *Nature* [61,92–94], *Naturwissenschaften* [84], indexed in Q1–Q2 *JCR* lists. Nevertheless, there are also some papers not included in *JCR* lists but in other repositories as *Communicative and Integrative Biology* [73], *Earth, Moon and Planets* [64], *Pathophysiology* [110] and *Star and Furrow* [71].

This analysis also raises the question of the extent to which the authors have a good basis in the physics behind all these phenomena, given that to date Moon has not been proved to affect plant biology regarding consolidated physics.

4. Next Steps from the Perspective of Science Teaching

We are concerned about the insidious spread of pseudo-scientific ideas, not only in the field of plant science (which determines many of the behaviours, habits and techniques of many farmers in rural areas) but into the broader population through both formal and informal education. As science educators, we are especially concerned about the widespread belief in pseudo-science throughout the general populace and especially in science teachers [111–114]. Solbes et al. [114] showed that 64.9% of a sample of 131 future science teachers agree or partially agree with the expression "The phase of the Moon can affect, to some extent, several factors such as health, the birth of children or certain agricultural tasks".

Given this worrying scenario, teachers must promote critical thinking as an essential part of citizenship development. Critical thinking implies being informed about issues or problems, not limiting oneself to the dominant discourses in the media, understanding alternative, well-argued positions and being able to analyse the evidence supporting each of them, studying the problem in its complexity, so that scientific, technical, social, economic, environmental, cultural and ethical dimensions are involved, etc. [115–117]. We believe that it is crucial for teachers to be aware of these beliefs in order to address them from a scientific perspective, as has been demanded for some time [118,119].

One way to approach pseudoscience is to involve students in the proper process of reasoning and knowledge building in science, and research-based teaching is postulated as a suitable teaching methodology to address the problem [120]. In the same way that Lie and Boker [121] analysed the perceptions of complementary therapies of medical students who claimed to have pseudo-scientific beliefs related to health, it would be of great interest to address these issues with agronomic students. In this line, a teaching–research sequence has been developed [122] with future science teachers, in which the strategy followed was to plant seeds of different plant species in each of the phases of the Moon and to measure their growth once a lunar cycle was completed. The participants specified the research question and the initial hypothesis. In addition, they established the experimental design as a whole, fixed the dependent, independent, and constant variables, the materials, the sequence of the sessions, etc. [123,124]. As a result of the proposal, still under analysis, it is expected that students will develop a critical attitude towards pseudoscience and improve their training in research methodologies. Didactic proposals of this type could help, not only in teacher training, but also in any scientific study, promoting critical thinking.

With this work, we wanted to draw attention to one of the many facets of current pseudo-scientific ideas, especially in agriculture. However, we want to emphasize that dismantling these ideas, which, as we have seen, lack in any scientific basis, should not be incompatible with knowing and preserving agricultural traditions that are an important part of an ethnographic and anthropological heritage, as some institutions such as the Food and Agriculture Organization of the United Nations (FAO) claim [125–128]. Many of these traditions (e.g., organic farming, traditional and seasonal crops) allow a harmonious and sustainable coexistence with their natural environments, compatible with the conservation of biodiversity, varieties of certain species.

Furthermore, this paper encourages new research on this long-lasting topic of the possible influence of the Moon on plants in order to clarify many aspects that still remain unanswered or which have not been approached. Considering that modern ecophysiology requires a good understanding of both the molecular aspects of plant processes and the environment, future studies will necessarily have to move to a higher level: scaling from physiology to the Globe [129], considering relationships between plant ecophysiological processes and those occurring at ecosystems but also including social aspects—as traditions, farmers' behaviours and protocols, etc.—that can determine the environment where plants grow.

This review opens the door to possible research that would help to complete the picture of the extent to which certain pseudo-scientific ideas have permeated different sectors of the population. In this sense, it would be interesting to carry out an in-depth analysis of what farmers, as well as students in careers related to the agriculture sector and plant biology, think about the relationship of the Moon with the growth and development of plants.

5. Conclusions

Science has widely established different evidences: (i) the Moon's gravity on the Earth cannot have any effect on the life cycle of plants due to the fact that it is 3.3×10^{-5} ms^{-2}, almost 300,000 times lower that the Earth's gravity; (ii) since all the oceans are communicated and we can consider their size being the size of the Earth, the Moon's influence on the tides is 10^{-6} ms^{-2}, but for a 2 m height plant such value is 3×10^{-13} ms^{-2} and, therefore, completely imperceptible; (iii) the Moon's illuminance cannot have any effect on plant life since it is, at best, 128,000 times lower than the minimum of sunlight on an average day; (iv) the rest of possible effects of the Moon on the Earth (e.g., magnetic field, polarization of light) are non-existent.

The logical consequence of such evidence is that none of these effects appear in physics and biology reference handbooks. However, many of these beliefs are deeply ingrained in both agricultural traditions and collective imagery. This shows that more research should be undertaken on the possible effects observed on plants and assigned to the Moon by the popular belief, addressing their causes, if any. It would also be interesting to address these issues in both compulsory education and formal higher agricultural education in order to address pseudo-scientific ideas and promote critical thinking.

Author Contributions: Conceptualization, O.M. and T.P.; methodology, O.M., J.S., J.C. and T.P.; formal analysis, O.M. and T.P.; investigation, O.M., J.S., J.C. and T.P.; data curation, O.M. and T.P.; writing—original draft preparation, O.M., J.S., J.C. and T.P.; writing—review and editing, O.M., J.S., J.C. and T.P.; supervision, O.M., T.P.; project administration, O.M., J.S. and T.P.; funding acquisition, O.M. and J.S. All authors have read and agreed to the published version of the manuscript.

Acknowledgments: A significant part of the research and bibliographic review was carried out by O.M. and T.P. in the different libraries of Harvard University—especially in the excellent library of the Arnold Arboretum of Harvard University. In this sense, we would like to express our gratitude to the Real Colegio Complutense (RCC) —Harvard, who granted two of the authors (O.M. and T.P.) for a stay at Harvard University during 2019.

References

1. Martínez, V.J.; Miralles, J.A.; Marco, E.; Galadí-Enríquez, D. *Astronomía Fundamental*; Universitat de València: Valencia, Spain, 2005.
2. Morgan, E. The moon and life on earth. *Earth Moon Planets* **2001**, *85–86*, 279–290.
3. Rackham, T. *Moon in Focus*; Pergamon: Oxford, UK, 1968.
4. Bakulin, P.I.; Kononovich, E.V.; Moroz, V.I. *Curso de Astronomía General*; Mir: Moscow, Russia, 1987.
5. Hewitt, P.G. *Conceptual Physics*, 9th ed.; Pearson Education: San Francisco, CA, USA, 2002.
6. Neumann, D. Timing in Tidal, Semilunar, and Lunar Rhythms. In *Annual, Lunar and Tidal Clocks: Patterns and Mechanisms of Nature's Enigmatic Rhythms*; Numata, H., Helm, B., Eds.; Springer: Tokyo, Japan, 2014; pp. 3–24.
7. Adushkin, V.V.; Riabova, S.A.; Spivak, A.A. Lunar–solar tide effects in the Earth's crust and atmosphere. *Izv. Phys. Solid Earth* **2017**, *53*, 565–580. [CrossRef]
8. Myers, D.E. Gravitational effects of the period of high tides and the new moon on lunacy. *Int. J. Emerg. Med.* **1995**, *13*, 529–532. [CrossRef]
9. RCA Corporation. *Electro-Optics Handbook*; RCA/Commercial Engineering: Harrison, NJ, USA, 1974.
10. Schlyter, P. (1997–2017) Radiometry and photometry in astronomy. Available online: http://www.stjarnhimlen.se/comp/radfaq.html#13 (accessed on 29 April 2020).
11. Hunt, R. *Basic Growth Analysis: Plant Growth Analysis for Beginners*; Unwin Hyman: London, UK, 2012.
12. Arteca, R.N. *Plant Growth Substances: Principles and Applications*; Springer Science & Business Media: Dordrecht, The Netherlands, 2013.
13. Evans, G.C. *The Quantitative Analysis of Plant Growth*; Blackwell Scientific Publications: Oxford, UK, 1972; Volume 1.
14. Larcher, W. *Physiological Plant Ecology: Ecophysiology and Stress Physiology of Functional Groups*; Springer-Verlag Berlin Heidelberg: Berlin, Germany, 2003.
15. Leopold, A.C. *Plant Growth and Development*; MCGraw-Hill Education: New York, NY, USA, 1964.
16. Srivastava, L.M. *Plant Growth and Development: Hormones and Environment*; Elsevier: San Diego, CA, USA, 2002.
17. Taiz, L.; Zeiger, E.; Møller, I.M.; Murphy, A. *Plant Physiology and Development*, 6th ed.; Sinauer Associates, Inc.: Sunderland, MA, USA, 2015.
18. Bulgari, R.; Franzoni, G.; Ferrante, A. Biostimulants Application in Horticultural Crops under Abiotic Stress Conditions. *Agronomy* **2019**, *9*, 306. [CrossRef]
19. Drobek, M.; Frąc, M.; Cybulska, J. Plant Biostimulants: Importance of the Quality and Yield of Horticultural Crops and the Improvement of Plant Tolerance to Abiotic Stress—A Review. *Agronomy* **2019**, *9*, 335. [CrossRef]
20. Strasburger, E.; Noll, F.; Schenck, H.; Schimper, A. *Tratado de Botánica*; Omega: Barcelona, Spain, 2004.
21. Bhatla, S.C.; Lal, M.A. *Plant Physiology, Development and Metabolism*; Springer: Singapore, 2018.

22. Brown, F.A., Jr.; Freeland, R.O.; Ralph, C.L. Persistent Rhythms of O_2-Consumption in Potatoes, Carrots and the Seaweed, Fucus. *Plant Physiol.* **1955**, *30*, 280. [CrossRef] [PubMed]
23. Ferris, T. *Coming of Age in the Milky Way*; Anchor Books: Morrow, NY, USA, 1988.
24. Solbes, J.; Palomar, R. ¿Por qué resulta tan difícil la comprensión de la astronomía a los estudiantes? *Didáctica Cienc. Exp. Soc.* **2011**, *25*, 187–211.
25. Böckler, C.G. *Donde Enmudecen las Conciencias: Crepúsculo y Aurora en Guatemala*; CIESAS: Mexico City, Mexico, 1986.
26. Phillips, S. *An Encyclopaedia of Plants in Myth, Legend, Magic and Lore*; The Crowood Press Ltd.: Marlborough, UK, 2012.
27. Navazio, J. *The Organic Seed Grower: A farmer's Guide to Vegetable Seed Production*; Chelsea Green Publishing Co.: London, UK, 2012.
28. Anglés Farrerons, J.M. *Influencia de la Luna en la Agricultura y Otros Temas de Especial Interés Para el Campesino y Gentes de la Ciudad*; Dilagro-Ediciones: Lleida, Spain, 1984.
29. Restrepo, J. *La Luna: El sol Nocturno en los Trópicos y su Influencia en la Agricultura*; (No. 630.2233 R436.); Servicio de Información Mesoamericano sobre Agricultura Sostenible: Managua, Nicaragua, 2004.
30. Tompkins, P.; Bird, C. *The Secret Life of Plants*; (No. QK50. T65I 1973.); Harper & Row: New York, NY, USA, 1973.
31. Galston, A.W.; Slayman, C.L. The Not-So-Secret Life of Plants: In which the historical and experimental myths about emotional communication between animal and vegetable are put to rest. *Am. Sci.* **1979**, *67*, 337–344.
32. Horowitz, K.A.; Lewis, D.C.; Gasteiger, E.L. "Plant primary perception": Electrophysiological unresponsiveness to brine shrimp killing. *Science* **1975**, *189*, 478–480. [CrossRef]
33. Bussagli, M. *Calendario Lunar de las Siembras y Labores Agrícolas (Pequeñas Joyas)*; Susaeta: Madrid, Spain, 2019.
34. Calendario Zaragozano. *Calendario Zaragozano. El Firmamento para toda España*; Castillo y Ocsiero, M., Ed.; Zaragozano: Madrid, Spain, 2019.
35. Geiger, P. (Ed.) *Farmers' Almanac for the Year 2020*; Almanac Publishing Company: Lewiston, ME, USA, 2019.
36. Gros, M. *Lunario 2020: Calendario Lunar para el Huerto y el Jardín Ecológico y También para Mantener la Salud*; Artús Porta Manresa: Manresa, Spain, 2019.
37. Leendertz, L. *The Almanac: A Seasonal Guide to 2020*; Mitchell Beazley: London, UK, 2019.
38. Trédoulat, T. *Réussir son Potager Avec la Lune*; Rustica éditions: Paris, France, 2020.
39. Littlewood, M. *A Guide to Gardening by the Moon*; Gaby Bartai: Glasgow, UK, 2009.
40. Descubre cómo las fases lunares pueden afectar a tus cultivos. Available online: https://www.todohusqvarna.com/blog/fases-lunares/ (accessed on 19 June 2020).
41. Chalker-Scott, L. The science behind biodynamic preparations: A literature review. *HortTechnology* **2013**, *23*, 814–819. [CrossRef]
42. Thun, M.; Thun, M.K. *Calendario de Agricultura Biodinámica 2020*; Editorial Rudolf Steiner: Madrid, Spain, 2019.
43. Steiner, R. Agriculture (English translation). 1958. Available online: http://wn.rsarchive.org/Biodynamics/GA327/English/BDA1958/Ag1958_index.html (accessed on 27 April 2020).
44. Brock, C.; Geier, U.; Greiner, R.; Olbrich-Majer, M.; Fritz, J. Research in biodynamic food and farming–a review. *Open Agric.* **2019**, *4*, 743–757. [CrossRef]
45. Demeter Association, Inc. Available online: https://www.demeter-usa.org/about-demeter/biodynamic-certification-marks.asp (accessed on 27 April 2020).
46. Kirchmann, H. Biological dynamic farming—An occult form of alternative agriculture? *J. Agric. Env. Ethics* **1994**, *7*, 173–187. [CrossRef]
47. Steiner, R. *A Modern Art of Education*; Anthroposophic Press: Great Barrington, MA, USA, 2004.
48. Beer, A.; Beer, P. *Kepler, Four Hundred Years*; Pergammon Press: Oxford, UK, 1975.
49. Tipler, P.A. *Physics for Scientist and Engineers*, 3rd ed.; Worth Publishers: New York, NY, USA, 1992.
50. Feynman, R.P.; Leighton, R.B.; Sands, M. *The Feynman Lectures on Physics*; Fondo Educativo Interamericano: Bogotá, Colombia, 1971; Volume 1.
51. Holton, G.; Brush, S. *Introduction to Concepts and Theories in Physical Science*; Addison Wesley: Reading, MA, USA, 1976.
52. Gettys, W.E.; Keller, F.J.; Skove, M.J. *Physics Classical and Modern*; McGraw-Hill: New York, NY, USA, 1989.
53. Giancoli, D.C. *Physics*; Prentice-Hall: Englewood Cliffs, NJ, USA, 1985.

54. Smith, F.G.; Thomson, J.H. *Optics*; Wiley: Hoboken, NJ, USA, 1989.
55. Fosket, D.E. *Plant Growth and Development: A Molecular Approach*; Academic Press Inc.: San Diego, CA, USA, 1994.
56. Raven, P.H.; Evert, R.F.; Eichhorn, S.E. *Biology of Plants*; W.H. Freeman and Company: New York, NY, USA, 2005.
57. Thomas, B.; Vince-Prue, D. *Photoperiodism in Plants*; Academic Press, Inc.: San Diego, CA, USA, 1996.
58. Salisbury, F.B. Plant adaptations to the light environment. In *Plant Production in the North*; Kaurin, A., Junttila, O., Nilsen, J., Eds.; Norwegian University Presss: Tromso, Norway, 1985; pp. 43–61.
59. Bünning, E.; Moser, I. Interference of moonlight with the photoperiodic measurement of time by plants, and their adaptive reaction. *Proc. Natl. Acad. Sci. USA* **1969**, *62*, 1018–1022. [CrossRef] [PubMed]
60. Kadman-Zahavi, A.; Peiper, D. Effects of moonlight on flower induction in *Pharbitis nil*, using a single dark period. *Ann. Bot.* **1987**, *60*, 621–623. [CrossRef]
61. Beeson, C.F.C. The moon and plant growth. *Nature* **1946**, *158*, 572. [CrossRef] [PubMed]
62. Brown, F.A., Jr.; Bennett, M.F.; Marguerite Webb, H. Persistent daily and tidal rhythms of O_2-consumption in fiddler crabs. *J. Cell Comp. Physiol.* **1954**, *44*, 477–505. [CrossRef]
63. Brown, F.A., Jr.; Webb, H.M.; Bennett, M.F.; Sandeen, M.I. Temperature-independence of the frequency of the endogenous tidal rhythm of *Uca*. *Physiol. Zool.* **1954**, *27*, 345–349. [CrossRef]
64. Schad, W. Lunar influence on plants. *Earth Moon Planets* **2001**, *85–86*, 405–409.
65. Covington, M.F.; Panda, S.; Liu, X.L.; Strayer, C.A.; Wagner, D.R.; Kay, S.A. ELF3 modulates resetting of the circadian clock in *Arabidopsis*. *Plant Cell* **2001**, *13*, 1305–1316. [CrossRef]
66. Van Norman, J.M.; Breakfield, N.W.; Benfey, P.N. Intercellular communication during plant development. *Plant Cell* **2011**, *23*, 855–864. [CrossRef]
67. Bitz, D.M.; Sargent, M.L. A failure to detect an influence of magnetic fields on the growth rate and circadian rhythm of *Neurospora crassa*. *Plant Physiol.* **1974**, *53*, 154–157. [CrossRef] [PubMed]
68. Mironov, V.L.; Kondratev, A.Y.; Mironova, A.V. Growth of *Sphagnum* is strongly rhythmic: Contribution of the seasonal, circalunar and third components. *Physiol. Plant* **2020**, *168*, 765–776. [CrossRef] [PubMed]
69. Spiess, H. Chronobiological Investigations of Crops Grown under Biodynamic Management. I. Experiments with Seeding Dates to Ascertain the Effects of Lunar Rhythms on the Growth of Winter Rye (Secale cereale cv. Nomaro). *Biol. Agric. Hortic.* **1990**, *7*, 165–178. [CrossRef]
70. Spiess, H. Chronobiological Investigations of Crops Grown under Biodynamic Management. II. Experiments with Seeding Dates to Ascertain the Effects of Lunar Rhythms on the Growth of Little Radish (*Raphanus sativus*, cv. Parat). *Biol. Agric. Hortic.* **1990**, *7*, 179–189. [CrossRef]
71. Thun, M. Nine years observation of cosmic influences on annual plants. *Star Furrow* **1964**, *22*.
72. Kollerstrom, N.; Staudenmaier, G. Evidence for lunar-sidereal rhythms in crop yield: A review. *Biol. Agric. Hortic.* **2001**, *19*, 247–259. [CrossRef]
73. Chaffey, N.; Volkmann, D.; Baluška, F. The botanical multiverse of Peter Barlow. *Comm. Integr. Biol.* **2019**, *12*, 14–30. [CrossRef]
74. Barlow, P.W.; Klingelé, E.; Klein, G.; Mikulecký Sen, M. Leaf movements of bean plants and lunar gravity. *Plant Signal. Behav.* **2008**, *3*, 1083–1090. [CrossRef]
75. Barlow, P.W. Leaf movements and their relationship with the lunisolar gravitational force. *Ann. Bot.* **2015**, *116*, 149–187. [CrossRef]
76. Fisahn, J.; Klingelé, E.; Barlow, P. Lunar gravity affects leaf movement of *Arabidopsis thaliana* in the International Space Station. *Planta* **2015**, *241*, 1509–1518. [CrossRef]
77. Zajączkowska, U.; Barlow, P.W. The effect of lunisolar tidal acceleration on stem elongation growth, nutations and leaf movements in peppermint (*Mentha* × *piperita* L.). *Plant Biol.* **2017**, *19*, 630–642. [CrossRef] [PubMed]
78. Barlow, P.W.; Mikulecký, M., Sr.; Střeštík, J. Tree-stem diameter fluctuates with the lunar tides and perhaps with geomagnetic activity. *Protoplasma* **2010**, *247*, 25–43. [CrossRef] [PubMed]
79. Barlow, P.W.; Fisahn, J. Lunisolar tidal force and the growth of plant roots, and some other of its effects on plant movements. *Ann. Bot.* **2012**, *110*, 301–318. [CrossRef] [PubMed]
80. Barlow, P.W.; Fisahn, J.; Yazdanbakhsh, N.; Moraes, T.A.; Khabarova, O.V.; Gallep, C.M. *Arabidopsis thaliana* root elongation growth is sensitive to lunisolar tidal acceleration and may also be weakly correlated with geomagnetic variations. *Ann. Bot.* **2013**, *111*, 859–872. [CrossRef] [PubMed]

81. Fisahn, J.; Yazdanbakhsh, N.; Klingele, E.; Barlow, P. *Arabidopsis thaliana* root growth kinetics and lunisolar tidal acceleration. *New Phytol.* **2012**, *195*, 346–355. [CrossRef] [PubMed]
82. Gallep, C.M.; Moraes, T.A.; Červinková, K.; Cifra, M.; Katsumata, M.; Barlow, P.W. Lunisolar tidal synchronism with biophoton emission during intercontinental wheat-seedling germination tests. *Plant Signal. Behav.* **2014**, *9*, e28671. [CrossRef]
83. Gallep, C.M.; Barlow, P.W.; Burgos, R.C.; van Wijk, E.P.R. Simultaneous and intercontinental tests show synchronism between the local gravimetric tide and the ultra-weak photon emission in seedlings of different plant species. *Protoplasma* **2017**, *254*, 315–325. [CrossRef]
84. Moraes, T.A.; Barlow, P.W.; Klingelé, E.; Gallep, C.M. Spontaneous ultra-weak light emissions from wheat seedlings are rhythmic and synchronized with the time profile of the local gravimetric tide. *Naturwissenschaften* **2012**, *99*, 465–472. [CrossRef]
85. Fisahn, J.; Klingelé, E.; Barlow, P. Lunisolar tidal force and its relationship to chlorophyll fluorescence in *Arabidopsis thaliana*. *Plant Signal. Behav.* **2015**, *10*, e1057367. [CrossRef]
86. Kronenberg, G.H.M.; Kendrick, R.E. The physiology of action. In *Photomorphogenesis in Plants*; Kendrick, R.E., Kronenberg, G.H.M., Eds.; Nijhoff Publ.: Dordrecht, The Netherlands, 1986; pp. 99–114.
87. Salisbury, F.B. *The Flowering Process*; Pergamon Press, Inc.: New York, NY, USA, 1963.
88. Kolisko, L. *The Moon and the Growth of Plants*; Bray-on-Thames, Anthroposophical Agricultural Foundation: London, UK, 1936.
89. Raven, J.A.; Cockell, C.S. Influence on photosynthesis of starlight, moonlight, planetlight, and light pollution (reflections on photosynthetically active radiation in the universe). *Astrobiology* **2006**, *6*, 668–675. [CrossRef]
90. Breitler, J.C.; Djerrab, D.; Leran, S.; Toniutti, L.; Guittin, C.; Severac, D.; Pratlong, M.; Dereeper, A.; Etienne, H.; Bertrand, B. Full moonlight-induced circadian clock entrainment in *Coffea arabica*. *BMC Plant Biol.* **2020**, *20*, 1–11. [CrossRef] [PubMed]
91. Rydin, C.; Bolinder, K. Moonlight pollination in the gymnosperm *Ephedra* (Gnetales). *Biol. Lett.* **2015**, *11*, 20140993. [CrossRef] [PubMed]
92. Semmens, E.S. Effect of Moonlight on the Germination of Seeds. *Nature* **1923**, *111*, 49–50. [CrossRef]
93. Semmens, E.S. Hydrolysis in Green Plants by Moonlight. *Nature* **1932**, *130*, 243. [CrossRef]
94. Semmens, E.S. Chemical Effects of Moonlight. *Nature* **1947**, *159*, 613. [CrossRef]
95. Macht, D.I. Concerning the influence of polarized light on the growth of seedlings. *J. Gen. Physiol.* **1926**, *10*, 41–52. [CrossRef]
96. Cronin, T.W.; Marshall, J. Patterns and properties of polarized light in air and water. *Philos. Trans. R. Soc. Lond. B Biol. Sci.* **2011**, *366*, 619–626. [CrossRef]
97. Nowinszky, L.; Szabó, S.; Tóth, G.; Ekk, I.; Kiss, M. The effect of the moon phases and of the intensity of polarized moonlight on the light-trap catches. *Zeitschrift Angewandte Entomologie* **1979**, *88*, 337–353. [CrossRef]
98. Kyba, C.C.; Ruhtz, T.; Fischer, J.; Hölker, F. Lunar skylight polarization signal polluted by urban lighting. *J. Geophys. Res.* **2011**, *116*, D24106. [CrossRef]
99. Begall, S.; Malkemper, E.P.; Cerveny, J.; Nemec, P.; Burda, H. Magnetic alignment in mammals and other animals. *Mammal. Biol.* **2013**, *78*, 10–20. [CrossRef]
100. Ritz, T.; Wiltschko, R.; Hore, P.J.; Rodgers, C.T.; Stapput, K.; Thalau, P.; Timmel, C.R.; Wiltschko, W. Magnetic compass of birds is based on a molecule with optimal directional sensitivity. *Biophys. J.* **2009**, *96*, 3451–3457. [CrossRef] [PubMed]
101. Yan, L.; Zhang, S.; Chen, P.; Liu, H.; Yin, H.; Li, H. Magnetotactic bacteria, magnetosomes and their application. *Microbiol. Res.* **2012**, *167*, 507–519. [CrossRef] [PubMed]
102. Belyavskaya, N.A. Biological effects due to weak magnetic field on plants. *Adv. Space Res.* **2004**, *34*, 1566–1574. [CrossRef]
103. Galland, P.; Pazur, A. Magnetoreception in plants. *J. Plant Res.* **2005**, *118*, 371–389. [CrossRef]
104. Maffei, M.E. Magnetic field effects on plant growth, development, and evolution. *Front. Plant Sci.* **2014**, *5*, 445. [CrossRef] [PubMed]
105. Nyakane, N.E.; Markus, E.D.; Sedibe, M.M. The effects of magnetic fields on plants growth: A comprehensive review. *Int. J. Food Eng.* **2019**, *5*, 79–87. [CrossRef]
106. Vian, A.; Davies, E.; Gendraud, M.; Bonnet, P. Plant responses to high frequency electromagnetic fields. *Biomed. Res. Int.* **2016**, *2016*, 1830262. [CrossRef]

107. Mighani, S.; Wang, H.; Shuster, D.L.; Borlina, C.S.; Nichols, C.I.; Weiss, B.P. The end of the lunar dynamo. *Sci. Adv.* **2020**, *6*, eaax0883. [CrossRef]
108. Tikoo, S.M.; Weiss, B.P.; Shuster, D.L.; Suavet, C.; Wang, H.; Grove, T.L. A two-billion-year history for the lunar dynamo. *Sci. Adv.* **2017**, *3*, e1700207. [CrossRef]
109. Mitchell, D.L.; Halekas, J.S.; Lin, R.P.; Frey, S.; Hood, L.L.; Acuña, M.H.; Binder, A. Global mapping of lunar crustal magnetic fields by Lunar Prospector. *Icarus* **2008**, *194*, 401–409. [CrossRef]
110. Bevington, M. Lunar biological effects and the magnetosphere. *Pathophysiology* **2015**, *22*, 211–222. [CrossRef] [PubMed]
111. Eve, R.A.; Dunn, D. Psychic powers, astrology and creationism in the classroom? Evidence of pseudoscientific beliefs among high school biology & life science teachers. *Am. Biol. Teach.* **1990**, *52*, 10–21. [CrossRef]
112. Happs, J.C. Challenging pseudoscientific and paranormal beliefs held by some pre-service primary teachers. *Res. Sci. Educ.* **1991**, *21*, 171–177. [CrossRef]
113. Kaplan, A.O. Research on the pseudo-scientific beliefs of preservice science teachers: A sample from astronomy-astrology. *J. Balt. Sci. Educ.* **2014**, *13*, 381–393.
114. Solbes, J.; Palomar, R.; Dominguez-Sales, M.C. To what extent do pseudosciences affect teachers? A look at the mindset of science teachers in training. *Mètode Sci. Stud. J. Ann. Rev.* **2018**, *8*, 188–195. [CrossRef]
115. Halpern, D. Teaching critical thinking for transfer across domains. *Am. Psychol.* **1998**, *53*, 449–455. [CrossRef]
116. Torres, N.; Solbes, J. Contribuciones de una intervención didáctica usando cuestiones sociocientíficas para desarrollar el pensamiento crítico. *Enseñanza Cienc.* **2016**, *34*, 43–65. [CrossRef]
117. Yager, R.E. Science and critical thinking. In *Teaching Critical Thinking: Reports from Across the Curriculum*; Clarke, J.H., Biddle, A.W., Eds.; Prentice Hall: Englewood Cliffs, NJ, USA, 1993.
118. Bates, J.; Culpepper, W. Using Pseudoscience to Teach Science: Encouraging Skepticism of Paranormal Powers in the Classroom. *J. Coll. Sci. Teach.* **1991**, *21*, 106–111. [CrossRef] [PubMed]
119. Wilson, J.A. Reducing pseudoscientific and paranormal beliefs in university students through a course in science and critical thinking. *Sci. Educ.* **2018**, *27*, 183–210. [CrossRef]
120. Chinn, C.A.; Malhotra, B.A. Epistemologically authentic inquiry in schools: A theoretical framework for evaluating inquiry tasks. *Sci. Educ.* **2002**, *86*, 175–218. [CrossRef]
121. Lie, D.; Boker, J. Development and validation of the CAM Health Belief Questionnaire (CHBQ) and CAM use and attitudes amongst medical students. *BMC Med Educ.* **2004**, *4*, 2. [CrossRef] [PubMed]
122. Pedaste, M.; Mäeots, M.; Siiman, L.A.; De Jong, T.; Van Riesen, S.A.; Kamp, E.T.; Constantinos, C.; Zacharias, M.; Tsourlidaki, E. Phases of inquiry-based learning: Definitions and the inquiry cycle. *Educ. Res. Rev.* **2015**, *14*, 47–61. [CrossRef]
123. Pina, T.; Mayoral, O.; Solbes, J. ¿Influye la Luna en el crecimiento de las plantas? Indagación para favorecer el pensamiento crítico. In *Propuestas de Educación Científica Basadas en la Indagación y Modelización en Contexto*; Solbes, J., Jiménez, M.R., Pina, T., Eds.; Tirant lo Blanch: Valencia, Spain, 2019; pp. 121–143.
124. Pina, T.; Mayoral, O.; Solbes, J. Do lunar phases influence the growth of plants? Scientific inquiry to encourage critical thinking in the classroom. In Proceedings of the ESERA conference, Bologna, Italy, 26–30 August 2019.
125. Dixon, J.A.; Gibbon, D.P.; Gulliver, A. *Farming Systems and Poverty: Improving Farmers' Livelihoods in a Changing World*; FAO: Rome, Italy; World Bank: Washington, DC, USA, 2001.
126. Kuhnlein, H.V.; Erasmus, B.; Spigelski, D. *Indigenous Peoples' Food Systems: The Many Dimensions of Culture, Diversity and Environment for Nutrition and Health*; FAO: Rome, Italy, 2009.
127. Kuhnlein, H.V. Biodiversity and sustainability of indigenous peoples' foods and diets. Sustainable diets and biodiversity. In *Sustainable Diets and Biodiversity: Directions and Solutions for Policy, Research and Action. Proceedings of the International Scientific Symposium, Biodiversity and Sustainable Diets United Against Hunger, Rome, Italy, 3–5 November 2010*; Burlingame, B., Dernini, S., Eds.; FAO Headquarters: Rome, Italy, 2012.
128. Sinclair, F.; Wezel, A.; Mbow, C.; Chomba, S.; Robiglio, V.; Harrison, R. *The Contribution of Agroecological Approaches to Realizing Climate-Resilient Agriculture*; GCA: Rotterdam, The Netherlands, 2019.
129. Lambers, H.; Chapin, F.S., III; Pons, T.L. *Plant Physiological Ecology*; Springer Science + Business Media: Berlin, Germany; LLC: New York, NY, USA, 2008.

8

The use of Proline in Screening for Tolerance to Drought and Salinity in Common Bean (*Phaseolus vulgaris* L.) Genotypes

Sugenith Arteaga [1], Lourdes Yabor [1,2], María José Díez [3], Jaime Prohens [3], Monica Boscaiu [4,*] and Oscar Vicente [3]

1 Institute for Plant Molecular and Cell Biology (IBMCP, UPV-CSIC), Universitat Politècnica de València, Camino de Vera s/n, 46022 Valencia, Spain; suarcas@alumni.upv.es (S.A.); lyabor@bioplantas.cu (L.Y.)

2 Permanent Address: Laboratory for Plant Breeding and Conservation of Genetic Resources, Bioplant Center, University of Ciego de Avila, Ciego de Ávila 69450, Cuba

3 Institute for the Conservation and Improvement of Valencian Agrodiversity (COMAV, UPV), Universitat Politècnica de València, Camino de Vera s/n, 46022 Valencia, Spain; mdiezni@btc.upv.es (M.J.D.); jprohens@btc.upv.es (J.P.); ovicente@upvnet.upv.es (O.V.)

4 Mediterranean Agroforestry Institute (IAM, UPV), Universitat Politècnica de València, Camino de Vera s/n, 46022 Valencia, Spain

* Correspondence: mobosnea@eaf.upv.es

Abstract: The selection of stress-resistant cultivars, to be used in breeding programmes aimed at enhancing the drought and salt tolerance of our major crops, is an urgent need for agriculture in a climate change scenario. In the present study, the responses to water deficit and salt stress treatments, regarding growth inhibition and leaf proline (Pro) contents, were analysed in 47 *Phaseolus vulgaris* genotypes of different origins. A two-way analysis of variance (ANOVA), Pearson moment correlations and principal component analyses (PCAs) were performed on all measured traits, to assess the general responses to stress of the investigated genotypes. For most analysed growth variables and Pro, the effects of cultivar, treatment and their interactions were highly significant ($p < 0.001$); the root morphological traits, stem diameter and the number of leaves were mostly due to uncontrolled variation, whereas the variation of fresh weight and water content of stems and leaves was clearly induced by stress. Under our experimental conditions, the average effects of salt stress on plant growth were relatively weaker than those of water deficit. In both cases, however, growth inhibition was mostly reflected in the stress-induced reduction of fresh weight and water contents of stems and leaves. Pro, on the other hand, was the only variable showing a negative correlation with all growth parameters, but particularly with those of stems and leaves mentioned above, as indicated by the Pearson correlation coefficients and the loading plots of the PCAs. Therefore, in common beans, higher stress-induced accumulation of Pro is unequivocally associated with a stronger inhibition of growth; that is, with a higher sensitivity to stress of the corresponding cultivar. We propose the use of Pro as a suitable biochemical marker for simple, rapid, large-scale screenings of bean genotypes, to exclude the most sensitive, those accumulating higher Pro concentrations in response to water or salt stress treatments.

Keywords: abiotic stress biomarkers; bean landraces; osmolytes; plant breeding; salt stress; salt stress tolerance; water deficit; water stress tolerance

---

1. Introduction

Drought and soil salinity are amongst the most restrictive environmental factors affecting agriculture worldwide. Even moderate degrees of water deficit or salt stress can lead to a reduction of

50–70% in average yields in most crops when compared with registered record yields [1–3]. Drought, brought about by the scarcity of rain, affects more than half of the agricultural land of our planet and is often linked to secondary salinisation of farmland due to intensive irrigation [4,5]. Cropland salinisation is becoming one of the major constrains for agriculture in many parts of the world, especially in arid and semi-arid regions. At the beginning of this century, it was estimated that around 20% of the irrigated lands were salinised [6], but this figure is increasing yearly, mainly due to anthropogenic alterations, such as irrigation with brackish water or the abusive and indiscriminate use of chemical fertilisers [4]. On the other hand, the scarcity of good-quality water for irrigation, mainly as a consequence of the effects of global warming, will mean more-significant crop losses in the near future, which will especially affect subsistence agriculture in developing countries [7]. Legumes are some of the most important crops, representing a significant component of the human diet. Globally, legumes complement cereal crops as the main sources of plant minerals and proteins [8]. Among the leguminous crops, *Phaseolus* L. is a large and diverse genus comprising about 70 American species [9], five of which have been domesticated (*Phaseolus vulgaris* L., *Phaseolus dumosus* Macfady, *Phaseolus coccineus* L., *Phaseolus acutifolius* A. Gray and *Phaseolus lunatus* L); moreover, a few additional species show signs of incipient domestication [10].

The common bean (*P. vulgaris*) is the most-consumed legume in human nutrition; it is an essential component of the diet, especially in developing countries, as a source of proteins, vitamins, minerals and fibre [8,11]. The species has a natural distribution area from northern Mexico to northwestern Argentina. It was domesticated independently in Central America and the Andes [12,13], but now it is cultivated practically all over the world. Beans from both origins were introduced to Spain in the 16th century [14–16], where they had to adapt to the new environmental conditions, which were very different from those in their native areas. The cropping system in small farms, spread in proximal areas, allowed the genetic flow between genotypes of Mesoamerican and Andean origin [17]. Due to centuries of bean cultivation, the Iberian Peninsula has become a secondary centre of diversification of this species [18].

Phaseolus vulgaris is not considered as very tolerant to water stress [19]; nevertheless, it is cultivated under diverse environmental conditions, including relatively dry areas [20,21]. In fact, globally, only a small percentage, around 7%, of the cropland planted with common bean receives adequate rainfall [11], and in some areas, drought causes yield losses of up to 80% [22] Like practically all cultivated plants, the bean is a glycophyte, sensitive to soil salinity even at electric conductivity values below 2 $dS{\cdot}m^{-1}$ [23]. However, just as there are cultivars that are more resistant to water stress, some respond better to high soil salinity [24,25].

As for other common crops, many bean genotypes no longer grown in the fields or cultivated only locally at a small scale (landraces, local varieties, heirlooms or minor commercial cultivars) are available from small farmers or germplasm banks and represent a rich source of genetic variability. Landraces appeared over time due to selection of traits specifically adapted to local conditions, often suboptimal or even highly stressful. Therefore, such genotypes are probably more competitive in low-input agriculture and represent a source of allelic richness that may enhance agronomic production under the foreseeable restrictive conditions imposed by climate change [26]. There is an increasing interest for the recovery of local landraces by consumers and markets, not only concerning global warming but also because of the commercial demand for local products, considered as tastier and healthier [27]. Unfortunately, many autochthonous varieties have been lost, and many others are at risk of extinction, due to genetic erosion. Screening this type of varieties for tolerance to stresses represents an interesting strategic path for the agriculture of the future.

The screening of a large number of genotypes would be greatly facilitated by identifying a suitable stress biomarker, easily quantified by simple, rapid and non-destructive assays, and unequivocally associated to the relative resistance of the cultivars to water deficit or salt stress. Proline (Pro), one of the commonest plant osmolytes [28,29], could be an appropriate candidate because a significant increase in Pro contents in response to water deficit, high salinity or other stressful conditions has been detected in beans—as in many other species. However, it is not yet clear whether Pro accumulation in *P. vulgaris* is associated with enhanced or reduced tolerance to stress since contradictory results are available in the literature. Some reports correlated higher Pro contents with a relatively higher stress tolerance when comparing different bean cultivars [30–37], whereas in other cases higher Pro concentrations were measured in the relatively more stress-sensitive cultivars [38–40]. All these studies were based on the comparison of a few genotypes. Only a wider analysis, based on a considerably higher number of cultivars, grown under the same experimental conditions and subjected to the same stress treatments, could establish whether responses to stress based on Pro accumulation are relevant, or not, for stress tolerance in *P. vulgaris*, and how Pro could be used as a reliable abiotic stress biomarker in this species.

Based on the ideas mentioned above, we have applied specific water deficit and salt stress treatments, under controlled greenhouse conditions, to a relatively large number of common bean cultivars, obtained from germplasm banks. The aims of this study were (i) to determine the overall response of the analysed genotypes to controlled water and salt stress treatments, (ii) to establish the role of Pro in bean stress responses, either as a mere stress biomarker or as an osmolyte directly involved on stress tolerance mechanisms and (iii) based on the results obtained, to propose Pro as a suitable biochemical marker for the rapid selection of bean cultivars with a (relatively) higher tolerance (or sensitivity) to drought or salinity.

2. Materials and Methods

2.1. Plant Material

The study included 47 accessions of common bean (*P. vulgaris*), from Spain (23), Colombia (19) and Cuba (5), provided by the Germplasm Bank of Universitat Politècnica de València (UPV), the International Center for Tropical Agriculture (CIAT) and the Bioplants Center, University of Ciego de Ávila, respectively.

Spanish genotypes are represented by local landraces, with geographic origins indicated in Table 1. Materials from Cuba are commercial varieties or experimental lines from INIFAT (Alexander Humboldt Institute for Basic Research in Tropical Agriculture) or IIHDL (Liliana Dimitrova Horticultural Research Institute, La Habana, Cuba), and those from Colombia are lines reported to be relatively resistant to drought and high temperatures.

Table 1. Origin of the analysed *Phaseolus vulgaris* accessions and duration of the applied stress treatments.

Abbreviation	Treatment (Weeks)	Genebank Code	Country	Origin	Cultivar Name
Sp 1	2	BGV000143	Spain	Lecina, Huesca	Judía amarilla de enrame
Sp 2	2	BGV001191	Spain	Velez Rubio, Almería	Judía
Sp 3	2	BGV001581	Spain	Mercado el Olivar, Palma de Mallorca	Judia de careta
Sp 4	2	BGV003176	Spain	Barlovento, Santa Cruz de Tenerife	Judia blanca mantecosa
Sp 5	2	BGV003616	Spain	La Bañeza, León	
Sp 6	2	BGV003941	Spain	AldeaNueva de Barbarroya, Toledo	Judía larguilla
Sp 7	2	BGV004159	Spain	Plascencia, Cáceres	
Sp 8	2	BGV011254	Spain	Las Presillas, Puente Viesgo, Cantabria	Garrafal oro
Sp 9	2	BGV013605	Spain	Campo, Huesca	Negra
Co 10	2	INB-39	Colombia	-	
Co 11	2	INB-40	Colombia	-	
Co 12	2	INB-42	Colombia	-	
Co 13	2	INB-43	Colombia	-	
Co 14	2	INB-48	Colombia	-	
Co 15	2	INB-48I	Colombia	-	
Cu 16	2	V-71	Cuba	INIFAT	Bolita 11 [a]
Cu 17	2	E-125	Cuba	IIHLD	E-125 [b]
Cu 18	2	Milagro VIII	Cuba	INIFAT	Milagro Villareño [a]
Sp 19	3	BGV001167	Spain	Chirivel, Almeria	Judia
Sp 20	3	BGV001169	Spain	Laujar de Andarax, Almeria	Judia mocha
Sp 21	3	BGV001182	Spain	Juviles, Granada	Alubias
Sp 22	3	BGV003610	Spain	Ponferrada, León	

Table 1. *Cont.*

Abbreviation	Treatment (Weeks)	Genebank Code	Country	Origin	Cultivar Name
Sp 23	3	BGV003614	Spain	La Bañeza, León	
Sp 24	3	BGV003618	Spain	La Bañeza, León	
Sp 25	3	BGV004161	Spain	Plasencia, Cáceres	
Sp 26	3	BGV004466	Spain	Bilbao, Vizcaya	Alubias pintas
Sp 27	3	BGV011235	Spain	Beranga, Hazas de Cesto, Cantabria	Carica
Sp 28	3	BGV013603	Spain	Beceite, Teruel	Judia de Franco
Sp 29	3	BGV013609	Spain	Centenero, Huesca	Judia Fartapobres
Sp 30	3	BGV014980	Spain	Alcorisa, Teruel	De tabilla ancha
Sp 31	3	BGV015856	Spain	Alicante	Habichuela del barco
Sp 32	3	BGV015859	Spain	Albarracín	Judia
Co 33	3	ALB-74	Colombia	-	
Co 34	3	INB-35	Colombia	-	
Co 35	3	INB-38	Colombia	-	
Co 36	3	INB-41	Colombia	-	
Co 37	3	INB-44	Colombia	-	
Co 38	3	INB-45	Colombia	-	
Co 39	3	INB-46	Colombia	-	
Co 40	3	INB-47	Colombia	-	
Co 41	3	SEF-9	Colombia	-	
Co 42	3	SEF-52	Colombia	-	
Co 43	3	SEF-53	Colombia	-	
Co 44	3	SEF-55	Colombia	-	
Co 45	3	SEF-56	Colombia	-	
Cu 46	3	V-13	Cuba	INIFAT	P 2240 [b]
Cu 47	3	V-51	Cuba	INIFAT	P 186 [b]

a: commercial varieties; b: experimental lines.

2.2. *Plant Growth and Stress Treatments*

The plants were obtained by seed germination. Several seeds of each genotype were germinated in trays with peat, perlite and vermiculite (2:1:1). When the first trifoliate true leaves were formed, the seedlings were transplanted to individual 1.6 L-pots with the same substrate in the greenhouse; Hoagland's nutrient solution [41] was used for irrigation. When the plants reached a height of at least 20 cm and had two to five true leaves, plants were selected for the treatments and placed in 55 × 40 cm plastic trays (10 pots per tray). Irrigation was performed twice a week by adding to each tray 1.5 L deionised water or a 150 mM NaCl solution, for the control and salt stress treatments, respectively. The water stress treatment was applied by completely withholding irrigation of the plants. Five individual plants (biological replicas) of each genotype were used per treatment. Treatments were stopped after two weeks for 18 genotypes when plants showed clear wilting and general decline symptoms in the water deficit treatment, but before plant mortality was observed; salt treatments of these cultivars were stopped at the same time. The remaining, relatively more resistant 29 genotypes were treated for an additional week. The two groups of plants were analysed independently. All treatments were carried out under controlled conditions in the greenhouse: long-day photoperiod (16 h of light), temperature set at 23 °C during the day and 17 °C at night. Once the treatments were finished, whole plants were harvested, collecting separately their roots, stems and leaves. Several growth parameters were measured in all plants: the diameter of the stem (SD), the length of the roots (RL) and stems (SL), the number of trifoliate leaves (Lno) and the fresh weight of roots (RFW), stems (SFW) and leaves (LFW).

Part of the fresh material of roots, stems and leaves was weighed (FW), placed at 65 °C in an oven for three days, and weighed again to determine the dry weight (DW). The water content percentage (WC%) of the three organs was calculated according to the formula:

$$WC\ (\%) = [(FW - DW)/Fw] \times 100 \tag{1}$$

2.3. *Quantification of Proline Contents*

Leaf Pro concentrations were quantified using dry plant material, according to the ninhydrin-acetic acid method [42]. Pro was extracted in a 3% (*w*/*v*) aqueous sulfosalicylic acid solution; the sample was mixed with the acid ninhydrin solution, incubated for 1 h at 95 °C, cooled on ice and extracted with toluene. Samples with known Pro amounts were assayed in parallel to obtain a standard curve.

The absorbance of the supernatants was read at 520 nm using toluene as a blank. Pro concentration was finally expressed as μmol g^{-1} DW.

2.4. Statistical Analysis

Plants from the two- and three-week treatments were analysed separately. A two-way analysis of variance (ANOVA) was performed for all determined traits, to check the effects of the 'cultivar' and 'treatment' factors, and the interaction between treatment and genotype. Pearson moment correlations were also performed for all measured parameters, and a principal component analysis (PCA) was used to check the similarity between the responses to the different types of stress within each cultivar, and the similarity between accessions. Data were analysed using Statgraphics Centurion v.16 software (Statpoint Technologies, Warrenton, VA, USA).

3. Results

3.1. Analysis of Variance of Registered Traits

Some cultivars (18) were apparently more sensitive to both salt and water stress and therefore treatments were stopped after two weeks. For the remaining genotypes (29), treatments were extended to three weeks (Table 1). All growth parameters and the leaf Pro concentration of control and stressed plants, for each cultivar, are summarised in Supplementary Table S1. Notwithstanding quantitative differences between genotypes, the overall picture is that plants of most cultivars were affected by both types of stress, water deficit and salinity, which inhibited growth as indicated by the general relative reduction observed in the measured morphological variables. Under the specific stress conditions applied in the experiments, in most cases, growth inhibition was more accentuated in the water-stressed plants than in the salt-stressed ones. Again for most cultivars, leaf Pro contents increased significantly in response to both types of stress. To assess the general responses to stress of the selected cultivars, a two-way ANOVA was performed considering the effect on each parameter of cultivar and treatment, and their interaction (Table 2).

Table 2. Two-way analysis of variance (ANOVA) of cultivar, treatment and their interactions for the parameters considered. Numbers represent percentages of the sum of squares at the 5% confidence level. Abbreviations: RL, root length; RFW, root fresh weight; RWC, root water content; SD, stem diameter; SL, stem length; SFW, stem fresh weight; SWC, stem water content; Lno, leaf number; LFW, leaf fresh weight; LWC, leaf water content; Pro, proline content. Asterisks indicate the degree of significance: ** $p < 0.01$; *** $p < 0.001$, ns = not significant.

Trait	Two Weeks				Three Weeks			
	Cultivar	Treatment	Interaction	Residual	Cultivar	Treatment	Interaction	Residual
RL	22.14 ***	18.42 ***	9.24 ns	50.19	32.90 ***	10.43 ***	21.52 ***	35.16
RFW	33.01 ***	17.20 ***	18.45 ***	31.34	24.88 ***	4.54 ***	15.05 **	55.53
RWC	4.34 ***	70.52 ***	14.75 ***	10.39	22.23 ***	52.05 ***	13.58 ***	12.15
SD	24.77 ***	27.08 ***	13.15 ***	35.00	14.17 ***	1.06 **	23.61 ***	61.16
SL	52.63 ***	11.77 ***	9.85 ***	25.76	42.71 ***	18.30 ***	18.17 ***	20.81
SFW	19.51 ***	47.29 ***	13.84 ***	19.36	26.50 ***	38.21 ***	15.10 ***	20.19
SWC	30.07 ***	33.12 ***	16.07 ***	20.74	31.64 ***	21.15 ***	32.13 ***	15.08
Lno	24.67 ***	32.12 ***	10.16 ***	33.05	14.69 ***	9.06 ***	21.14 ***	55.12
LFW	15.70 ***	53.55 ***	16.85 ***	13.89	32.54 ***	32.42 ***	23.8 ***	11.24
LWC	25.53 ***	37.59 ***	14.52 ***	22.36	32.50 ***	24.36 ***	20.21 ***	22.93
Pro	30.46 ***	28.92 ***	18.89 ***	21.72	40.52 ***	20.08 ***	15.61 ***	23.79

For most analysed variables, the effects of cultivar, treatment and their interactions were highly significant ($p < 0.001$). The only non-significant value was found in the two-week treatment and the trait 'root length', for the interaction cultivar × treatment. In plants subjected to the two-week treatment, relatively stronger contributions to the sum of squares were those of 'cultivar' for the variables root fresh weight (RFW) and stem length (SL), and 'treatment' for root water content (RWC), stem fresh weight

(SFW), leaf fresh weight (LFW) and leaf water content (LWC). For stem water content (SWC) and Pro, both factors, cultivar and treatment, contributed similarly to the sum of squares (SS). On the other hand, most of the variation observed for root length (RL), and stem diameter (SD) was due to uncontrolled variation, as shown by the higher SS percentage of the residual (Table 2).

The ANOVA of data obtained from the three-weeks-treated plants showed somewhat different results. The effect of 'cultivar' was the most substantial contributor to SS for the variables SL, LWC and Pro, and that of 'treatment' for RWC and SFW. The relative contributions of cultivar and treatment were similar for LFW, and those of cultivar and the interaction of both factors, for SWC. The most-significant contribution to variation of RL, RFW, SD and the number of leaves (Lno) is accounted for by the residual source of variation.

Disregarding the individual responses to water and salt stress of the selected bean genotypes, which vary quantitatively (Table S1), a general analysis was performed, including all cultivars and using the mean values calculated for all measured growth variables and Pro contents (Table 3). After the water stress treatments, either for two or three weeks, all morphological parameters determined in the stressed plants showed a significant decrease with respect to the corresponding values of the well-watered controls. The strongest reductions, down to less than 30% of the controls, were observed for root and leaf fresh weight. The effect of water deficit was relatively weaker regarding the reduction of root length and stem parameters (SD, SL and SWC), especially in the three-week treatments. Leaf Pro concentration, on the contrary, significantly increased in response to water stress, about 2.7-fold and 2.1-fold, as average, for the plants treated for two and three weeks, respectively (Table 3).

Table 3. Mean values and percentages with respect to the control (%) of traits measured in *Phaseolus vulgaris* cultivars after two and three weeks of control (C), water stress (WS) (withholding of irrigation) and salt stress (SS) (150 mM NaCl) treatments. Abbreviations: RL, root length; RFW, root fresh weight; RWC, root water content; SD, stem diameter; SL, stem length; SFW, stem fresh weight; SWC, stem water content; Lno, leaf number; LFW, leaf fresh weight; LWC, leaf water content; Pro, proline content. Different letters (lowercase for two-week and capital for three-week treatments) indicate significant differences between treatments for each trait, according to the Tukey test, at the 95% confidence level.

Trait	Two Weeks			Three Weeks		
	C	WS	SS	C	WS	SS
RL (cm)	36.04c	23.24a	30.39b	29.50C	21.14A	25.73B
%		64.48	84.32		71.66	87.22
RFW (g)	3.22b	0.44a	3.36b	2.78C	0.69A	1.85B
%		13.66	104.35		24.82	66.55
RWC (%)	85.70b	31.94a	82.08b	84.07B	42.11A	85.28B
%		37.27	95.78		50.09	101.44
SD (mm)	3.87c	2.91a	3.55b	3.89C	3.18A	3.45B
%		75.19	91.73		81.75	88.69
SL(cm)	148.63b	109.95a	115.75a	139.90B	95.43A	90.85A
%		73.98	77.88		68.21	64.94
SFW (g)	10.40c	2.82a	5.37b	9.00C	3.18A	4.63B
%		27.12	51.63		35.33	51.44
SWC (%)	82.14b	56.84a	78.31b	82.82B	64.06A	79.47B
%		69.20	95.34		77.35	95.96
Lno	12.43b	6.47a	7.42a	13.28B	7.35A	7.98A
%		52.05	59.69		55.35	60.09
LFW (g)	22.73c	2.57a	5.94b	18.21B	4.08A	5.40A
%		11.31	26.13		22.41	29.65
LWC	84.04c	38.39a	58.58b	81.62C	49.31A	55.48B
%		45.68	69.70		60.41	67.97
Pro (μmol g^{-1} DW)	31.67a	86.61b	82.74b	25.89A	53.57B	68.29C
%		273.48	261.26		206.91	263.77

Under the specific conditions of our experiments, salt stress had a smaller effect than water deficit on the average growth inhibition of the bean cultivars, reflected mostly in a sharp reduction

(>70%) of the leaf fresh weight with respect to the control, followed by that of stem fresh weight (about 50%). Other parameters, such as root and stem water content, stem diameter or root fresh weight (in the two-week treatment) did not change significantly or decreased only slightly in response to increased salinity. The mean values calculated for most growth variables were similar for both treatment times. Pro contents also rose significantly, about 2.6-fold over control values, in the salt-treated plants (Table 3).

3.2. Correlation Analysis

Pearson moment correlation between the analysed traits for salt and water stress are presented separately for plants from the two-week (Figure 1a) and three-week (Figure 1b) treatments. Correlations between all morphological variables were in most cases positive, for both stresses and the two treatment times, although the correlation coefficients varied widely, from $r < 0.1$ to $r > 0.9$. Considering specifically the two-week treatment, the strongest correlations ($r \geq 0.8$) for the salt stress treatment were found between root water content (RWC) and stem fresh weight (SFW) or water content (SWC); or between SFW, leaf fresh weight (LFW) and the number of leaves (Lno) (Figure 1a). Under conditions of water stress, the strongest positive correlations were also found between SFW, LFW and Lno; between water contents of roots, stems and leaves (RWC/SWC/LWC) or between SFW and stem diameter (SD) (Figure 1a). On the other hand, Pro contents showed negative correlations with all growth parameters (except for RWC in the salt stress treatment), most significantly with leaf water content, but also with LFW and stem growth parameters (SFW and SWC). Correlations followed a similar pattern for both types of stress but were weaker (lower '*r*' values) in the case of salt stress (Figure 1a).

Salt stress

	RL	RFW	RWC	SD	SL	SFW	SWC	Lno	LFW	LWC	Pro
RL		0.01	0.24	0.63	0.50	0.46	0.29	0.46	0.45	0.36	−0.16
RFW	0.50		0.11	0.57	0.50	0.78	0.38	0.48	0.19	0.13	−0.1
RWC	0.75	0.63		0.72	0.38	0.80	0.83	0.68	0.76	−0.08	0.16
SD	0.40	0.07	0.03		0.28	0.77	0.33	0.55	0.62	0.42	−0.42
SL	0.43	0.04	0.23	0.32		0.54	−0.03	0.44	0.51	0.12	−0.23
SFW	0.67	0.20	0.11	0.81	0.56		0.41	0.84	0.89	0.56	−0.54
SWC	0.62	0.57	−0.29	0.70	0.20	0.72		0.58	0.65	0.58	−0.4
LnO	0.68	−0.10	0.11	0.72	0.49	0.78	0.21		0.79	0.46	−0.41
LFW	0.63	0.76	0.18	0.71	0.56	0.92	0.40	0.81		0.65	−0.58
LWC	0.65	0.62	0.91	0.78	0.26	0.78	0.90	0.63	0.75		−0.60
Pro	−0.58	−0.57	−0.77	−0.77	−0.35	−0.74	−0.74	−0.6	−0.67	−0.80	

Water stress

(a)

Salt stress

	RL	RFW	RWC	SD	SL	SFW	SWC	Lno	LFW	LWC	Pro
RL		0.08	−0.04	0.40	0.34	0.23	−0.03	0.13	−0.22	0.13	−0.17
RFW	0.19		0.14	0.26	0.32	0.48	0.12	0.21	0.50	0.24	−0.15
RWC	0.45	0.39		0.47	0.27	0.56	0.67	0.28	0.40	0.05	0.14
SD	0.11	0.15	−0.09		0.16	0.60	0.50	0.35	0.48	0.48	−0.29
SL	0.24	0.26	−0.20	0.18		0.62	0.01	0.30	0.52	0.30	−0.22
SFW	0.49	0.38	−0.25	0.68	0.60		0.34	0.59	0.85	0.54	−0.34
SWC	0.43	0.27	−0.25	0.60	0.18	0.55		0.27	0.44	0.72	−0.31
LnO	0.26	0.13	−0.09	0.41	0.31	0.58	0.20		0.48	0.33	−0.18
LFW	0.37	0.58	−0.24	0.58	0.51	0.86	0.36	0.50		0.62	−0.47
LWC	0.39	0.36	0.08	0.56	0.31	0.63	0.83	0.36	0.62		−0.58
Pro	−0.30	−0.20	−0.45	−0.43	−0.10	−0.38	−0.45	−0.16	−0.41	−0.55	

Water stress

(b)

Figure 1. Heatmap of Pearson moment correlation coefficients (r) between the analysed traits in *Phaseolus vulgaris* cultivars submitted to two weeks (**a**) and three weeks (**b**) of water and salt stresses. Dark blue denotes high correlation ($r \to 1$), dark red high negative correlation ($r \to -1$). Abbreviations: RL, root length; RFW, root fresh weight; RWC, root water content; SD, stem diameter; SL, stem length; SFW, stem fresh weight; SWC, stem water content; Lno, leaf number; LFW, leaf fresh weight; LWC, leaf water content; Pro, proline content.

Correlations between the different measured variables, generally positive for growth parameters and negative between Pro contents and the rest of variables, were maintained, qualitatively, when comparing the two- and three-week treatments, and for both stresses, but with lower relative significance for the longer treatment time (Figure 1b).

3.3. *Principal Component Analysis (PCA)*

A PCA was performed, separately for the cultivars subjected to the two-week and three-week treatments, and including the mean values of all measured parameters and the three applied conditions (control, water stress and salt stress) (Table 4, Figure 2).

Table 4. Component weights in the PCA performed on cultivars subjected to two and three weeks of treatment. Abbreviations: RL, root length; RFW, root fresh weight; RWC, root water content; SD, stem diameter; SL, stem length; SFW, stem fresh weight; SWC, stem water content; Lno, leaf number; LFW, leaf fresh weight; LWC, leaf water content; Pro, proline.

Trait	Two Weeks		Three Weeks	
	Component 1	Component 2	Component 1	Component 2
RL	0.320	−0.054	0.243	0.109
RFW	0.222	0.329	0.282	−0.017
RWC	0.318	0.267	0.250	0.529
SL	0.201	−0.578	0.262	−0.470
SFW	0.388	−0.178	0.421	−0.217
SWC	0.325	0.436	0.319	0.491
Lno	0.322	−0.374	0.286	−0.282
LFW	0.370	−0.227	0.406	−0.264
LWC	0.354	0.260	0.384	0.218
Pro	−0.291	−0.053	−0.240	−0.030

The PCA corresponding to the two-week treatments detected two components with Eigenvalues higher than 1, which explained 70.1% of the total variability of data (56.5% and 13.6% for the first and second components, respectively). All growth parameters—most significantly the fresh weights of stems (SFW) and leaves (LFW), followed by the water contents of both organs (LWC and SWC)—were positively correlated with the first component, whereas the only one negatively correlated was Pro concentration in leaves. Regarding the second component, some morphological variables (especially SWC and RFW) were positively correlated, whereas for others (e.g., SL or Lno) the correlation was negative (Table 4, Figure 2a).

Two components with an Eigenvalue higher than one were also detected in the PCA corresponding to the three-week treatments, the first explaining 44.3% and the second 14.0% of the total variability; that is, together explaining 58.3% of the total variation. Correlations of the different variables followed similar patterns to those observed for the cultivars treated for two weeks, for example regarding the negative correlation of Pro with the first component, and the positive correlations of all growth variables, with SFW and LFW showing the highest significance (Table 4, Figure 2b).

The 18 cultivars from the shorter treatment period (Figure 3a) were dispersed onto the two axes of the scatterplot, indicating high variability in the selected genotypes. There was, however, good separation between the different treatments, not only when looking individually at each cultivar, but also considering the overall behaviour of all genotypes. Plants from the control (green symbols) and water stress (pink) treatments were clearly separated, with almost no overlapping between the two conditions. Those symbols (blue) corresponding to the salt stress treatments appear located in the scatterplot in-between the control and water stress samples, which was in agreement with the weaker effect (on average) of the salt treatments as compared to water deficit, under the specific conditions used in our experiments. The scatterplot corresponding to the 29 cultivars that were subjected to the more prolonged (three-week) treatment (Figure 3b) showed the same general picture, maybe with more overlapping of the water- and salt-stressed plants. More-significant dispersion of the scores was found, for both treatment times, in the controls indicating a high variability of morphological traits of the different cultivars. Under salt stress, the separation between scores was not so pronounced as under water stress, suggesting a more homogeneous general response of the bean genotypes to salinity than to drought, at least under the conditions of our experiments (Figure 3).

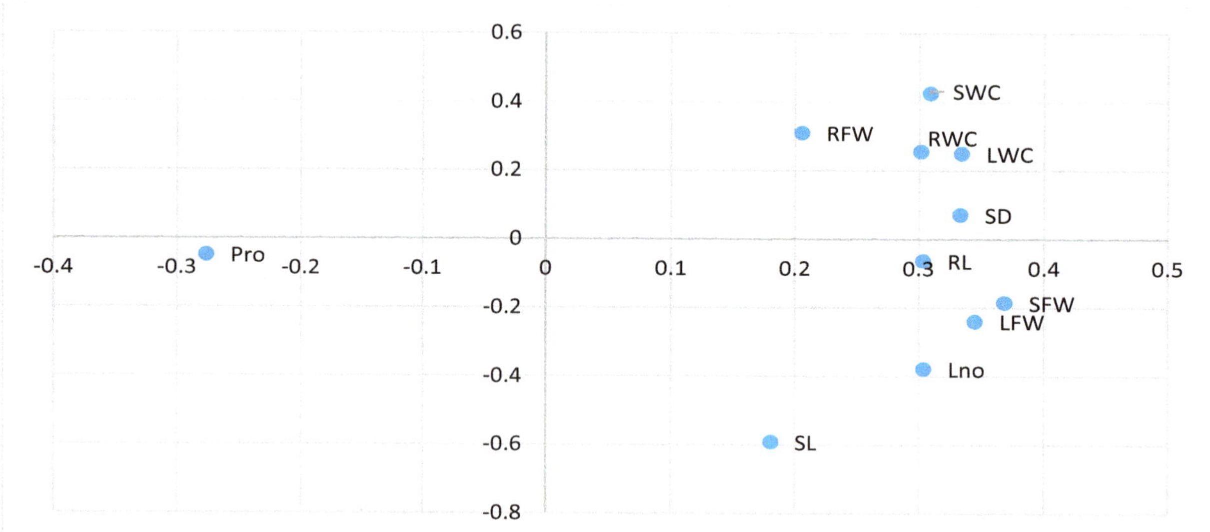

(a)

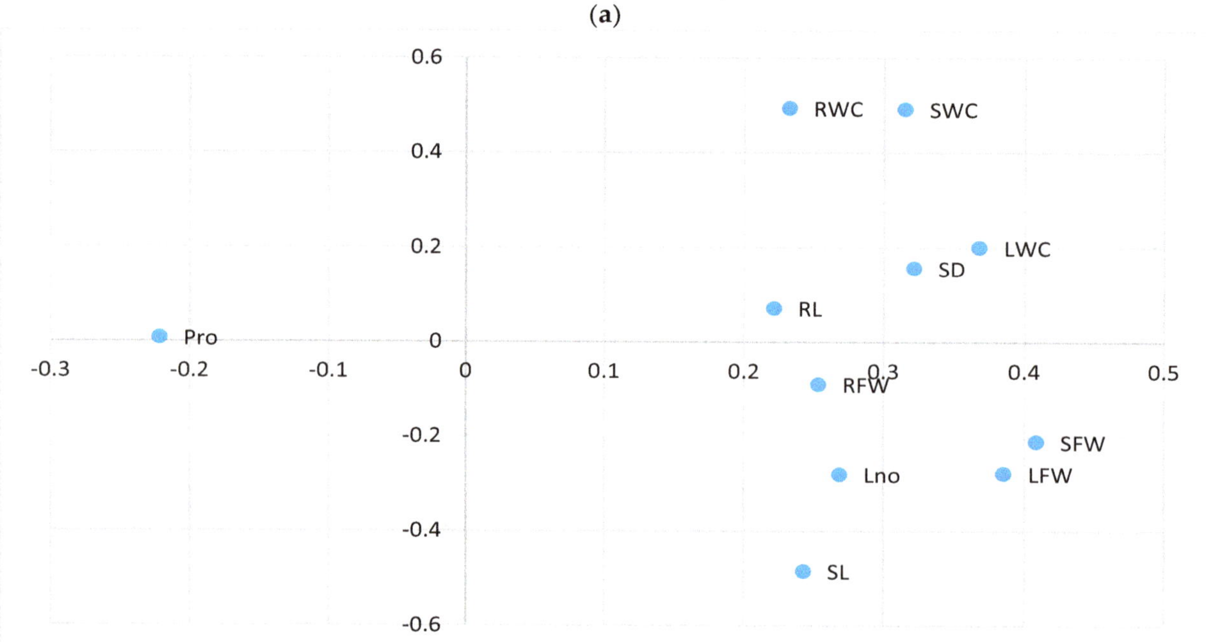

(b)

Figure 2. Loading plot of the principal component analysis (PCA) conducted with the analysed traits, in *P. vulgaris* cultivars subjected to control, water deficit and salt stress treatments. Two-week treatments (**a**); 56.5% and 13.6% of the total variability are explained by the first (x-axis) and the second (y-axis) components, respectively. Three-week treatments (**b**); 44.3% and 14.0% of the total variability are explained by the first (x-axis) and the second (y-axis) components, respectively. Abbreviations: RL, root length; RFW, root fresh weight; RWC, root water content; SD, stem diameter; SL, stem length; SFW, stem fresh weight; SWC, stem water content; Lno, leaf number; LFW, leaf fresh weight; LWC, leaf water content; Pro, proline.

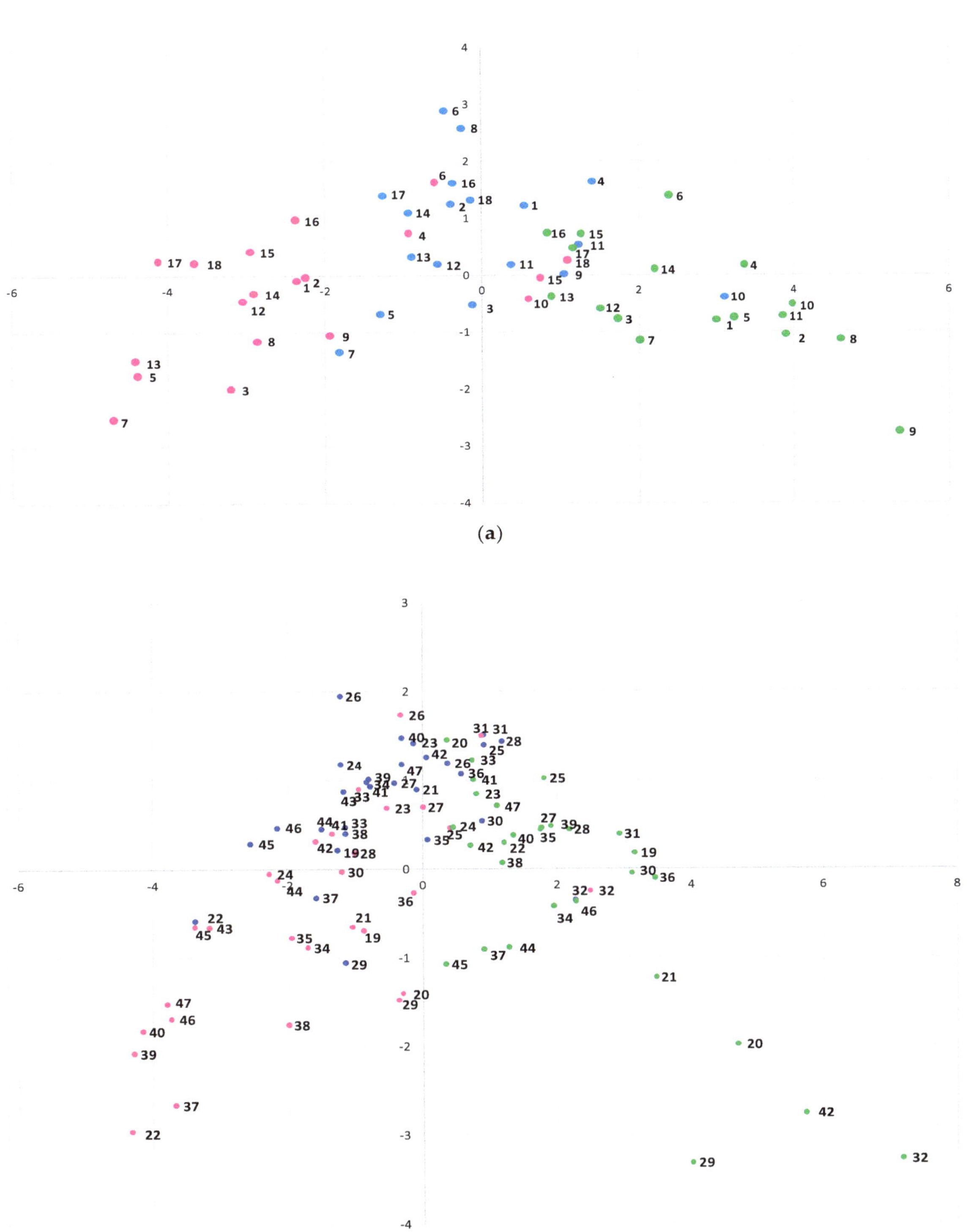

(**a**)

(**b**)

Figure 3. Scatter plot of the PCA scores. Plants treated for two (**a**) or three weeks (**b**); control (green), water deficit (pink) and salt stress (blue) treatments. (**a**) 1–9, cultivars from Spain; 10–15, from Colombia and 16–18, from Cuba and (**b**) 19–32, cultivars from Spain; 33–45, from Colombia and 46 and 47 from Cuba.

Based on the PCA scatter plot in the two-week trial (Figure 3a), we identified four accessions (7, 5, 13 and 17) with highly negative values for the first component (i.e., with high concentrations of Pro and low values for growth and water content parameters), both for the water deficit and salinity treatments; these cultivars can be considered as highly susceptible to both stresses. On the other hand, three accessions (25, 31 and 32) were detected in the three week scatter plot (Figure 3b), showing positive values for the first component (low Pro contents and limited growth inhibition), both for the drought and salt stress treatments, indicating that these accessions can be considered as the most tolerant to both stresses. Similarly, the relative position of other accessions along the x-axis should allow a ranking of their tolerance to water deficit and to salinity, within each group of cultivars (treated for two or three weeks).

4. Discussion

In the present study, responses to drought and salinity have been analysed in 47 *Phaseolus vulgaris* genotypes of different origins. Large variability was observed in the size and morphology of the plants of the different bean cultivars—as seen when comparing their growth parameters (Supplementary Table S1) individually and also by their dispersion in the PCA scatterplots (Figure 3)—making it difficult to determine, at first sight, the variables that are more relevant for assessing the relative degree of stress-induced growth inhibition and, therefore, for ranking the different cultivars according to their relative sensitivity or resistance to water deficit and salt stress. However, the statistical analyses performed with all experimental data provided a clear overall picture of the responses to stress of the *P. vulgaris* cultivars. Both 'cultivar' and 'treatment', as well as their interaction, had a highly-significant effect on (practically) all growth traits analysed, and on Pro contents, for the two- and three-week treatments of both water deficit and salt stress. In all cases, growth inhibition was mostly reflected in the stress-induced reduction of fresh weight and water contents of stems (SFW and SWC) and leaves (LFW and LWC), as reported in the same species [39,40] or other species of this genus [43]. These parameters are the growth variables most significantly correlated, positively, with the first principal component in the PCA. Pro, on the other hand, was the only variable showing a negative correlation with all growth variables, but particularly with those of stems and leaves mentioned above—as indicated by the Pearson correlation coefficients and the loading plots of the PCAs.

When comparing the stress tolerance of related taxa, for example, different cultivars of a particular crop, measurements of growth parameters are often complemented with the determination of several biochemical stress markers, associated with increased (or lower) tolerance; they include compatible solutes or osmolytes [44–47]. Proline (Pro) is a common osmolyte in plants, which accumulates in response to different types of abiotic stress, including drought and salinity, in a variety of plant species [28,48–50]. Besides its role in cellular osmotic adjustment, Pro has additional functions as 'osmoprotectant'; it directly stabilises sub-cellular structures, such as membranes and proteins, scavenges free radicals buffering redox potential, alleviates cellular acidosis and acts as a signalling molecule in the responses to stress [51,52]. Proline also plays essential roles in the absence of stress, being involved in many developmental processes; for example, Pro concentration increases during pollen and seed maturation. However, Pro can be toxic for certain tissues if it is partially catabolised to pyrroline-5-carboxylate (P5C), leading to apoptosis [53]. Considering the multiple functions of Pro, it is logical to assume that Pro accumulation would be associated with higher stress tolerance, and this has indeed been demonstrated for many plants, both wild species [54,55] and crops [43,56]. However, other comparative studies on related taxa, such as species of the same genus or cultivars or varieties of the same species, revealed higher Pro accumulation under stress in the less-tolerant genotypes [57,58]. There is some confusion, often found in the literature, between the concepts of 'stress responses' and 'stress tolerance'. Even though stress tolerance mechanisms are based on specific stress responses, not all responses are relevant for tolerance. On this line, Pro accumulation can be considered as a general 'response' to abiotic stress in many plant species, but Pro may or may not be involved in stress tolerance mechanisms, depending on the species.

Common bean is clearly a Pro accumulator species, as numerous reports have shown significant increases in Pro contents in *Phaseolus* plants in response to either salt stress [38,39,59] or water stress [40,60,61] treatments. Also, Pro appears to be a good bioindicator in other types of stress in beans, such as that induced by excess nitrogen dosage [62], herbicides [63] or heavy metals [64].

Moreover, exogenous application of Pro was shown to alleviate the salt stress deleterious effects in beans [65]. However, there are some contradictory data in the literature regarding the function of Pro in the mechanisms of stress tolerance in *Phaseolus*. Some published reports indicated higher Pro contents in more drought-tolerant [30,34–37] or salt-tolerant [31–33] cultivars than in less tolerant ones; that is, Pro accumulation correlates positively with the degree of stress resistance, suggesting a direct contribution to stress tolerance mechanisms. Other reports, on the contrary, showed that, under stress conditions, the less tolerant genotypes had a higher concentration of this osmolyte than the more resistant cultivars [38–40,66]; therefore, in this case, Pro is simply a marker of the level of stress affecting the plants, accumulating at higher concentrations in the more stressed—the more sensitive—cultivars, but is not directly involved in the mechanisms of tolerance. This was also the conclusion of previous work from our laboratory, comparing three commercial cultivars (two of *P. vulgaris* and one of *P. coccineus*) and one Spanish common bean landrace [39,40]. All these latter studies, based on the comparison of a few bean genotypes, generally some commercial cultivars, have been confirmed in the present work, using a much larger number of cultivars of different origins and an extensive statistical analysis of the experimental data.

Our results showed a strong negative correlation of Pro levels and growth variables, especially the fresh weight and water content of the aboveground organs of the plants; these are the most relevant parameters to evaluate the inhibition of growth induced under water deficit and high salinity conditions. Therefore, there is an unequivocal association of higher Pro contents with stronger growth inhibition; that is, with a higher sensitivity to stress of the bean cultivars.

5. Conclusions

Phaseolus vulgaris cannot be considered as drought- or salt-tolerant. It is even more sensitive to stress than many other crops such as barley or cowpea [67,68]. However, amongst the extremely high number of available genotypes of *P. vulgaris*, some will show a relatively higher resistance and could be used as parental lines in bean breeding programmes aimed at enhancing stress tolerance in this major crop. The identification of common bean accessions in the extremes of variation for susceptibility and tolerance to water deficit and salinity is of great interest for further studies on the physiological mechanisms of tolerance to both stresses. Also, the development of segregating generations after hybridisation between both types of materials can lead to the identification of genomic regions involved in tolerance to these stresses.

Proline concentrations in stressed plants can be determined by a simple and rapid spectrophotometric assay, requiring only small amounts of leaf material. From a practical point of view, our results support the use of Pro as a biochemical marker for the initial, large-scale screening of bean cultivars, to exclude the most sensitive, those accumulating higher Pro concentrations in response to water or salt stress.

Author Contributions: Conceptualization, M.B. and O.V.; methodology, S.A. and L.Y.; software, J.P.; validation, M.J.D. and J.P.; formal analysis, L.Y.; investigation, S.A. and L.Y.; resources, O.V.; data curation, S.A.; writing—original draft preparation, S.A. and M.B.; writing—review and editing, M.J.D., J.P. and O.V.; visualization, S.A.; supervision, M.B. and O.V.; project administration, M.B. and O.V.; funding acquisition, O.V. All authors have read and agreed to the published version of the manuscript.

Acknowledgments: We are indebted to Steve Beebe, from Alliance Biodiversity CIAT, for providing the Colombian bean cultivars and for his helpful comments on the manuscript.

References

1. Mantri, N.; Patade, V.; Penna, S.; Ford, R.; Pang, E. Abiotic stress responses in plants: Present and future. In *Abiotic Stress Responses in Plants: Metabolism, Productivity and Sustainability*; Ahmad, P., Prasad, M.N.V., Eds.; Springer: New York, NY, USA, 2012; pp. 1–19. [CrossRef]

2. Zörb, C.; Geilfus, C. M.; Dietz, K.J. Salinity and crop yield. *Plant Biol.* **2019**, *21*, 31–38. [CrossRef]
3. Osakabe, Y.; Osakabe, K.; Shinozaki, K.; Tran, L. Response of plants to water stress. *Front. Plant Sci.* **2014**, *5*, 86. [CrossRef] [PubMed]
4. Fita, A.; Rodríguez-Burruezo, A.; Boscaiu, M.; Prohens, J.; Vicente, O. Breeding and domesticating crops adapted to drought and salinity: A new paradigm for increasing food production. *Front. Plant Sci.* **2015**, *6*, 978. [CrossRef] [PubMed]
5. Shahid, S.A.; Zaman, M.; Heng, L. Soil salinity: Historical perspectives and a world overview of the problem. In *Guideline for Salinity Assessment, Mitigation and Adaptation Using Nuclear and Related Techniques*; Zaman, M., Shahid, S.A., Eds.; Springer: Cham, Switzerland, 2018; pp. 43–53. [CrossRef]
6. Flowers, T.; Flowers, S. Why does salinity pose such a difficult problem for plant breeders? *Agric. Water Manag.* **2005**, *78*, 15–24. [CrossRef]
7. Morton, J.F. The impact of climate change on smallholder and subsistence agriculture. *Proc. Natl. Acad. Sci. USA* **2007**, *104*, 19680–19685. [CrossRef]
8. Bellucci, E.; Bitocchi, E.; Rau, D.; Rodriguez, M.; Biagetti, E.; Giardini, A.; Attene, G.; Nanni, L.; Papa, R. Genomics of origin, domestication and evolution of *Phaseolus vulgaris*. In *Genomics of Plant Genetics Resources*; Tuberosa, R., Graner, A., Frison, E., Eds.; Springer: Dordrecht, The Netherlands, 2014; pp. 483–507. [CrossRef]
9. Freytag, G.F.; Debouck, D.G. *Taxonomy, Distribution, and Ecology of the genus Phaseolus (Leguminosae-Papilionoideae) in North America, Mexico and Central America*; Botanical Research Institute of Texas (BRIT): Forth Worth, TX, USA, 2002; pp. 1–298.
10. Delgado-Salinas, A.; Bibler, R.; Lavin, M. Phylogeny of the genus *Phaseolus* (Leguminosae): A recent diversification in an ancient landscape. *Syst. Bot.* **2006**, *31*, 779–791. [CrossRef]
11. Broughton, W.J.; Hernández, G.; Blair, M.; Beebe, S.; Gepts, P.; Vanderleyden, J. Beans (*Phaseolus* spp.)—Model food legumes. *Plant Soil* **2003**, *252*, 55–128. [CrossRef]
12. Gepts, P.; Debouck, D.G. Origin, domestication, and evolution of the common bean, *Phaseolus vulgaris*. In *Common Beans: Research for Crop Improvement*; van Schoonhoven, A., Voysest, O., Eds.; Cab Intern: Wallingford, UK, 1991; pp. 7–53.
13. Rendón-Anaya, M.; Montero-Vargas, J.M.; Saburido-Álvarez, S.; Vlasova, A.; Capella-Gutierrez, S.; Ordaz-Ortiz, J.J.; Aguilar, O.M.; Vianello-Brondani, R.P.; Santalla, M.; Delaye, L.; et al. Genomic history of the origin and domestication of common bean unveils its closest sister species. *Genome Biol.* **2017**, *18*, 60. [CrossRef]
14. Ortwin-Sauer, C. *The Early Spanish Men*; University of California Press: Berkeley, CA, USA, 1966.
15. Brucher, O.B.; Brucher, H. The South American wild bean (*Phaseolus aborigeneus* Burk.) as ancestor of the common bean. *Econ. Bot.* **1976**, *30*, 257–272. [CrossRef]
16. Debouck, D.G.; Smartt, J. Bean. In *Evolution of Crop Plants*, 2nd ed.; Smartt, J., Simmonds, N.W., Eds.; Longman Scientific and Technical: Harlow, UK, 1995; pp. 287–296.
17. Arteaga, S.; Yabor, L.; Torres, J.; Solbes, E.; Muñoz, E.; Díez, M.J.; Vicente, O.; Boscaiu, M. Morphological and agronomic characterization of Spanish landraces of *Phaseolus vulgaris* L. *Agriculture* **2019**, *9*, 149. [CrossRef]
18. Pinheiro, C.; Baeta, J.P.; Pereira, A.M.; Dominguez, H.; Ricardo, C. Mineral elements correlations in a Portuguese germplasm collection of *Phaseolus vulgaris*. Integrating Legume Biology for sustainable Agriculture. In Proceedings of the 6th European Conference on Grain Legumes, Lisbon, Portugal, 12–16 November 2007; pp. 125–126.
19. Molina, J.; Moda-Cirino, V.; Da Silva Fonseca, N.J.; Faria, R.; Destro, D. Response of common bean cultivars and lines to water stress. *Crop Breed. Appl. Biotechnol.* **2001**, *1*. [CrossRef]
20. Graham, P.; Ranalli, P. Common bean (*Phaseolus vulgaris* L.). *Field Crop Res.* **1997**, *53*, 131–146. [CrossRef]
21. Singh, S.P. Drought resistance in the race Durango dry bean landraces and cultivars. *Agron. J.* **2007**, *99*, 1219–1225. [CrossRef]
22. Cuellar-Ortiz, S.; Arrieta-Montiel, M.; Acosta-Gallegos, J.; Covar-Rubias, A. Relationship between carbohydrate partitioning and drought resistance in common bean. *Plant Cell Environ.* **2008**, *31*, 1399–1409. [CrossRef]
23. Maas, E.; Hoffman, G. Crop salt tolerance-current assessment. *J. Irrig. Drain. Eng.* **1977**, *103*, 115–134.
24. Gama, P.; Inanaga, S.; Tanaka, K.; Nakazawa, R. Physiological response of common bean (*Phaseolus vulgaris* L.) seedlings to salinity stress. *Afr. J. Biotechnol.* **2007**, *6*, 79–88.

25. Zhumabayeva, B.A.; Biotechnology, K.A.; Aytasheva, Z.G.; Dzhangalina, E.D.; Esen, A.; Lebedeva, L.P. Screening of domestic common bean cultivar for salt tolerance during in vitro cell cultivation. *Int. J. Biol.* **2019**, *12*, 94–102. [CrossRef]
26. Fess, T.L.; Kotcon, J.B.; Benedito, V.A. Crop breeding for low input agriculture: A sustainable response to feed a growing world population. *Sustainability* **2011**, *3*, 1742–1772. [CrossRef]
27. Hurtado, M.; Vilanova, S.; Plazas, M.; Gramazio, P.; Andújar, I.; Herraiz, F.J.; Prohens, J. Enhancing conservation and use of local vegetable landraces: The Almagro eggplant (*Solanum melongena* L.) case study. *Genet. Resour. Crop Evol.* **2014**, *61*, 787–795. [CrossRef]
28. Szabados, L.; Savouré, A. Proline: A multifunctional amino acid. *Trends Plant Sci.* **2010**, *15*, 89–97. [CrossRef]
29. Verslues, P.E.; Sharma, S. Proline metabolism and its implications for plant-environment interaction. *Arabidopsis Book* **2010**, *8*, e0140. [CrossRef] [PubMed]
30. Kapuya, J.A.; Barendse, G.W.M.; Linskens, H.F. Water stress tolerance and proline accumulation in *Phaseolus vulgaris* L. *Acta Bot. Neerl.* **1985**, *34*, 293–300. [CrossRef]
31. Misra, N.; Gupta, A.K. Effect of salt stress on proline metabolism in two high yielding genotypes of green gram. *Plant Sci.* **2005**, *169*, 331–339. [CrossRef]
32. Cárdenas-Avila, M.; Verde-Star, J.; Maiti, R.; Foroughbakhch, R.; Gámez-González, H.; Martínez-Lozano, S.; Núñez-González, M.; García Díaz, G.; Hernández-Piñero, J.; Morales-Vallarta, M. Variability in accumulation of free proline on in vitro calli of four bean (*Phaseolus vulgaris* L.) cultivars exposed to salinity and induced moisture stress. *Phyton* **2006**, *75*, 103–108.
33. Kaymakonova, M.; Stoeva, N. Physiological responses of bean plants (*Phaseolus vulg.* L.) to salt stress. *Gen. Appl. Plant Physiol.* **2008**, *34*, 177–188.
34. Herrera Flores, T.S.; Ortíz Cereceres, J.; Delgado Alvarado, A.; Acosta Galleros, J.A. Growth and, proline and carbohydrate content of bean seedlings subjected to drought stress. *Rev. Mexicana Cienc. Agric.* **2012**, *3*, 713–725.
35. Ghanbari, A.A.; Mousavi, S.H.; Mousapou Gorji, A.; Rao, I. Effects of water stress on leaves and seeds of bean (*Phaseolus vulgaris* L.). *Turk. J. Field Crops* **2013**, *18*, 73–77.
36. Kusvuran, S.; Dasgan, H.Y. Effects of drought stress on physiological and biochemical changes in *Phaseolus vulgaris* L. *Legume Res.* **2017**, *40*, 55–62.
37. Wang, Q.; Ang, Q.; Lin, F.; Wei, S.H.; Meng, X.X.; Yin, Z.G.; Guo, Y.F.; Yang, G.D. Effects of drought stress on endogenous hormones and osmotic regulatory substances of common bean (*Phaseolus vulgaris* L.) at seedling stage. *Appl. Ecol. Environ. Res.* **2019**, *17*, 4447–4457. [CrossRef]
38. Jiménez-Bremont, J.F.; Becerra-Flora, A.; Hernández-Lucero, E.; Rodríguez-Kessler, M.; Acosta-Gallegos, J.; Ramírez Pimentel, J. Proline accumulation in two bean cultivars under salt stress and the effect of polyamines and ornithine. *Biol. Plant.* **2006**, *50*, 763–766. [CrossRef]
39. Al Hassan, M.; Morosan, M.; López-Gresa, M.P.; Prohens, J.; Vicente, O.; Boscaiu, M. Salinity-induced variation in biochemical markers provides insight into the mechanisms of salt tolerance in common (*Phaseolus vulgaris*) and runner (*P. coccineus*) beans. *Int. J. Mol. Sci.* **2016**, *17*, 1582. [CrossRef] [PubMed]
40. Morosan, M.; Al Hassan, M.; Naranjo, M.; Lopez-Gresa, M.P.; Vicente, O. Comparative analysis of drought responses in *Phaseolus vulgaris* (common bean) and *P. coccineus* (runner bean) cultivar. *EuroBiotech J.* **2017**, *1*, 247–252. [CrossRef]
41. Hoagland, D.; Arnon, D. The water-culture method for growing plants without soil. *Circ. Califor. Agric. Exp. Stat.* **1950**, *347*, 32–63.
42. Bates, L.S.; Waldren, R.P.; Teare, I.D. Rapid determination of free proline for water stress studies. *Plant Soil* **1973**, *39*, 205–207. [CrossRef]
43. Arteaga, S.; Al Hassan, M.; Wijesinghe, C.; Yabor, L.; Llinares, J.; Boscaiu, M.; Vicente, O. Screening for Salt Tolerance in Four Local Varieties of *Phaseolus lunatus* from Spain. *Agriculture* **2018**, *8*, 201. [CrossRef]
44. Andrade, E.; Ribeiro, V.; Azvedo, C.; Chiorato, A.; Williams, T.; Carbonell, S. Biochemical indicators of drought tolerance in the common bean (*Phaseolus vulgaris* L.). *Euphytica* **2016**, *210*, 277–289. [CrossRef]
45. Bacha, H.; Tekaya, M.; Drine, S.; Guasami, F.; Touil, L.; Enneb, H.; Triki, T.; Cheour, F.; Ferchichi, A. Impact of salt stress on morpho-physiological and biochemical parameters of *Solanum lycopersicum* cv. Microtom leaves. *S. Afr. J. Bot.* **2017**, *108*, 364–369. [CrossRef]

46. Şen, A.; Öztürk, İ.; Yaycılı, O.; Alıkamanoglu, S. Drought tolerance in irradiated wheat studied by genetic and biochemical markers. *J. Plant Growth Regul.* **2017**, *36*, 669–676. [CrossRef]
47. Koźmińska, A.; Wiszniewska, A.; Hanus-Fajerska, E.; Boscaiu, M.; Al Hassan, M.; Halecki, W.; Vicente, O. Identification of salt and drought biochemical stress markers in several *Silene vulgaris* populations. *Sustainability* **2019**, *11*, 800. [CrossRef]
48. Verbruggen, N.; Hermans, C. Proline accumulation in plants: A review. *Amino Acids* **2008**, *35*, 753–759. [CrossRef]
49. Grigore, M.N.; Boscaiu, M.; Vicente, O. Assessment of the relevance of osmolyte biosynthesis for salt tolerance of halophytes under natural conditions. *Eur. J. Plant Sci. Biotechnol.* **2011**, *5*, 12–19.
50. Parvaiz, A.S.; Satyawati, S. Salt stress and phyto-biochemical responses of plants—A review. *Plant Soil Environ.* **2008**, *54*, 89–99. [CrossRef]
51. Hayat, S.; Hayat, Q.; Alyemeni, M.N.; Wani, A.S.; Pichtel, J.; Ahmad, A. Role of proline under changing environments: A review. *Plant Signal. Behav.* **2012**, *7*, 1456–1466. [CrossRef]
52. Rana, V.; Ram, S.; Nehra, K. Proline biosynthesis and its role in abiotic stress. *Int. J. Agric. Res. Innov. Technol.* **2017**, *6*.
53. Kavi Kishor, P.; Sreenivasulu, N. Is proline accumulation per se correlated with stress tolerance or is proline homeostasis a more critical issue? *Plant Cell Environ.* **2014**, *37*, 300–311. [CrossRef] [PubMed]
54. Al Hassan, M.; López-Gresa, M.P.; Boscaiu, M.; Vicente, O. Stress tolerance mechanisms in *Juncus*: Responses to salinity and drought in three *Juncus* species adapted to different natural environments. *Funct. Plant Biol.* **2016**, *43*, 949–960. [CrossRef]
55. Al Hassan, M.; Pacurar, A.; López-Gresa, M.P.; Donat-Torres, M.; Llinares, J.; Boscaiu, M.; Vicente, O. Effects of salt stress on three ecologically distinct *Plantago* species. *PLoS ONE* **2016**, *11*, e0160236. [CrossRef]
56. Plazas, M.; Nguyen, H.; González-Orenga, S.; Fita, A.; Vicente, O.; Prohens, J.; Boscaiu, M. Comparative analysis of the responses to water stress in eggplant (*Solanum melongena*) cultivars. *Plant Physiol. Biochem.* **2019**, *143*, 72–82. [CrossRef]
57. Chen, Z.; Cuin, T.; Zhou, M.; Twomei, A.; Naidu, B.; Shabala, S. Compatible solute accumulation and stress-mitigating effects in barley genotypes contrasting in their salt tolerance. *J. Exp. Bot.* **2007**, *58*, 4245–4255. [CrossRef]
58. Koźmińska, A.; Al Hassan, M.; Hanus-Fajerska, E.; Naranjo, M.A.; Vicente, O.; Boscaiu, M. Comparative analysis of water deficit and salt tolerance mechanisms in Silene. *S. Afr. J. Bot.* **2018**, *117*, 193–206. [CrossRef]
59. Nagesh, B.; Devaraj, V. High temperature and salt stress response in French bean (*Phaseolus vulgaris*). *Austr. J. Crop Sci.* **2008**, *2*, 40–42.
60. Ashraf, M.; Iram, A. Drought stress induced changes in some organic substances in nodules and other plant parts of two potential legumes differing in salt tolerance. *Flora* **2005**, *200*, 535–546. [CrossRef]
61. Rosales, M.A.; Ocampo, O.; Rodríguez-Valentín, R.; Olvera-Carrillo, Y.; Acosta-Gallegos, J.; Covarrubias, A.A. Physiological analysis of common bean (*Phaseolus vulgaris* L.) cultivars uncovers characteristics related to terminal drought resistance. *Plant Physiol. Biochem.* **2012**, *56*, 24–34. [CrossRef] [PubMed]
62. Sánchez, E.; Ruiz, J.M.; López-Lefebre, L.R.; Rivero, R.M.; García, P.C.; Romero, L. Proline metabolism in response to highest nitrogen dosages in green bean plants (*Phaseolus vulgaris* L. cv Strike). *J. Plant Physiol.* **2001**, *158*, 593–598.
63. Mackay, C.E.; Hall, C.; Hofstra, G.; Fletcher, R.A. Uniconazole-induced changes in abscisic acid, total amino acids, and proline in *Phaseolus vulgaris*. *Pestic. Biochem. Phys.* **1990**, *37*, 74–82. [CrossRef]
64. Zengin, F.K.; Munzuroglu, O. Effects of some heavy metals on content of chlorophyll, proline and some antioxidant chemicals in bean (*Phaseolus vulgaris* L.) seedlings. *Acta Biol. Cracov. Bot.* **2005**, *47*, 157–164.
65. Abdelhamid, M.T.; Rady, M.M.; Osman, A.S.; Abdalla, M.A. Exogenous application of proline alleviates salt-induced oxidative stress in *Phaseolus vulgaris* L. plants. *J. Hortic. Sci. Biotech.* **2013**, *88*, 439–446. [CrossRef]
66. Domínguez, A.; Yunel Pérez, Y.; Alemán, S.; Sosa, M.; Fuentes, L.; Darias, R.; Demey, J.; Rea, R.; Sosa, D. Respuesta de cultivares de *Phaseolus vulgaris* L. al estrés por sequía. *Biot. Veg.* **2014**, *14*, 29–36.
67. Gürel, F.; Öztürk, Z.N.; Uçarlı, C.; Rosellini, D. Barley genes as tools to confer abiotic stress tolerance in crops. *Front Plant Sci.* **2016**, *7*, 1137. [CrossRef]
68. Yoshida, J.; Tomooka, N.; Khaing, T.Y.; Sunil Shantha, P.G.; Naito, H.; Matsuda, Y.; Ehara, H. Unique responses of three highly salt-tolerant wild *Vigna* species against salt stress. *Plant Prod. Sci.* **2020**, *23*, 114–128. [CrossRef]

9

Transcriptomic Profiling of Pomegranate Provides Insights into Salt Tolerance

Cuiyu Liu [1,2], Yujie Zhao [1,2], Xueqing Zhao [1,2], Jinping Wang [1,2], Mengmeng Gu [3,*] and Zhaohe Yuan [1,2,*]

1 Co-Innovation Center for Sustainable Forestry in Southern China, Nanjing Forestry University, Nanjing 210037, China; liucuiyu88@gmail.com (C.L.); z1184985369@163.com (Y.Z.); zhaoxq402@163.com (X.Z.); wangjp0304@163.com (J.W.)

2 College of Forestry, Nanjing Forestry University, Nanjing 210037, China

3 Department of Horticultural Sciences, Texas A&M AgriLife Extension Service, College Station, TX 77843-2134, USA

* Correspondence: mgu@tamu.edu (M.G.); zhyuan88@hotmail.com (Z.Y.)

Abstract: Pomegranate (*Punica granatum* L.) is widely grown in arid and semi-arid soils, with constant soil salinization. To elucidate its molecular responses to salt stress on mRNA levels, we constructed 18 cDNA libraries of pomegranate roots and leaves from 0 (controls), 3, and 6 days after 200 mM NaCl treatment. In total, we obtained 34,047 genes by mapping to genome, and then identified 2255 DEGs (differentially expressed genes), including 1080 up-regulated and 1175 down-regulated genes. We found that the expression pattern of most DEGs were tissue-specific and time-specific. Among root DEGs, genes associated with cell wall organization and transmembrane transport were suppressed, and most of metabolism-related genes were over-represented. In leaves, 41.29% of DEGs were first suppressed and then recovered, including ions/metal ions binding-related genes. Also, ion transport and oxidation-reduction process were restricted. We found many DEGs involved in ABA, Ca^{2+}-related and MAPK signal transduction pathways, such as ABA-receptors, Ca^{2+}-sensors, MAPK cascades, TFs, and downstream functional genes coding for HSPs, LEAs, AQPs and PODs. Fifteen genes were selected to confirm the RNA-seq data using qRT-PCR. Our study not only illuminated pomegranate molecular responses to salinity, but also provided references for selecting salt-tolerant genes in pomegranate breeding processes.

Keywords: pomegranate; salt stress; transcriptome; tissue-specific; signaling transduction pathways; transcription factors

---

1. Introduction

Soil salinization is defined as the excess or deposition of salt ions in land, which may interfere with plant growth. With the aggravation of soil salinization, it has become a considerable threat to healthy and sustainable development of worldwide agriculture [1]. Approximate 20% of the global cultivated lands and 50% of the irrigated lands are affected by salinity [2]. Plants exposed to saline conditions mainly suffer from osmotic stress, ion toxicity, and nutrient deficiency [3,4]. Consequently, all of the significant processes involved in plant growth and development, such as photosynthesis, protein synthesis, energy conversion and ion balance, could be affected by salinity [5]. The effects of salinity depend, not only on species, genotypes, and the age of plants, but also on the duration and intensity of stress [6]. Meanwhile, plants can adapt to saline environment with following strategies: (1) Efficiently controlling the uptake, transport, and compartmentalization of toxic ions;

(2) synthesis of osmoregulation substances and activation of antioxidant enzymes; (3) formation of unique morphological structures, such as succulent leaves, salt glands and bladders [7,8].

Plant salt-tolerance is a quantitative trait controlled by multiple genes [4]. Using transcriptomic analysis, researchers can identify the salt-related genes with differential and temporal expression patterns in plants, and provide a clear picture of transcripts in response to salt stress. These salt-related genes are identified and categorized into two types according to the functions of proteins [4]. The first type is the effector code the functional proteins, which are directly involved in the physiological and biochemical responses to salt stress in plants. These proteins include the superoxide dismutase (SOD) [9], ascorbate peroxidase (APX) [10], high-affinity potassium transporter (HKT) [11], Na^+/H^+ antiporter (NHX) [12], aquaporin (AQP) [13], late embryogenesis abundant (LEA) [14], and H^+-ATPase (VHA) [15], etc. The second type is regulator that involved in regulating the expressions of genes and the signal transduction pathways, such as transcription factors (TFs) and various kinases [16]. These well-characterized TFs include Apetala2/ethylene response factor (AP2/ERF), dehydration-responsive element-binding protein/C-repeat binding factor (DREB/CBF), WRKY, NAC, basic leucine zipper (bZIP), MYB, and basic helix-loop-helix (bHLH) family members [4]. These genes regulate the expressions of downstream genes via various ways, then may influence the plants salt-tolerance eventually [17]. Under salt stress, many signaling transduction pathways are stimulated in plants, such as Ca^{2+}-related, abscisic acid (ABA), and mitogen-activated protein kinase (MAPKs), as well as the crosstalk networks among them, which play crucial roles in responses to salt stress [4,18].

Pomegranate (*Punica granatum* L.) is an emerging commercial fruit tree of the Lythraceae family [19]. In recent years, pomegranate is increasingly popular with extensive usage of its fruits and products, and it is considered a 'super fruit' with high nutritional and medicinal values [20]. The species is widely grown in arid and semiarid regions where the availability and irrigation of saline water are significant issues [21]. Thus, it is important to explore pomegranate potential salt tolerance. Considerable research has been done on pomegranate physiological and biochemical responses to salinity, especially on its growth, ion balance, osmoregulation, and the scavenging of reactive oxygen species (ROS) [22–24]. In this study, we used roots and leaves of a pomegranate cultivar 'Taishanhong' with whole genomic sequence released to perform deep transcriptome sequencing and then demonstrate a global representation of potential candidate genes under salt stress. We aim to unravel the fundamental molecular mechanisms in pomegranate underlying the responses to salt stress.

2. Materials and Methods

2.1. Plant Growth and Stress Treatments

Uniform rooted cuttings of pomegranate ('Taishanhong') were obtained from the Pomegranate Repository of Nanjing Forestry University, China. They were grown in plastic pots (2.5 L) containing medium (1:1 by volume of perlite and peat, 1.5 ± 0.1 Kg) in a climate controlled chamber for six months (14 h light 26 °C/10 h dark 22 °C), and fertigated weekly with $\frac{1}{2}$ Hoagland's solution [25].

At the beginning of experiment, forty-five plants were randomly selected and divided into 3 groups, fifteen plants per group. The roots and leaves of the first group of plants were harvested as control samples (T0R and T0L). The The other two groups of plants were fertigated once with 500 mL of $\frac{1}{2}$ Hoagland's solution containing 200 mM NaCl. A saucer was placed under each container to collect leachate solution during the experiment period. The roots and leaves of the second group of plants on day 3 (T1R and T1L) and the third group of plants on day 6 (T2R and T2L) were harvested later. There was not visible difference between the treated and untreated plant (Figure S1). Samples from five plants in the same group were pooled together as a replicate due to the small amount of biomass. All samples were immediately frozen in liquid nitrogen and then stored at −80 °C.

2.2. RNA Preparation, cDNA Library Construction and Sequencing

Total RNA was extracted from pomegranate roots and leaves using the Total RNA Kit (Tiangen, Beijing, China) according to the manufacturer's instructions. RNA purity, concentration and integrity were checked using the NanoPhotometer® spectrophotometer (IMPLEN, Westlake Village, CA, USA), Qubit® RNA Assay Kit in Qubit® 2.0 Flurometer (Life Technologies, Carlsbad, CA, USA) and the RNA Nano 6000 Assay Kit of the Agilent Bioanalyzer 2100 (Agilent Technologies, Santa Clara, CA, USA), respectively.

A total amount of 3 μg RNA per sample was used as input material for the RNA sample preparations. Sequencing libraries were generated using NEB Next® Ultra™ RNA Library Prep Kit for Illumina® (NEB, Ipswich, MA, USA) following manufacturer's recommendations and index codes were added to attribute sequences to each sample. The RNA samples were concentrated using magnetic oligo (dT) beads and then broken into short fragments using an RNA Fragmentation buffer (NEB Next First Strand Synthesis Reaction Buffer (5X), Ambion, Austin, TX, USA). The first-strand cDNA was synthesized using random hexamer primer and M-MuLV Reverse Transcriptase (RNase H-). Second-strand cDNA was synthesized subsequently using DNA Polymerase I and RNase H. After adenylation of 3'ends of DNA fragments, NEB Next Adaptor with hairpin loop structure were ligated to prepare for hybridization. The cDNA fragments with suitable lengths (150~200 bp) were purified with AMPure XP system (Beckman Coulter, Beverly, MA, USA) to construct the final cDNA libraries. Then 3 μL USER Enzyme (NEB, USA) was used with size-selected, adaptor-ligated cDNA at 37 °C for 15 min followed by 5 min at 95 °C before PCR. Next, the selected cDNA fragments were enriched via PCR with Phusion High-Fidelity DNA polymerase, Universal PCR primers and Index (X) Primer. PCR products were purified with AMPure XP system and the cDNA libraries were assessed on the Agilent Bioanalyzer 2100 system. Finally, the clustering of the index-coded samples was performed on a cBot Cluster Generation System using TruSeq PE Cluster Kit v3-cBot-HS (Illumina, Inc., San Diego, CA, USA) according to the manufacturer's instructions. After cluster generation, the library preparations were sequenced on an Illumina HiSeq 4000 platform (Illumina, Inc., San Diego, CA, USA) and 150 bp paired-end reads were generated.

2.3. Sequence Assembly

Low quality reads and reads containing adapter and ploy-N were filtered by NGS QC ToolKit [26] from raw reads. At the same time, Q20, Q30, GC-content and sequence duplication level of the clean data were calculated. All the downstream analyses were based on clean data with high quality. The clean data from all 18 libraries were separately mapped to the pomegranate genome assembly (ASM220158v1) using HISAT2 software [27]. StringTie [28] was used to construct and identify both known and novel transcripts from HISAT2 alignment results. StringTie was also used to count the reads numbers mapped to each gene. After that, FPKM (Fragments Per Kilobase of exon per million fragments Mapped) were calculated based on the length of the gene and reads count mapped to this gene [27].

2.4. Identification and Functional Annotation of DEGs

All genes were compared against various protein databases by BLASTX, including the Nr (NCBI non-redundant protein sequences); Pfam (Protein family); KOG/COG (Clusters of Orthologous Groups of proteins), and Swiss-Prot (A manually annotated and reviewed protein sequence database) with an E-value cut-off of 10^{-5}. Then genes with the best BLAST hit (the highest score) were chosen along with their protein functional annotations. DESeq and Q-value were employed to evaluate the differential expression genes (DEGs) between controls and treatments. The false discovery rate (FDR) and $\log_2$FC (log of fold change) were calculated for all genes, and only transcripts with FDR < 0.05 and$|\log_2(\text{fold change})| \geq 1$ were considered as DEGs. To annotate the DEGs with gene ontology (GO) terms, the Nr BLAST results were imported into the Blast2GO program [29]. To examine the expression patterns of DEGs, the expression data from roots and leaves (T0, T1 and T2) were normalized to 0, $log_2(T1/T0)$, $log_2(T2/T0)$, and then clustered by Short Time-series Expression Miner software (STEM) separately, using a FDR correction method and p-value ≤ 0.05 as the cutoff [30]. The results of GO annotations in each pattern were enriched and refined using TBtools v0.6652 (Toolbox for Biologists, https://github.com/CJ-Chen/TBtools). We used KOBAS software [31] to test the enrichment of DEGs in KEGG pathways.

2.5. Validation of RNA-Seq by qRT-PCR

We selected 15 DEGs as experimental validation by quantitative real-time PCR (qRT-PCR), which were performed with three biological and three technical replicates for each cDNA sample. The primers for these genes were designed with NCBI primer-BLAST (Table S1). Reverse transcription was conducted with the GoScriptTM Reverse Transcription System (Promega, Madison, WI, USA). We conducted qRT-PCR in a 7500 fast Real-Time PCR system (Applied Biosystems, CA, MA, USA), and analyzed results with the $\Delta\Delta$Ct method, and *Pgactin* (F: ATCCTCCGTCTTGACCTTG, R: TGTCCGTCAGGCAACTCAT) gene was used as a reference gene. Each reaction was carried out in a final volume of 20 μL, containing 7.5 μL of ddH_2O, 10 μL of SYBR Green PCR master mix, 0.5 μL of each gene-specific primer and 2 μL of diluted cDNA. The PCR thermal cycling conditions were as follows: 95 °C for 10 min; 40 cycles of 95 °C for 5 s, 60 °C for 30 s and 72 °C for 30 s. Data were collected during the extension step: 95 °C for 15 s, 60 °C for 1 min, 95 °C for 30 s and 60 °C for 15 s.

3. Results

3.1. Overview of Sequencing and Mapping

To obtain a global overview of the salt-induced changes at whole-transcriptomic scale in pomegranate, we constructed 18 cDNA libraries from the roots and leaves of 0 (controls), 3 and 6 days after 200 mM NaCl treatment. A high correlation ($R^2 > 0.80$) between biological replicates was observed for all treatments (Figure S2), which indicated that the biological replicates were reliable in this study. In totally, sequencing libraries yielded 519.47 million reads with 150 bp for both paired ends. After adapter removal and refining, we obtained 155.84 Gb of clean data, and Q30 percentage were all above 90.08% (Table 1). The ratios of reads mapping to the pomegranate genome were high, with values ranging from 93.61% to 95.33%. Then, the unique-mapped reads were 91.89–93.79% and the mapped sequences in genome exon regions were 62.88–69.32% (Table 1). Finally, we assembled 29,226 transcripts from the 18 cDNA libraries of pomegranate roots and leaves under salt stress. To splice the genes more completely and accurately, the StringTie software was employed to reconstruct the transcripts, and then retrieved 34,047 assemblies, including 5396 novel transcripts. Finally, 26,444 genes (including 2421 new genes) were assembled from annotated genes using 7 different databases (see methods), and can be used as a reference of pomegranate genome annotation (Table S2).

Table 1. Summary of RNA-Seq results and the alignments in pomegranate genome.

Tissues	Samples ID	Read Number (M)	Base Number (M)	GC Content (%)	Q30 (%)	Mapped Reads (%)	Unique Alignments (%)	Mapped to Exonic (%)
Root (CK)	T01	31,493,153	9,447,945,900	50.43	91.10	94.06	92.27	66.54
	T02	28,438,357	8,531,507,100	50.10	91.78	94.53	92.76	66.84
	T03	26,939,697	8,081,909,100	50.03	91.27	94.99	93.33	62.88
Leaf (CK)	T04	33,023,098	9,906,929,400	52.08	91.22	95.19	93.59	69.32
	T05	27,023,343	8,107,002,900	51.60	90.70	94.88	93.34	68.26
	T06	25,709,176	7,712,752,800	51.48	91.84	95.21	93.54	68.57
Root (3 d)	T07	31,433,566	9,430,069,800	49.95	91.16	94.46	92.67	66.13
	T08	28,144,078	8,443,223,400	50.16	91.23	94.12	92.32	67.30
	T09	26,542,339	7,962,701,700	50.51	91.32	94.63	92.88	67.69

Table 1. *Cont.*

Tissues	Samples ID	Read Number (M)	Base Number (M)	GC Content (%)	Q30 (%)	Mapped Reads (%)	Unique Alignments (%)	Mapped to Exonic (%)
Leaf (3 d)	T10	28,771,523	8,631,456,900	51.83	91.24	94.97	93.36	68.38
	T11	29,759,918	8,927,975,400	51.32	91.46	95.33	93.79	68.04
	T12	31,628,291	9,488,487,300	51.18	90.08	94.91	93.03	67.80
Root (6 d)	T13	31,761,056	9,528,316,800	50.56	91.91	93.73	91.89	66.87
	T14	29,807,151	8,942,145,300	50.07	91.40	93.89	92.01	66.11
	T15	25,298,243	7,589,472,900	50.29	91.46	93.61	91.89	66.33
Leaf (6 d)	T16	30,075,932	9,022,779,600	51.41	90.84	95.35	93.79	68.50
	T17	27,643,746	8,293,123,800	51.37	91.18	95.18	93.63	68.38
	T18	25,980,948	7,794,284,400	50.88	91.38	94.32	92.66	67.57
Total	—	519,473,615	155,842,084,500	—	—	—	—	—

Reads Number: total Number of paid-end reads in Clean Data; Base Number: total number of bases in Clean Data; Q30%: the ratio of nucleotides with quality value ≥ 30; GC content: The ratio of guanidine and cytosine nucleotides; Mapped ratio: The percentage of Mapped Reads in proportion of Clean reads.

3.2. Identification and Annotation of DEGs

A total of 2255 DEGs were identified from salt treatments with FDR < 0.05 and$|\log_2(\text{fold change})|\geq 1$, including 1080 up-regulated and 1175 down-regulated genes, and more DEGs were found in roots (1623 genes) than in leaves (632 genes) (Table S2). To provide insights into the underlying functions of pomegranate transcripts under salt stress, DEGs were annotated after they were compared to well-studied sequences in the GO, Nr, Swiss-Prot, KEGG, and KOG databases (Table S2). In this study, most of DEGs in T1R were distributed in T subgroup (Signal transduction mechanisms), E subgroup (Amino acid transport and metabolism) and G subgroup (Carbohydrate transport and metabolism) (Figure 1a). DEGs of T2R and T1L samples were mostly concentrated in R subgroup (General function prediction only), Q subgroup (Secondary metabolites biosynthesis, transport and cabalisms), and K subgroup (Transcription) (Figure 1b,c). The DEGs in T2L were distributed in O subgroup (Posttranslational modification, protein turnover and chaperones), P (Inorganic ion transport and metabolisms) and Q subgroup (Secondary metabolites biosynthesis, transport and cabalisms) (Figure 1d). These subgroups were close to the metabolism, such as DNA transcription, cell division, protein modification and energy conversion.

The DEGs of roots and leaves were also analyzed by KEGG pathways enrichment. These pathways were mainly classified into 5 categories (Figure S3) and showed the top 20 pathways (Figure 2). The majority of pathways in pomegranate roots and leaves were associated with metabolisms, such as global and overview maps, amino acid metabolism, carbohydrate metabolism, lipid metabolism, and energy metabolism, etc. (Figure S3). The metabolism-related genes were significantly enriched in metabolic pathways (ko01100), biosynthesis of secondary metabolites (ko01110), cysteine and methionine metabolism (ko00270), phenylpropanoid biosynthesis (ko00940), starch and sucrose metabolism (ko00500), amino sugar and nucleotide sugar metabolism (ko005200) and protein processing in endoplasmic reticulum (ko04141) pathways. Also, there were some DEGs involved in environmental adaptation (plant-pathogen interaction, ko04626) and signal transduction (plant hormone signal transduction, ko04075) pathways (Figure 2; Figure S3). Genes clustered in genetic information processing were enriched in folding, sorting and degradation, replication and repair, transcription and translation, including protein processing in endoplasmic reticulum (ko04141), DNA replication (ko03030), RNA polymerase (ko03020), RNA transport (ko03013) and RNA degradation (ko03018). In addition, genes involved in the photosynthetic pathways, such as carotenoid biosynthesis (ko00906), porphyrin and chlorophyll metabolism (ko00860) and photosynthesis-antenna proteins (ko00196) were suppressed by salinity.

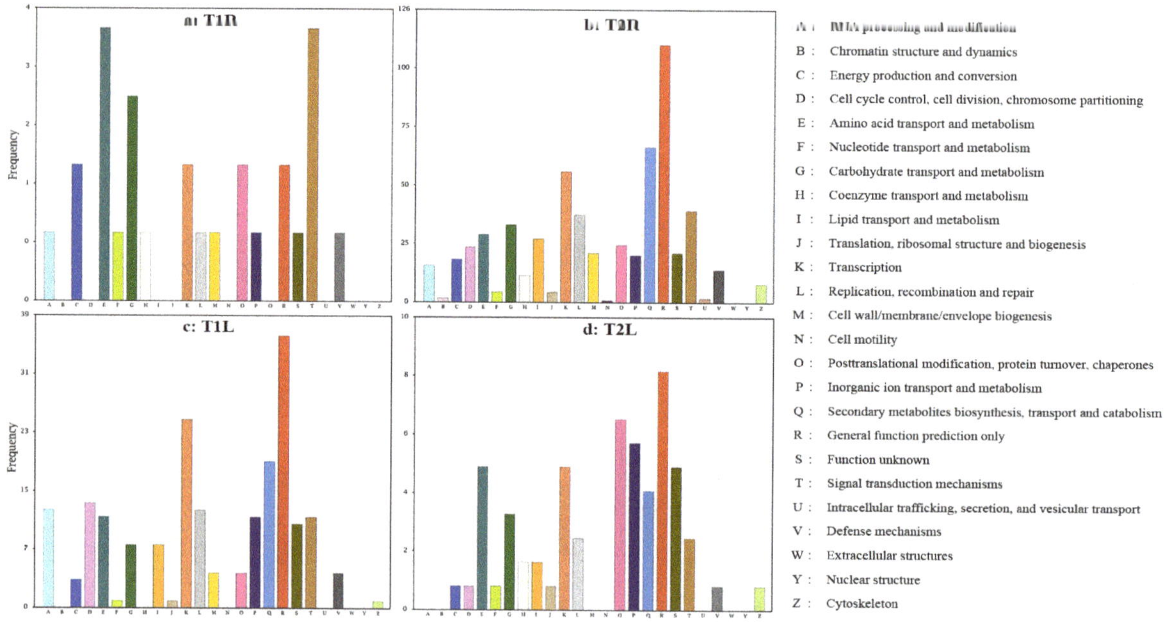

Figure 1. COG classification of differentially expressed genes (DEGs) in pomegranate roots and leaves. DEGs enrichments in T1R (**a**), T2R (**b**), T1L (**c**) and T2L (**d**).

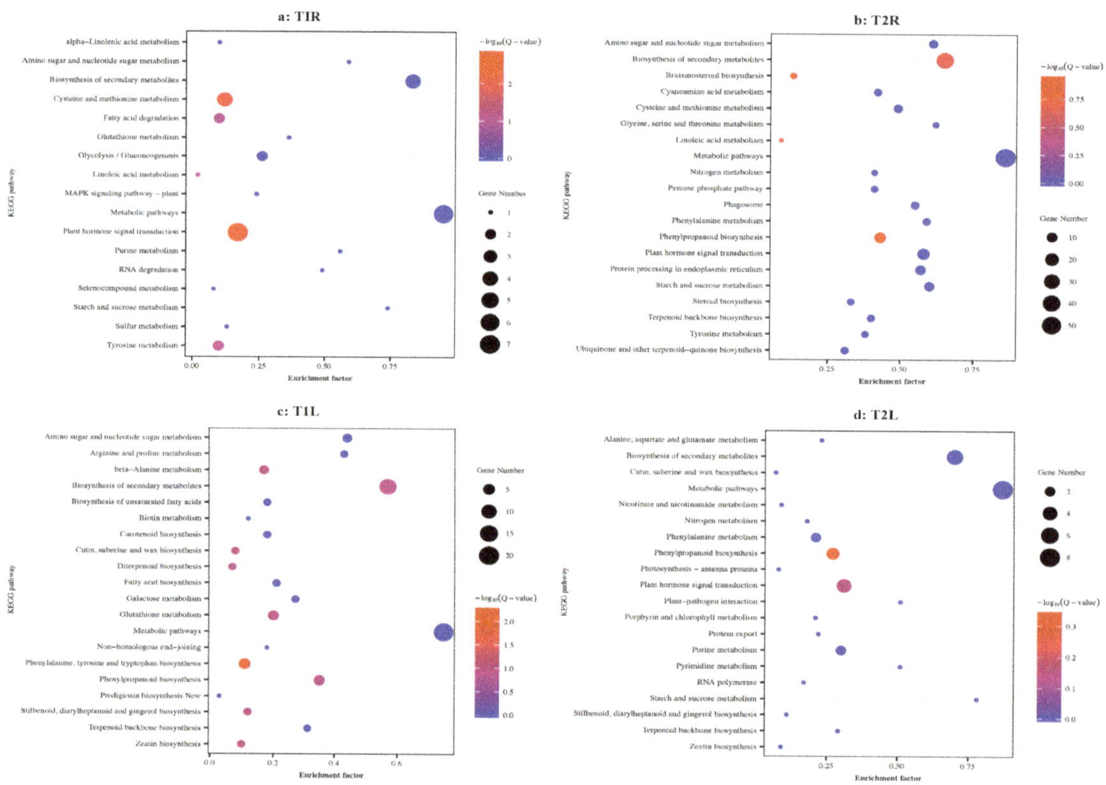

Figure 2. The top 20 KEGG pathway of DEGs in pomegranate root and leaf. The horizontal axis showed an enrichment factor, the smaller the enrichment factor is the more significant enrichment of DEGs in this pathway. While the vertical axis illustrated the KEGG pathway, and the color illustrated the -log(Q-value), red color is more reliable and significance enrichment in this pathway. The size of black spots showed the number of DEGs enriched into each pathway. DEGs enrichments in T1R (**a**), T2R (**b**), T1L (**c**), and T2L (**d**).

3.3. The Expressional Patterns of DEGs

Numbers of DEGs increased in roots with the duration of salt-stress, which were opposite in leaves. The genes were both with little overlap in roots and leaves at two points of stress time (Figure 3a). Only a small portion (2.0%) of DEGs (20 up-regulated and 24 down-regulated) shared the common expression tendency between roots and leaves, and even less (1.2%) DEGs showed an utterly opposite tendency between two tissues (3) genes were up-regulated in leaves and down-regulated in roots, 23 genes were down-regulated in leaves but up-regulated in roots) (Figure 3b). The remaining majority of DEGs were exclusively up-regulated or down-regulated in either tissue. Almost 85.9% of up-regulated and 53.4% of down-regulated DEGs were activated in roots after 6-days salt stress, most of down-regulated genes in T1L recovered in T2L (Figure 3c,d), and only 22 (2.0%) genes were suppressed in leaves at two points of time.

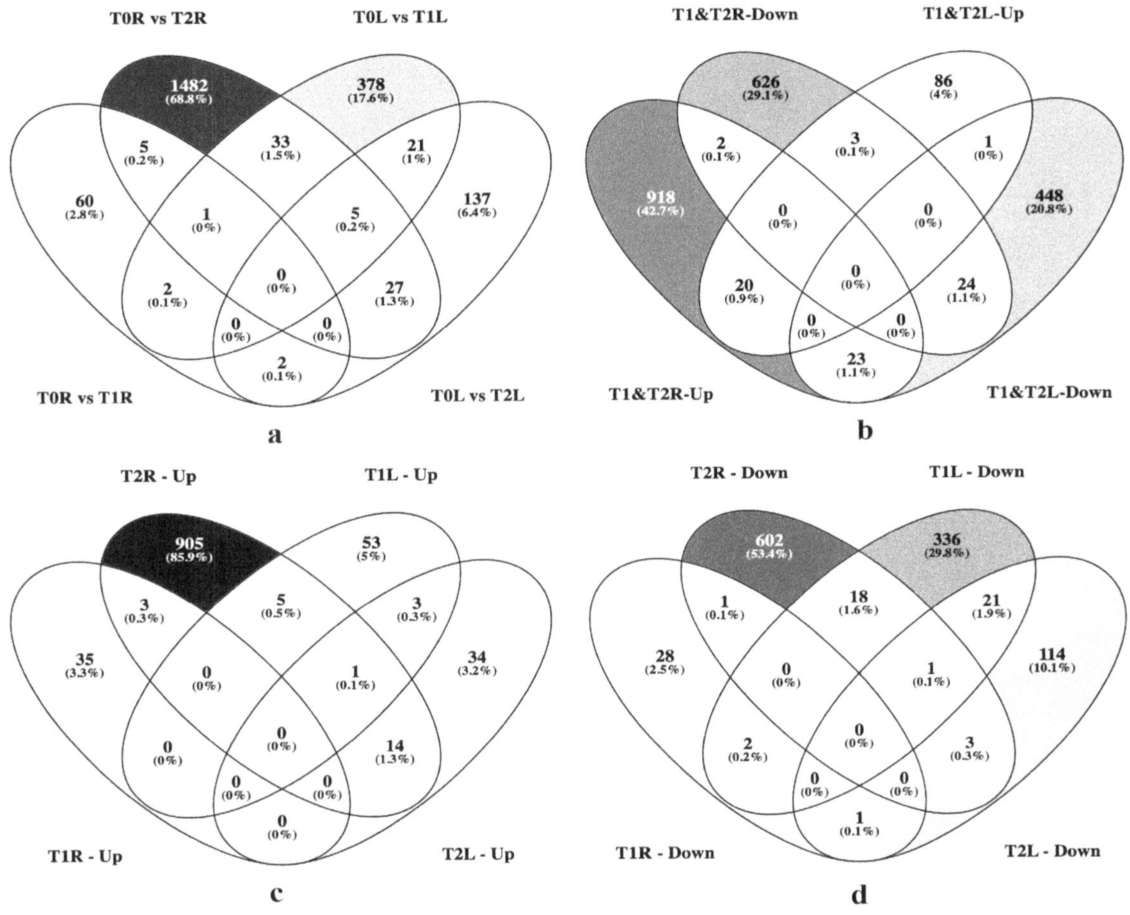

Figure 3. The numbers of up-regulated and down-regulated genes in treatments (T1R, T2R, T1L, and T2L), comparing to the controls (T0R and T0L). (**a**) All DEGs in T1R, T2R, T1L, and T2L; (**b**) Up- or down-regulated genes in leaves and roots; (**c**) Up-regulated genes in T1R, T2R, T1L, and T2L; and (**d**) Down-regulated genes in T1R, T2R, T1L, and T2L.

The STEM was employed to analyze the expressional patterns of DEGs. All the 1,623 DEGs in roots were clustered into 8 profiles, whereby 1,340 DEGs were significantly clustered into 4 profiles with p-value $\leq 2 \times 10^{-5}$ (Figure 4a). The two up-regulated patterns were Profile 4 (31.55%, 512 DEGs) and Profile 7 (15.16%, 246 DEGs), and two down-regulated patterns were Profile 3 (28.59%, 464 DEGs) and Profiles 0 (7.27%, 118 DEGs). All the 632 DEGs in leaf samples were also clustered into 8 profiles and then 2 enriched profiles, including one first down-regulated and then the up-regulated pattern of Profile 2 (41.29%, 261 DEGs, p-value $= 3 \times 10^{-32}$), and one down-regulated pattern of Profile 1 (16.61%, 105 DEGs, p-value $= 6 \times 10^{-17}$) (Figure 4b).

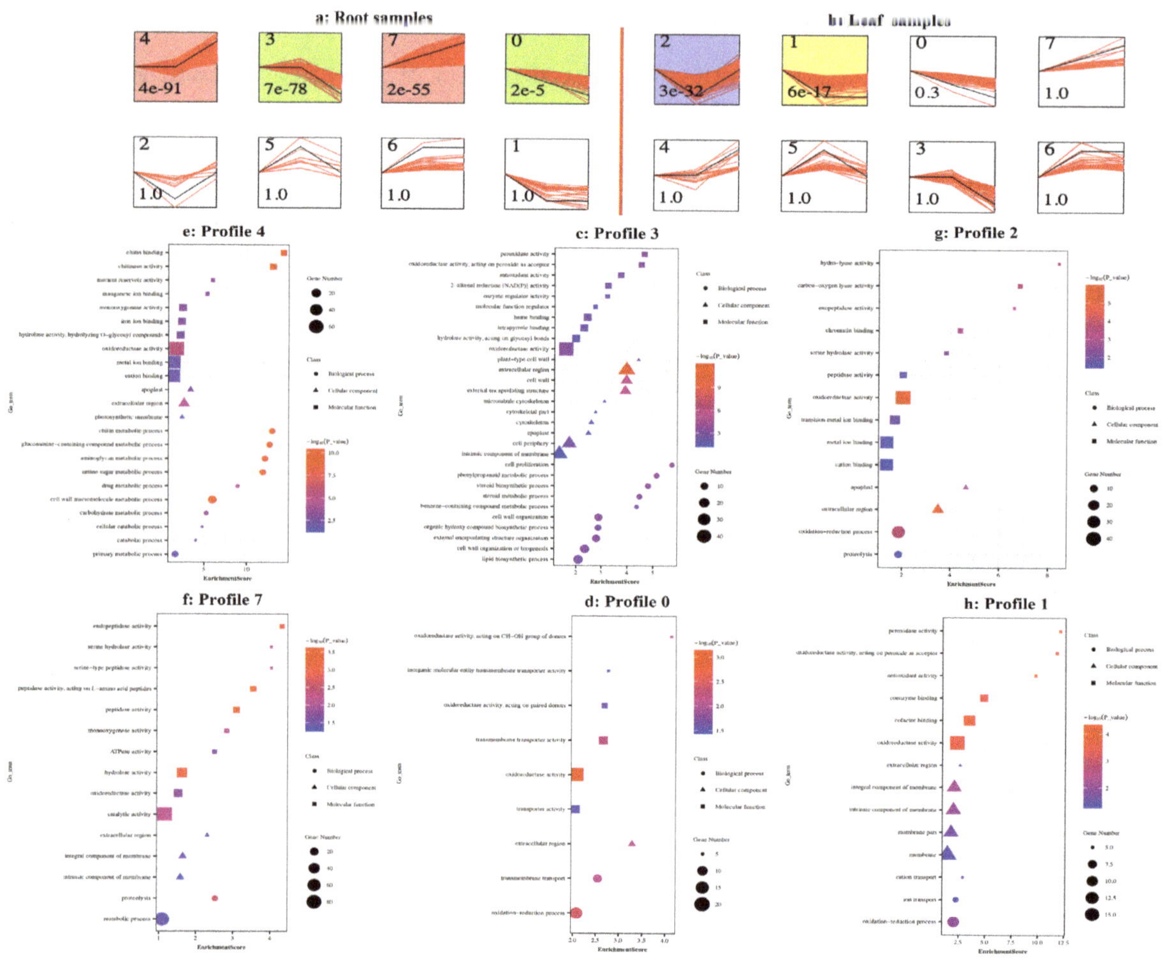

Figure 4. The expression patterns of DEGs in pomegranate roots and leaves under NaCl stress using STEM. The P-value in roots; (**a**) and leaves (**b**) are shown. The trends of each pattern showed in black lines and the clustered genes are shown in red lines. GO enrichments of Profile 4 (**c**), Profile 7 (**d**) Profile 3 (**e**) and Profile 0 (**f**) in root libraries (T0R, T1R and T2R); and Profile 2 (**g**) and Profile 1 (**h**) in leaf libraries (T0L, T1Land T2L).

Next, DEGs within these profiles were subjected to GO-term enrichment analysis. They were classified into three main categories, including cellular component, biological process, and molecular function, and the top 30 sub-categories (if more than 30, p-value < 0.05) of each category were shown (Figure 4). In roots, genes in up-regulated pattern of Profile 4 were just inducted in T2R. They were enriched in chitinase activity, chitin binding, oxidoreductase activity, metal ion, and cation binding under molecular function; chitin, glucosamine-containing compound, aminoglycan, amino sugar and cell wall macromolecule metabolic process under biological process (Figure 4c; Table S3). Gens were continuously suppressed with the extension of stress in Profile 7, and they were enriched in endopeptidase activity, peptidase activity, hydrolase activity, and catalytic activity under molecular function and proteolysis and metabolic process under biological process (Figure 4d; Table S3). The top subcategories of down-regulated pattern (Profile 3) were extracellular region (GO:0005576), cell wall (GO:0005618) and external encapsulating structure (GO:0030312) under cellular component, and lipid biosynthetic process (GO:0008610), cell wall organization (GO:0071555), and external encapsulating structure organization (GO:0045229) under biological process. (Figure 4e; Table S4). Many DEGs coding cell wall components were mostly suppressed in T2R samples, including four extensins, two pectin acetylesterase, three polygalacturonase, and three xyloglucan endotransglucosylase/hydrolase proteins. At the same time, transmembrane transport (GO:0055085) and oxidation-reduction process (GO:0055114) in Profile 0 were inhibited under salt stress. Many genes were identified in these biological process, such as adenine/guanine permease, aquaporin, cationic amino acid transporter, sugar carrier, and zinc transporter (Figure 4f; Table S4). Profile 2 showed a first down- and then up-regulated pattern in leaves, there were many subcategories, such as proteolysis and oxidation-reduction process under biological process, oxidoreductase activity, metal ion, and cation binding under molecular function. The top subcategories in down-regulated pattern of Profile 1 were coenzyme binding, cofactor binding and oxidoreductase activity under molecular function, and oxidation-reduction process, cation and ion transport under biological process. We identified genes coding ion transporters, such as cation/calcium exchanger, zinc transporter, ammonium transporter 1 member, copper-transporting ATPase, potassium transporter (Figure 4h; Table S5).

3.4. ABA Signaling Pathway

The ABA signaling pathway mainly consists of three protein classes: ABA receptors (PYR/PYL/RCAR), type 2C protein phosphatase (PP2Cs) and sucrose non-fermenting1-related protein kinase 2 (SnRK2s). In our study, three ABA receptors, *PYLs*, were significantly down-regulated (Figure 5), while 13 *PP2Cs* have different expressional patterns in pomegranate roots and leaves under salt stress. Among these genes, seven *PP2Cs* were up-regulated in roots or leaves, five *PP2Cs* down-regulated, and another gene up-regulated in roots and down-regulated in leaves (Figure 5). In contrast, just one *SnRK2* was identified among the DEGs, and it was up-regulated in leaves, which allowed the accumulation of phosphorylated downstream ABFs and activation of ABA-response genes.

a

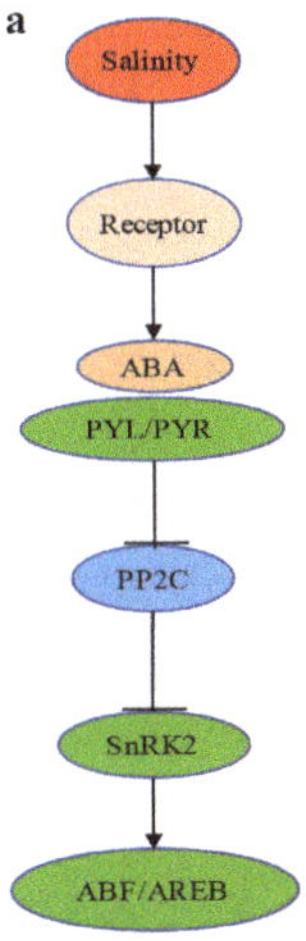

b

Name	GeneID	Gene_annotation
PYL	CDL15_Pgr005008	Abscisic acid receptor PYL3
	CDL15_Pgr022187	Abscisic acid receptor PYL4
	CDL15_Pgr025279	Abscisic acid receptor PYL4
PP2C	CDL15_Pgr001936	Probable protein phosphatase 2C 38
	CDL15_Pgr009612	Probable protein phosphatase 2C 40
	CDL15_Pgr010336	Protein phosphatase 2C 3
	CDL15_Pgr010390	Probable protein phosphatase 2C 33
	CDL15_Pgr014465	Probable protein phosphatase 2C 12
	CDL15_Pgr017165	Probable protein phosphatase 2C 15
	CDL15_Pgr021717	Probable protein phosphatase 2C 52
	CDL15_Pgr021723	Probable protein phosphatase 2C 5
	CDL15_Pgr022500	Probable protein phosphatase 2C 75
	CDL15_Pgr023042	Protein phosphatase 2C 57
	CDL15_Pgr023064	Protein phosphatase 2C 32
	CDL15_Pgr028018	Probable protein phosphatase 2C 8
	MSTRG.23006	Probable protein phosphatase 2C 25
SnRK2	CDL15_Pgr021460	Serine/threonine-protein kinase SAPK2

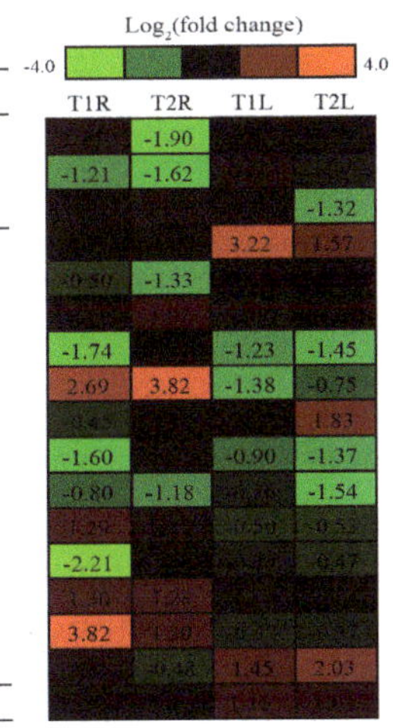

Figure 5. DEGs involved in ABA signal transduction pathway in pomegranate under salt stress. (**a**) ABA signal transduction pathway; (**b**) DEGs annotations and expressional levels. The color represents with $\log_2$(fold change), the DEGs with $\log_2FC \geq 1$ were up-regulated (red) and $\log_2FC \leq 1$ were down-regulated (green) (FDR < 0.05).

3.5. Ca^{2+}-Related Signaling Pathways

Thirty DEGs involved in Ca^{2+}-related signaling pathway were identified, these transcripts coded function proteins included three Ca^{2+}-ATPases (ACAs), four cation/H^+ antiporters (CAXs), two cation/calcium exchangers (CCXs), four glutamate receptor (GLRs), ten calcium-binding proteins (CaM/CMLs), two CBL-interacting protein kinases (CIPKs) and five calcium-dependent protein kinases (CDPKs) (Figure 6). Expectedly, two *ACAs* were up-regulated in roots (T2R), and one was down-regulated in leaves (Figure 6). Ten *CAXs* expressed differently, including two *CAXs* located on vacuoles were both up-regulated in T2R. Two *CIPKs* were significantly up-regulated in leaves, but no obvious change was observed in roots. The majority of *CaM/CMLs* were up-regulated in roots and down-regulated in leaves. The *CDPKs* were up- or down-regulated in roots and/or leaves. Also, two DEGs, that were involved in the MAPK signaling pathway were identified, and both were down-regulated in leaves (Figure 6).

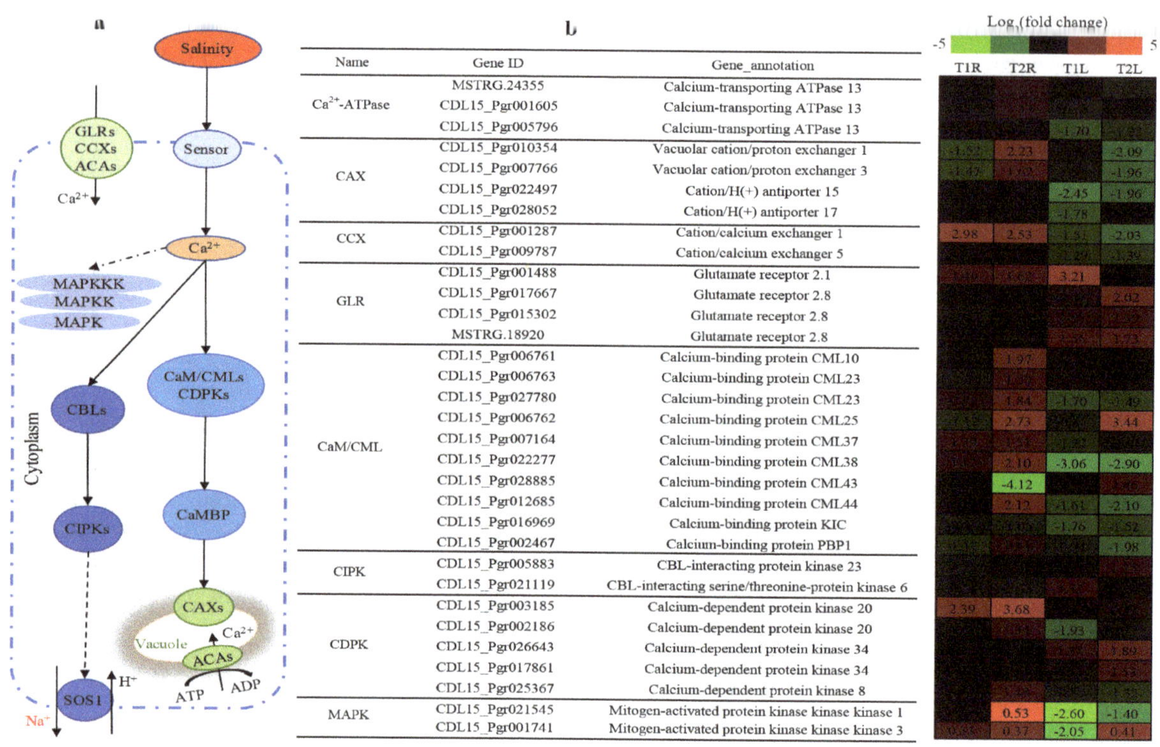

Figure 6. The DEGs involved in the Ca^{2+}-related signal transduction pathways. (**a**) Ca^{2+}-related signal transduction pathway; (**b**) DEGs annotations and expressional levels. The color represents with log_2(fold change), the DEGs with $log_2FC \geq 1$ were up-regulated (red) and $log_2FC \leq 1$ were down-regulated (green) (FDR < 0.05).

3.6. The Transcription Factors (TFs)

For identification of the TFs in pomegranate transcriptome, all the mapped genes were analyzed by BLAST against the Plant Transcription Factor Database (PlantTFDB, http://planttfdb.cbi.pku.edu.cn). A total of 1346 TFs that were classified into 55 putative families (Table S6). Among these TFs, twenty-seven TF families, including 151 genes expressed differently under salt stress compared to controls. The most abundant of differential expression TFs included NAC, ERF, MYB-related, C2H2, MYB, bHLH, GRAS, WRKY, LBD, B3 and bZIP family genes (Table 2; Table S6).

Under NaCl stress, we found that 12 of 19 *NAC* genes were up-regulated, and 10 of 19 genes were down-regulated in leaves or roots of pomegranate plants. Interestingly, we also observed that 18 *PgNACs* significantly changed in T2R when compared to controls, and 4 genes were up-regulated or down-regulated in roots, but reversed in leaves (Table 2; Table S6). These results suggested that many *NACs* were involved in salt stress, but there were different potential responding mechanisms of NAC domain genes to salinity. Thirteen *MYB* genes were induced by salt treatment in pomegranate (Table 2). Among these genes, six in roots and four in leaves were up-regulated, and four in roots and eight in leaves were down-regulated. Moreover, sixteen *MYB-related* genes were detected in our RNA-Seq analyses, thirteen genes were down-regulated, and seven genes were up-regulated by salt stress (Table 2; Table S6). But so far, little was reported that the MYB-related type proteins are related with the responses to salt stress.

The AP2/ERF superfamily is divided into three families: the AP2 family proteins containing two repeated AP2/ERF domains, the ERF family proteins, containing a single AP2/ERF domain, and the RAV family proteins containing a B3 domain. There were 19 *PgERFs*, and 3 *PgAP2s* significantly expressed in treatments when compared to controls (Table 2). Fifteen *ERFs* were down-regulated, and all of *AP2* were repressed by salinity. There were 9 up-regulated and 8 down-regulated *C2H2*, three up-regulated and nine down-regulated *bHLH*, five up-regulated and three down-regulated *WRKY* identified in pomegranate roots and leaves. Remarkably, nine *GRASs* were up-regulated in the root, but the no-significant change in leaves (Table 2; Table S6).

Table 2. The different expressional TFs in pomegranate under NaCl stress.

TFs	Total No.	DEGs	T1R		T2R		T1L		T2L	
			Up	Down	Up	Down	Up	Down	Up	Down
NAC	97	19	1	1	12	6	1	2	1	4
ERF	114	19	1	1	8	5	1	7		9
MYB_related	91	16	-	-	4	3	1	11	3	2
C2H2	91	13	-	-	8	3	-	1		2
MYB	79	13	-	1	6	4	-	8	4	2
bHLH	102	11	-	1	2	5	1	4	1	1
GRAS	50	9	1	-	9	-	-	-	-	-
LBD	38	7	2	2	2	2	-	2	-	1
WRKY	66	7	-	-	5		-	2	1	3
B3	49	3	-	-	1	1	-	1		1
bZIP	45	3	-	-	1	2	-	1	1	1
AP2	8	3	-	-	-	3	-	-	-	-
G2-like	40	3	-	-	1	1	-		-	1
HD-ZIP	36	3	-	-	-	2	-	-	-	1
HSF	21	3	-	-	3	-	-	-	-	-
Others	419	19	-	-	8	3	-	6	1	3
Total	1346	151	5	6	70	40	6	46	13	31

3.7. qRT-PCR Validation

To confirm the reliability of the expression levels obtained from the RNA-Seq, fifteen DEGs, including five *PgERFs*, three *PgMYBs* and seven *PgNACs* were selected for qRT-PCR assays. The results showed that the expression level of each transcript closely corresponded to the transcript level estimated from the sequence data with $R^2 \geq 0.85$ (Figure 7), which implies reproducibility and accuracy of the RNA-Seq results.

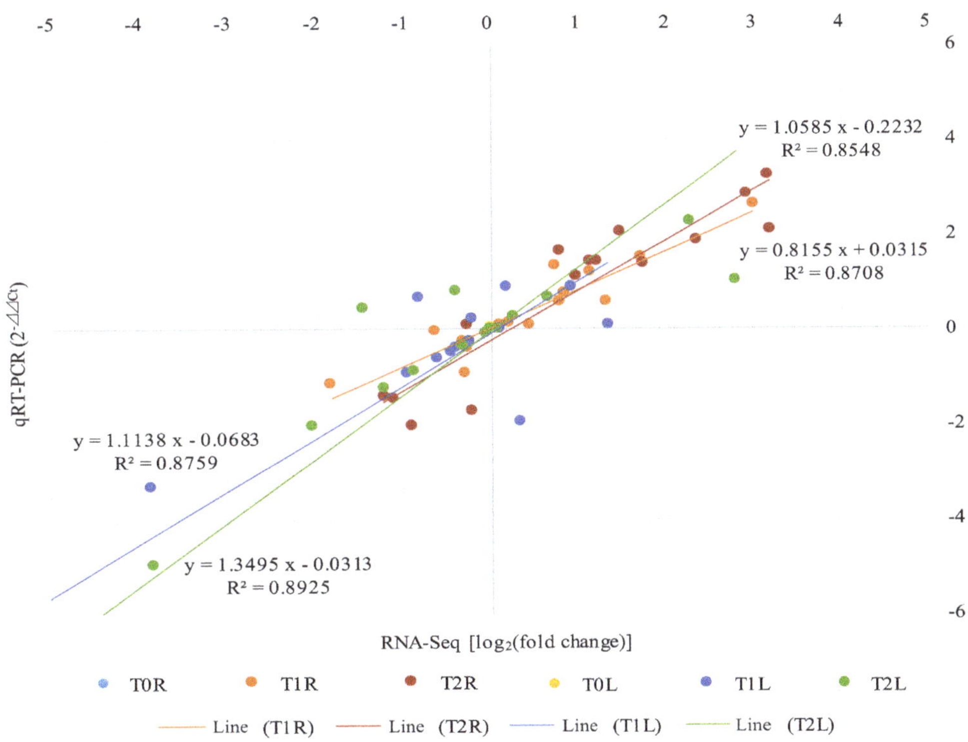

Figure 7. The relationship between qRT-PCR and RNA-Seq date. The color circle is gene expression level. The lines represent co-relationship with dependent linear equations.

4. Discussion

RNA-sequencing techniques have proven to be beneficial and economical for scanning the transcribed genes, both in model and non-model plant species. In this study, we used the RNA-Seq approach as a powerful tool to elucidate the molecular responses to salt stress in pomegranate. Finally, we reconstructed the transcripts and identified 5396 novel genes from the 18 cDNA libraries. There were significant differences between the percentage of total mapped reads and the mapped reads in exonic regions. Theoretically, the ratio of sequencing reads that produced from mature mRNA mapped into exonic regions could be 100%. However, a portion of 30.68–37.12% reads were also mapped into intronic and intergenic regions for the following reasons: (1) The coding and non-coding sequences in the reference genome might have been annotated incorrectly; (2) the new alternative splicing (SNPs) and novel genes resulted from the alternative mRNA splicing and transcripts reconstructing; (3) variations exist between sequencing and reference genome genotypes, which are attributed to the gaps and diversity of their sequences [32,33].

Under salt stress, plants firstly experience osmotic stress due to a disorder of water uptake. Then a process of gradual recovery due to the partially or completely re-established uptake of water occurs within days after the salt treatment [34,35]. We identified 2255 DEGs through the 18 mRNA libraries. From the third to sixth day after treatment, there were significant differences between pomegranate roots and leaves, with little overlap of DEGs responding salinity (Figure 3). More salt-related genes were up-regulated in roots, while the majority of suppressed genes recovered with salt-treating process in leaves. In contrast, the osmotic adjustment of leaves started only after roots had reached new water equilibrium [36]. These tissue-specific and time-specific differences between roots and leaves have also been reported in *Arabidopsis thaliana* (L.) [37], *Populus euphratica* (Oliv.) [36], and *Millettia pinnata* (L.) [38]. Many DEGs were involved in down-regulated patterns in roots, such as cell wall organization, transmembrane transport and oxidation-reduction process, but the proteolysis and metabolic process were over-expressed. Otherwise, most of DEGs involved in oxidation-reduction process and ion transport in leaves were suppressed (Figure 4). The cell wall plays an important role in protecting plant from salt toxicity [39]. In our study, extensin, pectin acetylesterase, polygalacturonase, and xyloglucan endotransglucosylase/hydrolase proteins were identified in roots (Table S4). These proteins are important components of the cell wall, which affect plant growth by mediating cell enlargement and expansion [39]. Many DEGs involved in transmembrane transport code for various channels, carriers and pumps, and ion transporters, such as aquaporin, cationic amino acid transporter, sugar carrier, and zinc transporter (Table S4). They worked together to transport coenzymes, amino acids, carbohydrates, and ions under salt stress. The suppressed genes coding proteins, such as dehydrogenase, cytochrome P450s and peroxidase involved in oxidation-reduction process, play essential roles in responding to salinity [40,41] (Tables S4 and S5). These results indicated that salinity affected pomegranate growth and development by inhibiting the cell division and transmembrane transport, as well as slowing down the redox reactions.

On the other hand, over-represented genes in roots were enriched in metabolic process, carbohydrate metabolic process, and catabolic process, etc. These metabolism-related genes in roots were also enriched in KEGG pathways, such as metabolic pathways (ko01100), biosynthesis of secondary metabolites (ko01110), phenylpropanoid biosynthesis (ko00940), starch and sucrose metabolism (ko00500), amino sugar and nucleotide sugar metabolism (ko005200) (Figure 2; Table S3). The carbohydrates such as glycan, starch, sucrose, amino sugar have been demonstrated as osmolytes and energy resouces in plant responses to salt stress, especially in halophyte species [42]. The accelerated metabolisms in pomegranate might contribute to its adaption responses to salt stress. Similar expression patterns of metabolism-related genes were observed in *Oryza sativa* (L.) [43], *Thellungiella halophila* (C. A. Mey.) [42] and *Helianthus tuberosus* (L.) [44] under salt stress. Metal ions/cations, such as K^+, Ca^{2+}, Mg^{2+}, Fe^{2+}, Cu^{2+}, and Zn^{2+}, act as cofactors participating in the catalytic activity of the enzymes they bind to [44]. The up-regulated DEGs involved in metal ion/cation binding might accelerate the catalytic activity. At the same time, the ion transport process was inhibited in roots and leaves (Tables S3 and S5). The restriction of ion transport indicated that the excess of Na^+ inhibited the uptake of mineral ions [4]. The results correspond to the physiological responses in other pomegranate cultivars under salt stress [45]. We suspected that pomegranate could cope with the toxic ions to some extent via decreasing uptake of ions from soil and increasing utilization of ions in cells. The toxic ions access into the leaf cells with the transpiration, but the detrimental effect is a slow process taking weeks or months, eventually cause salt toxicity in leaves [46]. Plants were only exposed to salinity stress just for six days in our research, so further study is needed to reveal the long-term effect of salinity.

ABA is a critical hormone, which regulates plant growth, development, as well as responses to environmental stresses such as salinity, drought, heat, cold, wound, and pathogen [47,48]. Once ABA is produced, ABA-bound receptors bind ABA, inhibit *PP2Cs*, such as *ABI1* and *ABI2*, thereby activate *SnRK2s* [49]. The *SnRK2s* are involved in plant responses to abiotic stresses, which can phosphorylate the ABA-responsive element binding factor (ABF) and trigger the expression of ABA-responsive genes [50]. In our study, three ABA receptors, *PYLs*, were significantly down-regulated in roots or leaves (Figure 5), which is consistent with results in tea (*Camellia sinensis* L.) [51] and grape (*Vitis vinifera* L.) [52]. The down-regulation of *PYLs* can reduce plant sensitivity to ABA in response to salt stress and help plants adapt to salinity. The expressional differences of *PP2Cs* between roots and leaves in pomegranate are similar to other plants, which indicate that *PP2Cs* may participate in response to salinity via different strategies [52,53]. *PYR/PYLs* may regulate *SnRK2s* directly or indirectly, however, whether *SnRK2s* responding to various abiotic stresses in an ABA-independent or ABA-dependent pathway needs further investigation.

Ca^{2+} referred as a second messenger plays a very important role in many stress-related signaling transduction pathways. Under various abiotic stresses, over-accumulation of ions induce a temporary fluctuations in plant cytosolic ($[Ca^{2+}]_{cyt}$) levels [54]. The genes involved in CIPKs and CDPKs pathways, such as *ACAs*, *CAXs*, *GLRs*, *CaM/CMLs*, *CBLs* etc., participate in transporting and binding Ca^{2+}, sensing and relaying signals in plant cell under salinity stress [55,56]. In our study, we found two *ACAs* and two *CAXs* located on vacuoles were up-regulated in T2R, and the majority of *CaM/CMLs* were up-regulated in roots but down-regulated in leaves (Figure 6). Previous reports revealed a central role for CaMs in the regulation of Ca^{2+} channels and pumps, like CNGCs and ACAs [57,58]. The negative effect of CaM on CNGC activity provides a direct feedback pathway to restrict the influx of Ca^{2+} into plant cells [58]. The CaM stimulates the activity of these *ACAs* by preventing their auto-inhibition [58]. Collectively, under salt stress, pomegranate plants restrain the excess influx of Ca^{2+} into root cells under salinity. Meanwhile, it coped with the excess Ca^{2+} via removing Ca^{2+} from the cytosol by *ACAs* and *CAXs* in plasma membrane, including efflux of excess Ca^{2+} into the outer rhizosphere and/or the influx into vacuoles [59,60].

Specifically, transcription factors, such as NAC, MYB, AP2/ERF, bHLH, WRKY, GRAS family genes, regulated the expression of downstream genes to cope with various stresses [17]. Our study indicated that the TFs of pomegranate participated in the response to salinity with different patterns. For example, most of *NAC* and *C2H2* genes were up-regulated in roots, while most of *MYB* and *MYB-related* genes were down-regulated in leaves (Table 2; Table S6). Interestingly, nine *GRASs* were all up-regulated in T2R, which is mainly due to that these proteins play roles in plant development, including root development, axillary shoot development, and maintenance of the shoot apical meristem [61]. We also found 57 DEGs coding for salt-inducted effectors, including two *SODs*, two *APXs*, nineteen *PODs*, five *LEAs*, five *AQPs*, ten *HSFs*, three *AKTs*, and one *HKT* (Table S7). Two *APXs* were up-regulated and two *SODs* were down-regulated in leaves. Fourteen and nine *PODs* were suppressed in roots and leaves, respectively. Most of *AQPs* (4 of 5) were down-regulated, while all of *LEAs* (5 of 5) and majority of *HSFs* (9 of 10) were up-regulated in pomegranate tissues under salt stress. *LEAs* and *HSPs* were reported to prevent protein denaturation and maintain cell membrane fluidity under stress [62]. Then the DEGs of *LEA* and *HSF* in pomegranate contributed to mitigating salt stress. The expressions of three *AKTs* and one *HKT* were significantly down-regulated, which suggested that the influxes of K^+ and Na^+ in the cell were restricted [63].

Briefly, when pomegranate plants are exposed to high salinity, Na^+ enters the cell and simultaneously generates stress signals. The messengers Ca^{2+}, ROSs (reactive oxygen species), and ABA transduced the signals, and then the cascades, including Ca^{2+}-sensors, MAPK cascades and TFs, involved in secondary and tertiary regulatory networks were activated (Figure S4). TFs regulated the expression of downstream functional genes, such as *HSPs*, *LEAs*, *AQPs*, *SODs*, and *APXs*, to cope with salt stress.

5. Conclusions

In summary, our research sheds light on the pomegranate molecular response mechanisms to salinity. The differentially expressed genes could also provide references for selecting salt-tolerant breeding materials in pomegranate breeding processes.

Supplementary Materials
Figure S1: The images of untreated and treated plants. Figure S2: Pearson correlation of all samples, Figure S3: Pathways classification into cellular process, environmental information processing, genetic processing, metabolism and organismal system based on the KEGG analysis, Figure S4: Signaling networks in pomegranate under salt stress, Table S1: The primers used for qPCR validation of 15 DEGs, Table S2: Total numbers of DEGs assigned to databases, Table S3: DEGs involved in up-regulated patterns of pomegranate roots, Table S4: DEGs involved in down-regulated patterns of pomegranate roots, Table S5: DEGs involved in expressional patterns of pomegranate leaves, Table S6: The numbers of differently expressed TFs in pomegranate under NaCl stress, Table S7: The salt-inducted DEGs in pomegranate under NaCl stress.

Author Contributions: Conceptualization, C.L. and Z.Y.; methodology, C.L. and Y.Z.; formal analysis, C.L., Y.Z., J.W., and X.Z.; investigation, C.L., Y.Z., X.Z., and J.W.; writing—original draft preparation, C.L.; writing—review and editing, C.L., X.Z., M.G., and Z.Y.; supervision, M.G. and Z.Y.; Funding acquisition, Z.Y. All authors have read and agreed to the published version of the manuscript.

References

1. Li, J.; Pu, L.; Zhu, M.; Zhang, R. The present situation and hot issues in the salt-affected soil research. *Acta Geogr. Sin.* **2012**, *67*, 1233–1245.
2. Zhu, J.K. Plant salt tolerance. *Trends Plant Sci.* **2001**, *6*, 66–71. [CrossRef]
3. Pandey, P.; Ramegowda, V.; Senthil-Kumar, M. Shared and unique responses of plants to multiple individual stresses and stress combinations: Physiological and molecular mechanisms. *Front. Plant Sci.* **2015**, *6*, 723. [CrossRef]
4. Munns, R.; Tester, M. Mechanisms of salinity tolerance. *Annu. Rev. Plant Biol.* **2008**, *59*, 651–681. [CrossRef]
5. Zhang, H.; Han, B.; Wang, T.; Chen, S.; Li, H.; Zhang, Y.; Dai, S. Mechanisms of plant salt response: Insights from proteomics. *J. Proteome Res.* **2011**, *11*, 49–67. [CrossRef]
6. Bui, E.N. Soil salinity: A neglected factor in plant ecology and biogeography. *J. Arid Environ.* **2013**, *92*, 14–25. [CrossRef]
7. Apse, M.P.; Blumwald, E. Na^+ transport in plants. *FEBS Lett.* **2007**, *581*, 2247–2254. [CrossRef]
8. Munns, R.; James, R.A.; Gilliham, M.; Flowers, T.J.; Colmer, T.D. Tissue tolerance: An essential but elusive trait for salt-tolerant crops. *Funct. Plant Biol.* **2016**, *43*, 1103–1113. [CrossRef]
9. Attia, H.; Karray, N.; Msilini, N.; Lachaâl, M. Effect of salt stress on gene expression of superoxide dismutases and copper chaperone in *Arabidopsis thaliana*. *Biol. Plant.* **2011**, *55*, 159–163. [CrossRef]
10. Shafi, A.; Chauhan, R.; Gill, T.; Swarnkar, M.K.; Sreenivasulu, Y.; Kumar, S.; Kumar, N.; Shankar, R.; Ahuja, P.S.; Singh, A.K. Expression of *SOD* and *APX* genes positively regulates secondary cell wall biosynthesis and promotes plant growth and yield in *Arabidopsis* under salt stress. *Plant Mol. Biol.* **2015**, *87*, 615–631. [CrossRef]
11. Kumar, S.; Beena, A.S.; Awana, M.; Singh, A. Salt-induced tissue-specific cytosine methylation downregulates expression of *HKT* genes in contrasting wheat (*Triticum aestivum* L.) genotypes. *DNA Cell Biol.* **2017**, *36*, 283–294. [CrossRef] [PubMed]

12. Yokoi, S.; Quintero, F.J.; Cubero, B.; Ruiz, M.T.; Bressan, R.A.; Hasegawa, P.M.; Pardo, J.M. Differential expression and function of *Arabidopsis thaliana NHX* Na^+/H^+ antiporters in the salt stress response. *Plant J.* **2010**, *30*, 529–539. [CrossRef] [PubMed]
13. Hu, W.; Yuan, Q.; Wang, Y.; Cai, R.; Deng, X.; Wang, J.; Zhou, S.; Chen, M.; Chen, L.; Huang, C. Overexpression of a wheat aquaporin gene, *TaAQP8*, enhances salt stress tolerance in transgenic tobacco. *Plant Cell Physiol.* **2012**, *53*, 2127–2141. [CrossRef] [PubMed]
14. Duan, J.; Cai, W. *OsLEA3-2*, an abiotic stress induced gene of rice plays a key role in salt and drought tolerance. *PLoS ONE* **2012**, *7*, e45117. [CrossRef]
15. Zhou, A.; Liu, E.; Ma, H.; Feng, S.; Gong, S.; Wang, J. NaCl-induced expression of *AtVHA-c5* gene in the roots plays a role in response of Arabidopsis to salt stress. *Plant Cell Rep.* **2018**, *37*, 443–452. [CrossRef]
16. Hasegawa, P.M.; Bressan, R.A.; Zhu, J.K.; Bohnert, H.J. Plant cellular and molecular responses to high salinity. *Annu. Rev. Plant Physiol. Plant Mol. Biol.* **2000**, *51*, 463–499. [CrossRef]
17. Deinlein, U.; Stephan, A.B.; Horie, T.; Luo, W.; Xu, G.; Schroeder, J.I. Plant salt-tolerance mechanisms. *Trends Plant Sci.* **2014**, *19*, 371–379. [CrossRef]
18. Guo, S.M.; Tan, Y.; Chu, H.J.; Sun, M.X.; Xing, J.C. Transcriptome sequencing revealed molecular mechanisms underlying tolerance of *Suaeda salsa* to saline stress. *PLoS ONE* **2019**, *14*, e0219979. [CrossRef]
19. Yuan, Z.; Fang, Y.; Zhang, T.; Fei, Z.; Han, F.; Liu, C.; Liu, M.; Xiao, W.; Zhang, W.; Wu, S. The pomegranate (*Punica granatum* L.) genome provides insights into fruit quality and ovule developmental biology. *Plant Biotechnol. J.* **2018**, *16*, 1363–1374. [CrossRef]
20. Silva, J.A.T.D.; Rana, T.S.; Narzary, D.; Verma, N.; Meshram, D.T.; Ranade, S.A. Pomegranate biology and biotechnology: A review. *Sci. Hortic.* **2013**, *160*, 85–107. [CrossRef]
21. Holl, D.; Hatib, K.; Bar-Ya'Akov, I. Pomegranate: Botany, horticulture, breeding. *Hortic. Rev.* **2009**, *35*, 127–191.
22. Karimi, H.R.; Hasanpour, Z. Effects of salinity and water stress on growth and macro nutrients concentration of pomegranate (*Punica granatum* L.). *J. Plant Nutr.* **2014**, *37*, 1937–1951. [CrossRef]
23. Okhovatianardakani, A.R.; Mehrabanian, M.; Dehghani, F.; Akbarzadeh, A. Salt tolerance evaluation and relative comparison in cuttings of different pomegranate cultivars. *Plant Soil Environ.* **2010**, *56*, 176–185. [CrossRef]
24. Liu, C.; Yan, M.; Huang, X.; Yuan, Z. Effects of salt stress on growth and physiological characteristics of pomegranate (*Punica granatum* L.) cuttings. *Pak. J. Bot.* **2018**, *50*, 457–464.
25. Feng, Z.T.; Deng, Y.Q.; Fan, H.; Sun, Q.J.; Sui, N.; Wang, B.S. Effects of NaCl stress on the growth and photosynthetic characteristics of *Ulmus pumila* L. seedlings in sand culture. *Photosynthetica* **2014**, *52*, 313–320. [CrossRef]
26. Patel, R.K.; Jain, M. NGS QC Toolkit: A toolkit for quality control of next generation sequencing data. *PLoS ONE* **2012**, *7*, e30619. [CrossRef]
27. Kim, D.; Pertea, G.; Trapnell, C.; Pimentel, H.; Kelley, R.; Salzberg, S.L. TopHat2: Accurate alignment of transcriptomes in the presence of insertions, deletions and gene fusions. *Genome Biol.* **2013**, *14*, R36. [CrossRef]
28. Pertea, M.; Kim, D.; Pertea, G.M.; Leek, J.T.; Salzberg, S.L. Transcript-level expression analysis of RNA-seq experiments with HISAT, StringTie and Ballgown. *Nat. Protoc.* **2016**, *11*, 1650–1667. [CrossRef]
29. Götz, S.; Garcíagómez, J.M.; Terol, J.; Williams, T.D.; Nagaraj, S.H.; Nueda, M.J.; Robles, M.; Talón, M.; Dopazo, J.; Conesa, A. High-throughput functional annotation and data mining with the Blast2GO suite. *Nucleic Acids Res.* **2008**, *36*, 3420–3435. [CrossRef]
30. Ernst, J.; Nau, G.J.; Bar-Joseph, Z. Clustering short time series gene expression data. *Bioinformatics* **2005**, *21*, i159–i168. [CrossRef]
31. Mao, X.; Cai, T.; Olyarchuk, J.G.; Wei, L. Automated genome annotation and pathway identification using the KEGG Orthology (KO) as a controlled vocabulary. *Bioinformatics* **2005**, *21*, 3787–3793. [CrossRef] [PubMed]
32. Wang, J.; Zhu, J.; Zhang, Y.; Fan, F.; Li, W.; Wang, F.; Zhong, W.; Wang, C.; Yang, J. Comparative transcriptome analysis reveals molecular response to salinity stress of salt-tolerant and sensitive genotypes of indica rice at seedling stage. *Sci. Rep.* **2018**, *8*, 2085. [CrossRef] [PubMed]
33. Zhou, Y.; Yang, P.; Cui, F.; Zhang, F.; Luo, X.; Xie, J. Transcriptome analysis of salt stress responsiveness in the seedlings of Dongxiang wild rice (*Oryza rufipogon* Griff.). *PLoS ONE* **2016**, *11*, e0146242. [CrossRef]
34. Munns, R.; James, R.A.; Läuchli, A. Approaches to increasing the salt tolerance of wheat and other cereals. *J. Exp. Bot.* **2006**, *57*, 1025–1043. [CrossRef]

35. Munns, R. Genes and salt tolerance. Bringing them together. *New Phytol.* **2005**, *167*, 645–663. [CrossRef]
36. Monika, B.; Mikael, B.; Basia, V.; Atef, A.O.; Payam, F.; Dennis, J.; Ottow, E.A.; Cullmann, A.D.; Joachim, S.; Jaakko, K.R. Linking the salt transcriptome with physiological responses of a salt-resistant Populus species as a strategy to identify genes important for stress acclimation. *Plant Physiol.* **2010**, *154*, 1697–1709.
37. Ma, S.; Gong, Q.; Bohnert, H.J. Dissecting salt stress pathways. *J. Exp. Bot.* **2006**, *57*, 1097–1107. [CrossRef]
38. Huang, J.; Xiang, L.; Hao, Y.; Chen, S.; Zhang, W.; Huang, R.; Zheng, Y. Transcriptome characterization and sequencing-based identification of salt-responsive genes in *Millettia pinnata*, a semi-mangrove plant. *DNA Res.* **2012**, *19*, 195–207. [CrossRef]
39. Le Gall, H.; Philippe, F.; Domon, J.-M.; Gillet, F.; Pelloux, J.; Rayon, C. Cell wall metabolism in response to abiotic stress. *Plants* **2015**, *4*, 112–166. [CrossRef]
40. Fatehi, F.; Hosseinzadeh, A.; Alizadeh, H.; Brimavandi, T.; Struik, P.C. The proteome response of salt-resistant and salt-sensitive barley genotypes to long-term salinity stress. *Mol. Biol. Rep.* **2012**, *39*, 6387–6397. [CrossRef]
41. Bushman, B.S.; Amundsen, K.L.; Warnke, S.E.; Robins, J.G.; Johnson, P.G. Transcriptome profiling of Kentucky bluegrass (*Poa pratensis* L.) accessions in response to salt stress. *BMC Genom.* **2016**, *17*, 48. [CrossRef] [PubMed]
42. Wang, X.; Chang, L.; Wang, B.; Wang, D.; Li, P.; Wang, L.; Yi, X.; Huang, Q.; Peng, M.; Guo, A. Comparative proteomics of *Thellungiella halophila* leaves from plants subjected to salinity reveals the importance of chloroplastic starch and soluble sugars in halophyte salt tolerance. *Mol. Cell. Proteom.* **2013**, *12*, 2174–2195. [CrossRef] [PubMed]
43. Boriboonkaset, T.; Theerawitaya, C.; Yamada, N.; Pichakum, A.; Supaibulwatana, K.; Cha-um, S.; Takabe, T.; Kirdmanee, C. Regulation of some carbohydrate metabolism-related genes, starch and soluble sugar contents, photosynthetic activities and yield attributes of two contrasting rice genotypes subjected to salt stress. *Protoplasma* **2013**, *250*, 1157–1167. [CrossRef] [PubMed]
44. Zhang, A.; Han, D.; Wang, Y.; Mu, H.; Zhang, T.; Yan, X.; Pang, Q. Transcriptomic and proteomic feature of salt stress-regulated network in Jerusalem artichoke (*Helianthus tuberosus* L.) root based on de novo assembly sequencing analysis. *Planta* **2018**, *247*, 715–732. [CrossRef] [PubMed]
45. Sun, Y.; Niu, G.; Masabni, J.G.; Ganjegunte, G. Relative salt tolerance of 22 pomegranate (*Punica granatum*) cultivars. *HortScience* **2018**, *53*, 1513–1519. [CrossRef]
46. Aroca, R.; Porcel, R.; Ruiz-Lozano, J.M. Regulation of root water uptake under abiotic stress conditions. *J. Exp. Bot.* **2011**, *63*, 43–57. [CrossRef] [PubMed]
47. Finkelstein, R.R.; Gampala, S.S.; Rock, C.D. Abscisic acid signaling in seeds and seedlings. *Plant Cell* **2002**, *14*, S15–S45. [CrossRef]
48. Shinozaki, K.; Yamaguchi-Shinozaki, K. Molecular responses to dehydration and low temperature: Differences and cross-talk between two stress signaling pathways. *Curr. Opin. Plant Biol.* **2000**, *3*, 217–223. [CrossRef]
49. Fan, W.; Zhao, M.; Li, S.; Bai, X.; Li, J.; Meng, H.; Mu, Z. Contrasting transcriptional responses of *PYR1/PYL/RCAR* ABA receptors to ABA or dehydration stress between maize seedling leaves and roots. *BMC Plant Biol.* **2016**, *16*, 99. [CrossRef]
50. Zhang, H.; Li, W.; Mao, X.; Jing, R.; Jia, H. Differential activation of the wheat SnRK2 family by abiotic stresses. *Front. Plant Sci.* **2016**, *7*, 420. [CrossRef]
51. Wan, S.; Wang, W.; Zhou, T.; Zhang, Y.; Chen, J.; Xiao, B.; Yang, Y.; Yu, Y. Transcriptomic analysis reveals the molecular mechanisms of *Camellia sinensis* in response to salt stress. *Plant Growth Regul.* **2018**, *84*, 481–492. [CrossRef]
52. Boneh, U.; Biton, I.; Zheng, C.; Schwartz, A.; Ben-Ari, G. Characterization of potential ABA receptors in *Vitis vinifera*. *Plant Cell Rep.* **2012**, *31*, 311–321. [CrossRef] [PubMed]
53. Cao, J.; Min, J.; Peng, L.; Chu, Z. Genome-wide identification and evolutionary analyses of the PP2C gene family with their expression profiling in response to multiple stresses in *Brachypodium distachyon*. *BMC Genom.* **2016**, *17*, 175. [CrossRef] [PubMed]
54. McCormack, E.; Tsai, Y.-C.; Braam, J. Handling calcium signaling: Arabidopsis *CaMs* and *CMLs*. *Trends Plant Sci.* **2005**, *10*, 383–389. [CrossRef] [PubMed]
55. Dang, Z.H.; Zheng, L.L.; Jia, W.; Zhe, G.; Wu, S.B.; Zhi, Q.; Wang, Y.C. Transcriptomic profiling of the salt-stress response in the wild recretohalophyte *Reaumuria trigyna*. *BMC Genom.* **2013**, *14*, 29. [CrossRef] [PubMed]
56. Yang, Y.; Zhang, C.; Tang, R.; Xu, H.; Lan, W.; Zhao, F.; Luan, S. Calcineurin B-Like proteins CBL4 and CBL10 mediate two independent salt tolerance pathways in Arabidopsis. *Int. J. Mol. Sci.* **2019**, *20*, 2421. [CrossRef]

57. Virdi, A.S.; Singh, S.; Singh, P. Abiotic stress responses in plants: Roles of calmodulin-regulated proteins. *Front. Plant Sci.* **2015**, *6*, 809. [CrossRef]
58. Cheval, C.; Aldon, D.; Galaud, J.-P.; Ranty, B. Calcium/calmodulin-mediated regulation of plant immunity. *Biochim. Biophys. Acta Mol. Cell Res.* **2013**, *1833*, 1766–1771. [CrossRef]
59. Apse, M.P.; Sottosanto, J.B.; Blumwald, E. Vacuolar cation/H^+ exchange, ion homeostasis, and leaf development are altered in a T-DNA insertional mutant of *AtNHX1*, the Arabidopsis vacuolar Na^+/H^+ antiporter. *Plant J.* **2010**, *36*, 229–239. [CrossRef]
60. Huda, K.M.; Banu, M.S.; Tuteja, R.; Tuteja, N. Global calcium transducer P-type Ca^{2+}-ATPases open new avenues for agriculture by regulating stress signalling. *J. Exp. Bot.* **2013**, *64*, 3099–3109. [CrossRef]
61. Bolle, C. The role of GRAS proteins in plant signal transduction and development. *Planta* **2004**, *218*, 683–692. [CrossRef] [PubMed]
62. Hoekstra, F.A.; Golovina, E.A.; Buitink, J. Mechanisms of plant desiccation tolerance. *Trends Plant Sci.* **2001**, *6*, 431–438. [CrossRef]
63. Xu, M.; Chen, C.; Cai, H.; Wu, L. Overexpression of *PeHKT1*; 1 Improves Salt Tolerance in Populus. *Genes* **2018**, *9*, 475. [CrossRef] [PubMed]

10

Physiological Responses of Selected Vegetable Crop Species to Water Stress

Eszter Nemeskéri * and Lajos Helyes

Institute of Horticulture, Szent István University, H-2100 Gödöllő, Hungary
* Correspondence: Nemeskeri.Eszter@mkk.szie.hu

Abstract: The frequency of drought periods influences the yield potential of crops under field conditions. The change in morphology and anatomy of plants has been tested during drought stress under controlled conditions but the change in physiological processes has not been adequately studied in separate studies but needs to be reviewed collectively. This review presents the responses of green peas, snap beans, tomatoes and sweet corn to water stress based on their stomatal behaviour, canopy temperature, chlorophyll fluorescence and the chlorophyll content of leaves. These stress markers can be used for screening the drought tolerance of genotypes, the irrigation schedules or prediction of yield.

Keywords: vegetable crops; stomatal conductance; canopy temperature; chlorophyll fluorescence; SPAD; water stress

---

1. Introduction

As a result of climate change, the increasing atmospheric CO_2 enhances the photosynthesis capacity and improves water use efficiency therefore the amount of yield will increase in most of vegetable crops, however its advantage cannot be shown under limited water and nitrogen deficiency. The high temperature during reproductive growth is harmful for many important vegetable crops, such as tomatoes, peppers, beans and sweet corn, and yield reduction will probably occur [1]. The frequency of drought periods decreases vegetable yield and quality, however soluble solid content of produce may be increased by water deficiency in some crops [2,3]. Nevertheless, the occurrence of excess precipitation causes waterlogging in soils, the symptoms of which are similar to water deficit. Soil waterlogging impedes the oxygen supply and respiration of roots, water uptake and hydraulic conductance which results in stomatal closure [4,5]. Under these conditions the stomatal closure results in a reduction of net photosynthesis which is due to the decrease in stomatal conductance, chlorophyll fluorescence and chlorophyll content of leaves [6]. Excess water causes a decline in grain filling and grain weight of corn leading to decreased yield [7]. However, water stress commonly refers to water deficits not excess water.

The selection of the vegetable crops grown under field conditions for the investigation was based on their production in the world and Europe. During the last twenty years from 1997 to 2017 the growing area of tomatoes increased intensively, that of green peas increased moderately while the growing area of snap beans and sweet corn increased slowly in the world. During this time in Europe the growing area of tomatoes and snap beans decreased from 650.4 to 496.2 thousand ha while that of green peas increased slightly from 208 to 212 thousand ha and sweet corn's increased intensively (from 50.5 to 110.1 thousand ha) (FAOSTAT 2017 [8]. In Hungary, the production of green peas and sweet corn is performed in large field growing areas (19.5 thousand hectares and 34.5 thousand hectares, respectively) while snap beans are grown in smaller ones (1.6 thousand hectares) (FAOSTAT 2017) [8]. The other aspect of the selection was the sensitivity of plant species to water stress.

Corn, soybeans, beans and peas are considered to be moderately water stress sensitive while tomatoes belong to the extremely drought sensitive group [9,10]. The responses of plant species significantly depend on the intensity and duration of stress and their stages of development. The spring-sown green pea utilizes the precipitation well (if there is any) and requires a low temperature during vegetative growth but during the flowering and seed development periods it is sensitive to water deficiency. The crops require a warm temperature, even though they have different ripening times, snap beans have short (60 days), sweet corn has medium (75–90 days) and tomatoes have long ripening times (110–130 days), their generative stages of development coincided with dry June and July, thus they require irrigation. Irrigation scheduling and the amount of irrigation water are determined by the water stress tolerance and water use of the plant varieties. The evaluation of drought tolerance in field conditions is difficult because low soil moisture and high air temperature stress generally occur together, and it is difficult to evaluate the responses separately. Drought under field conditions promotes the evapotranspiration and affects the photosynthesis, which leads to reduced yield [11]. Use of remote-sensing methods makes the measurement of physiological responses of varieties to various strong water stresses easy. These non-destructive methods help the breeder to select drought tolerant genotypes and the growers to measure the water deficit of plants and decide the time of irrigation.

The selection for water stress tolerance in traditional breeding is based on the suitability of performance under a series of environmental conditions using extensive statistical methods. This progress could be improved by the introduction of traits which contribute to the prediction of yield in the drought-prone environments. In this study, the effect of water stress on the plants and those physiological traits which influenced the yield are mainly demonstrated. Information was gathered on the physiological responses of selected vegetable species to drought to analyse their use in breeding for high and stable yield.

2. Water Stress during Growth of Vegetable Crops

Sensitivity of plants to water stress such as snap beans and green peas differs with the stages of development. During the early stages of vegetative growth most crops are less sensitive to water scarcity [2,3,12,13], but during the generative stage the water deficiency results in changes of many physiological traits [2,14–16], causing the disturbance of fertility and reduction of yield. During flowering of legume plants water stress increases the ratio of flower drop [17], decreases the pod numbers and seed abortion in the pods [18,19] and increases the ratio of curved pods [20]. Under water deficiency, bean plants produce shorter shoots and smaller leaves and decrease the length of pods [21]. Semi-leafless pea varieties have reduced leaf area that is presumed to have a low water use and they have higher water use efficiency (WUE) than traditional varieties with normal leaves [22]. In sweet corn, ear differentiation begins at the six- or eight-leaf stage growth when the water deficiency decreases the length of ears and the numbers of ear rows [23], but during tasselling the water deficiency causes significant yield reduction [24,25]. Tomatoes are most sensitive to water deficiency at fruit setting and intensive fruit development periods [3], when the increasing water stress resulted in a 25 to 50% decrease in the yield [10,26–29]. During early flowering of tomatoes, water scarcity causes flower shedding and lack of fertilization [30], and during fruit setting, plants with small sized fruits are produced [10,31].

The effect of water stress on morphology and anatomy of plants has been studied by several researchers under controlled conditions [6,32–35], however, the changes in physiological responses have been less investigated under field conditions. The physiological characteristics that have an important role in the defence against drought can be measured by remote sensing techniques using non-destructive methods in open field conditions. The leaf photosynthetic activity of plants can be monitored with measurement of chlorophyll content using a portable chlorophyll meter and chlorophyll fluorescence while the measurement of stomatal conductance indicates the severity of water stress [2,3,14]. Spectral vegetation indices such as the green normalized vegetation index

(GNDVI) and the normalized differential vegetation index (NDVI) were used for monitoring the growth of the plant to detect the water stress and for yield prediction [36–39]. Crop water stress index is determined by an infrared thermometry technique to indicate the change in canopy temperature of plants under water stress conditions. More physiological indices such as leaf water potential, relative water content, turgor potential, osmotic adjustment, difference between canopy and air temperature can also be used as a screening tools for testing the water stress tolerance of genotypes [40]. Studies have focused on the identification of drought tolerance-related traits using Quantitative Trait Locus (QTL)s and Marker Assisted Selection (MAS) techniques [41–44], however, the identification of the most relevant loci controlling drought tolerance and drought-related traits could be achieved by the integration of molecular genetics with physiology [45].

3. Drought Tolerance

Adaptive mechanisms promoting the survival of plants have been grouped into three categories; drought escape, drought avoidance and drought tolerance [46]. Drought escape is the ability of plants to accomplish their life cycle before the development of soil and plant water deficit. The varieties with early flowering and short maturity are able to escape drought [47], however, they are not drought tolerant in every case. The varieties with moderate drought sensitivity developed different defence strategies to avoid short- and long-term water stress which prevents the water loss in their cells and tissues. The essential defence mechanism against drought operating in the plants is the maintenance of the water status and the reduction of tissue water loss (Figure 1).

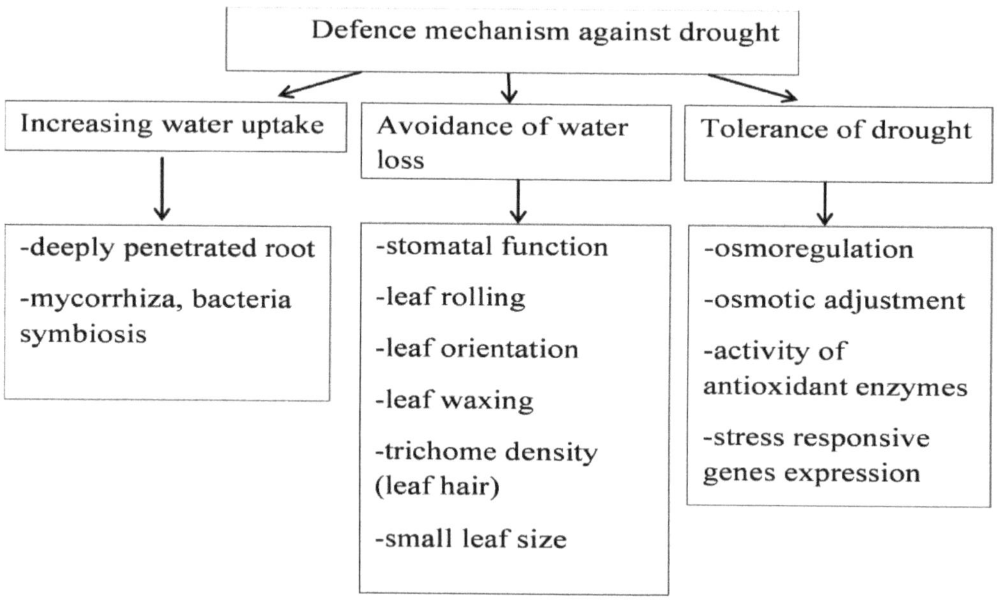

Figure 1. Defence mechanism against drought (Leonardis et al. [46]).

A well-developed deeply penetrating root system provides the water uptake and maintenance of water circulation inside the plant despite the low soil-moisture content. Nevertheless, in dry soil the lives of microorganisms are retarded when the activity of mycorrhiza living symbiotically with root nodules of legumes is low, which results in a decrease in the nitrogen uptake [48,49], therefore the growth of the plant is retarded. Long-term drought of soil accelerates the senescence of root nodules and production of reactive oxygen species (ROS) [50,51], therefore the nodule weight, root and shoot weight are decreased [52]. Water stress results in a change in the proportion of root weight as the ratio of root to shoot increases [53]. Under permanent low moisture content of soil, a 27–42% decrease in leaf weight and 12–27% decrease in specific leaf area of snap bean varieties were found [54]. Tomatoes are able to survive prolonged periods of low soil water content by the development of a deep root system [28,55]. In dry years, tomatoes inoculated with mycorrhiza easily endured the water scarcity, for example larger weight fruits and higher yield were produced by deficit irrigation than under non-irrigated conditions [56].

4. Reduction of Water Loss

Drought avoidance is the ability of plants to maintain high tissue water potential despite the deficiency of soil moisture. The mechanisms developed for the reduction of water loss are related to the duration of water stress.

During short-term water deficiency the leaf movement, deep penetrating roots with strong suction force and partial or total stomatal closure provide a decrease in the water loss. Leaf movement of plants not only protects from the photodamage caused by high irradiation but reduces the effective leaf area for transpiration [57]. Paraheliotropic movement of leaves occurs mainly in beans while leaf rolling is typical for maize. Fernandez and Castrillo [58] found that the extent of leaf rolling is linearly correlated with the water potential. During leaf rolling of maize the transpiration, stomatal conductance, intracellular CO_2 concentration and net photosynthetic rate decreased [59]. Pastenes et al. [60] found that the degree of paraheliotropic leaf movement was larger in the water stressed plants because of lower water potential, however, it also occurred in the water supplied crops. Deep, thick and dense roots intensively promote the use of available water and the optimal development of aboveground parts. During short-term water stress (<7 nap), abscisic acid (ABA) is produced in the roots then it is transported into the leaves where ABA induces the stomatal closure and thus decreases the water loss [61,62]. Partial or total stomatal closure restricts the transpiration therefore the water and nutrient uptake is decreased, which results in a decrease in photosynthesis and growth of plants [63]. Stomatal responses of legume species are different; under water deficiency, beans have a rapid and complete stomatal closure generating the stomatal conductivity and photosynthesis decreases significantly, whereas in cowpeas (*Vigna unguiculata*), the stomata remain partially open and have a lower decrease in their net photosynthetic rate under the same conditions [64]. Under moderate water deficit conditions, the growth of snap beans was already retarded, and their leaf area decreased while the leaf area index (LAI) of sweet corn did not change [3,15]. Nevertheless, water scarcity did not influence the leaf area of tomatoes [65] but heat and water stress up to 6 days already significantly decreased the weight of shoots and roots of tomato seedlings under a controlled environment [66].

During long-term water deficiency, plants try to prevent the dehydration of cells of vegetative and generative organs with some morphological and physiological changes. Trichome density (leaf hairs) on the leaf protects the tissues from sunlight injury, decreases the water loss by evaporation and enhances the transpiration resistance [67]. Under water stress conditions, a lower number of trichomes was found only on the basal zone of leaves on both surfaces in comparison with irrigated plants [68]. However, according to Nobel [69], the length of trichomes can be more important than their frequency. The epicuticular wax layer of the leaf controls the water flow across the cuticle and protects from high radiation and prevents damage caused by UV light. Water stress induced the increase in the wax layer on the leaf surface of peas and the wax-rich varieties had significantly lower canopy temperature [70].

Drought tolerance is the ability of the plants to endure the long-term moisture deficit and survive the water loss. When the morphological changes seem to be insufficient to avoid the water deficiency, biochemical processes of plants are activated to maintain the osmotic adjustment and the structure of cell membranes in order to avoid cell dehydration. Decreasing the water potential of leaves induces the accumulation of different osmotic compounds such as sugars, amino acids and quaternary ammonium compounds. The osmotic pressure of cells is increased by the accumulation of osmotic compounds because water movement into the cells and tissues provides the maintenance of turgor [71]. It was found that peas and castor beans exposed to water deficit accumulate a significant amount of soluble sugars and proline [72,73], and the raffinose and sucrose level of leaves are significantly increased by water stress during flowering of snap beans [74]. Action of enzymatic and non-enzymatic antioxidants is intensified to alleviate the oxidative damages in the tissues. Concerted operation of numerous water soluble antioxidant compounds (ACW) contributes to the adaptation of plants to environmental stresses. The level of ACW in the leaves is influenced by stomatal closure because it is related to ascorbic acid redox potential of guard cells [75]. In snap bean genotypes that have a high ACW level in leaves during the flowering and pod development periods, this provides a defence against water deficiency [74].

4.1. Regulation of Water Circulation in Plants under Drought

Many physiological processes are activated to mitigate the water loss of plants (Table 1). Transpiration is restrained as a result of stomatal closure and by decreasing leaf area. Stomata play an important role in the regulation of transpiration and CO_2 uptake. Use of light energy gathered by photosynthesis determines the growth and biomass production of plants. In these processes, the stomatal characteristics such as stomatal size, number and ratio of stomata on abaxial and adaxial surfaces significantly affect the C assimilation and water use efficiency (WUE) [76,77]. The higher stomatal density on the abaxial surface of the leaf is related to a higher water use efficiency [78], while those existing on upper epidermis (adaxial surface) of the leaf influenced the water use of plants [15]. Nevertheless, the number of stomata on both epidermis of leaves changes significantly depending on the variety and water supply.

Table 1. Physiological traits relevant for response to drought.

Physiological Traits	Effect Relevant for Yield	Alteration under Stress	References
Size and density of stomata	relation to leaf water potential and water consumption	increase/decrease depending on species	Hardy et al. [79], Nemeskéri et al. [14,15]
Leaf temperature	relation to transpiration	increase	Helyes et al. [31], DeJonge et al. [80]
Stomatal conductance	correlation with water consumption, decrease in individual yield	decrease in diffusion of CO_2, stomatal resistance increases	Jones [81], Nemeskéri et al. [14,15]
Photosynthetic capacity	modulation of activity enzymes of Calvin cycle	reduction under stress	Lawlor and Cornic [82]
Change in chlorophyll fluorescence	alteration of quantum yield of PSII photosystem	decrease in Fv/Fm under severe drought	Flagella et al. [83], Pol et al. [84], Yordanov et al. [85]
Chlorophyll content of leaf	decrease in photosynthesis	decrease under stress, relative chlorophyll content (SPAD value) can increase	Nankishore and Farrell [32], Nemeskéri et al. [16]
Reduced growth rate		leaf area reduces, biomass decreases	Ghanbari et al. [86], Guida et al. [87]

4.1.1. Stomatal Characteristics

More stomata (134–195 stomata mm^{-2}) were observed on the abaxial surface of tomato leaves but it was significantly less (40–62 mm^{-2}) on the adaxial surface of leaves [76]. A significant difference can be demonstrated in stomatal density of leaves between snap beans, green peas and sweet corn grown under non-irrigated and deficit irrigated (50% water deficiency) conditions (Table 2). On the basis of 3 year experiments, on the lower epidermis of leaves the stomata density was significantly higher for snap beans under moderate and severe water stress and it was similarly high for sweet corn only in severe water deficiency, but no difference could be shown for green peas in comparison with the optimal water supplied plants [14,15,88]. On the upper epidermis of leaves more and larger sized stomata can be found for snap beans exposed to drought while there were fewer similar sized stomata for the green peas compared to the irrigated plants (Table 2). However, under water scarcity, significant differences in stomata number and size can be detected between the varieties. Under non-irrigated conditions, the size of stomata on the upper (adaxial) surface of leaves of green-podded bean varieties was smaller by 5–12%, but more of them were found than on optimal water supplied plants. Nevertheless, yellow-podded snap bean varieties had 13–18% larger sized stomata on the adaxial surface of leaves of plants exposed to water deficiency in comparison with the irrigated plants [15]. A larger stomatal density was observed for late ripening green pea varieties [14] and late ripening sweet corn hybrids under water scarcity [88] than for the short duration ones. Nevertheless, the distribution and size of stomata can be different on both areas and surfaces of the same leaf. Various number and sized stomata were detected on different areas of leaves of tomatoes; on the abaxial surface of leaves and their apical and middle areas, larger sized (32–34 μm) and more stomata were found than that on the adaxial surface. The stomata on the apical areas of leaves responded sensitively to water deficiency in that they showed fewer and larger sized stomata on the adaxial surface of leaves than for well-watered plants [68]. A significant correlation between the stomatal density and stomatal conductance ($r^2 = 0.958$) was established in tomatoes. According to this correlation, 130 stomata mm^{-2} was associated with high stomatal conductance (0.1 mol H_2O m^{-2} s^{-1}) [76]. Others [89] found that the relationship between stomatal density and WUE was positive and the size of stomata correlated negatively with the WUE for grass peas.

Table 2. Size and density of stomata measured during generative stages of vegetable crops under different water supplies Source: modified from Nemeskéri et al. [14,15,88].

Species	Water Supply	Lower Epidermis		Upper Epidermis	
		Stomata mm^{-2} *	Size of Stomata μ	Stomata mm^{-2}	Size of Stomata μ
Snap bean	I0	387.79	23.72	104.81	30.51
	DI	374.17	-	93.41	-
	WI	331.22	24.90	78.61	29.64
	average	364.39	24.31	92.28	30.08
Green pea	I0	214.29	25.82	165.70	25.79
	DI	214.65	25.48	170.86	24.68
	WI	214.74	24.35	194.72	25.21
	average	214.56	25.22	177.10	25.23
Sweet corn	I0	145.61	-	95.23	-
	DI	140.79	-	94.98	-
	WI	136.13	-	93.73	-
	average	140.84	50.04	94.65	53.22

* Based on average of years [14,15,88], μ = micron, I0 = non-irrigation, DI = deficit irrigation, WI = optimal water supply.

4.1.2. Canopy Temperature-Transpiration

Under high photosynthetically active radiation (PAR) water deficit combined with high temperature results in an increase in leaf temperature and air temperature oscillation (±3–4 °C) due to the opening and closing of stomata [53]. Stomata closure triggers the decrease in the transpiration which contributes to the increase in canopy temperature of plants. One of the tasks of transpiration is to keep the temperature of plants at a favourable level for life processes. Decreasing transpiration causes the temperature of plants to increase. If soil water content is sufficient for the plant stand, the difference between canopy temperature and air temperature is zero or negative, but if the plants suffer from water stress this value is positive. An increase of 1 °C in canopy temperature related to a 10% decrease in the transpiration [31]. Size and stomatal density of genotypes are different thereby the transpiration varies in intensity which correlates with the difference of the canopy temperature of plants. Changes in canopy temperature have often been used to signal water stress [90] to evaluate the drought tolerance of bean genotypes and the difference in canopy temperature and air temperature was used for the real time irrigation [91–94]. During the daytime the canopy temperature rises along with the daily air temperature and radiation as the available soil water changes. The lowest value of crop water stress index (CWSI) of maize was measured at 10:00 and 11:00 and it was the largest between 12:00 and 13:00 [95]. Under water deficiency, the canopy temperature of both snap beans and tomatoes was higher than the air temperature from 09:00 to 15:00 however, that of tomatoes was higher than the air temperature only at 12:00 and 13:00 [96]. Under water stress conditions, between 09:00 and 15:00, the canopy temperature of snap beans was higher by 3.8 °C than the air temperature while it was lower by 1.6 °C in well-watered plants [93]. When the amount of available moisture in the soil for the plants decreases, then the transpiration is limited depending on the air temperature, which results in an increasing canopy temperature. Under moderate water deficiency, at 25–50% available soil water the canopy temperature of snap beans almost coincided with the air temperature (Figure 2a) that denoted the need for irrigation [93]. Nevertheless, the available soil water below 25% was not able to satisfy the water demands of plants. In this case the cooling of the canopy was not shown by transpiration and the temperature on the foliage surface was higher than the air temperature by 2.5 °C on average, indicating the plants suffered from water stress (Figure 2b) [93].

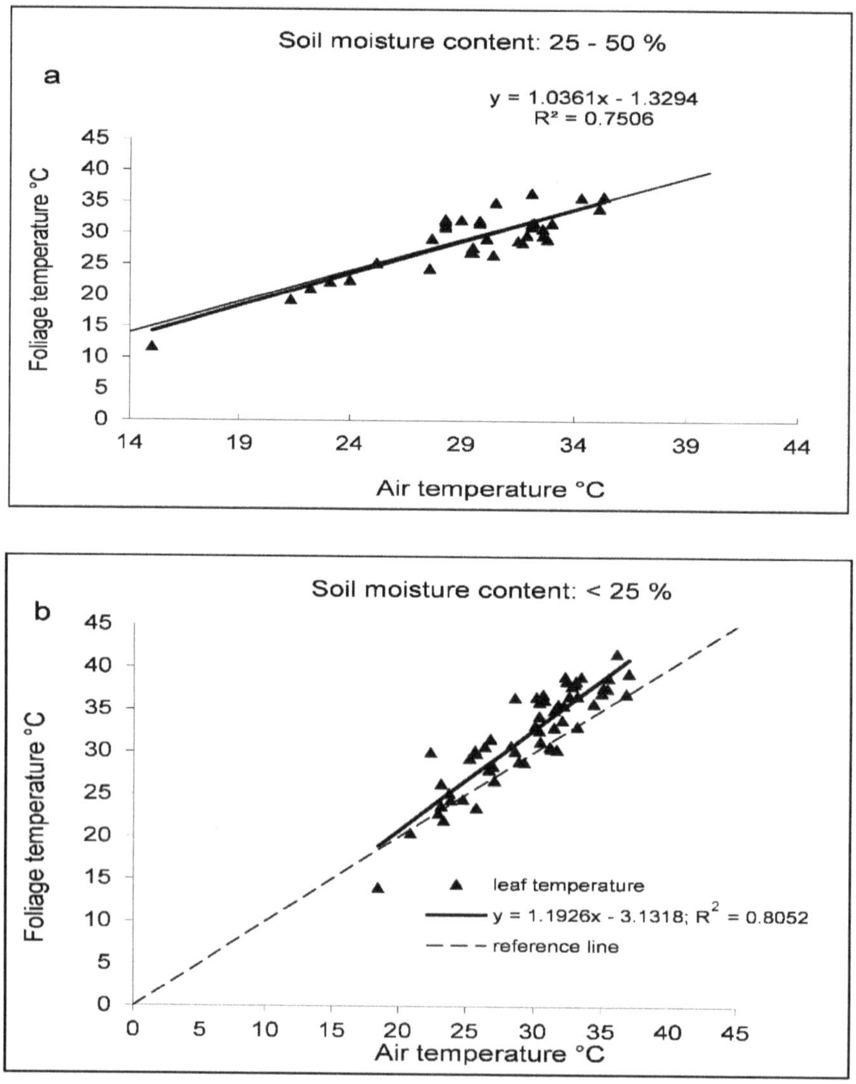

Figure 2. Relationship between air and canopy temperature for snap beans under water deficit (**a**) and severe water stress (**b**). The thick line shows the increase in leaf temperature compared to air temperature (broken line) Source: Helyes et al. [93].

Tomatoes seemed to better use deep soil moisture with deep, strong suction force roots than the shallow rooted snap beans. Under water stress conditions, the canopy temperature of tomatoes was only higher than the air temperature by 1.8 °C, while it was significantly lower (0.6 °C) under optimal water supply conditions [92]. Air temperature had a small impact on the canopy temperature of tomatoes grown under regular irrigation and cut-off stand (i.e., irrigation was stopped 30 days before harvest) (r^2 = 0.60; r^2 = 0.55), however, the canopy temperature of water stressed plants increased with rising air temperature (r^2 = 0.59) (Figure 3) [31]. A close correlation between canopy temperature and leaf water potential of maize was established [80] and the lowest CWSI values were measured between 10:00 and 11:00 and the highest ones between 12:00 and 13:00 [95].

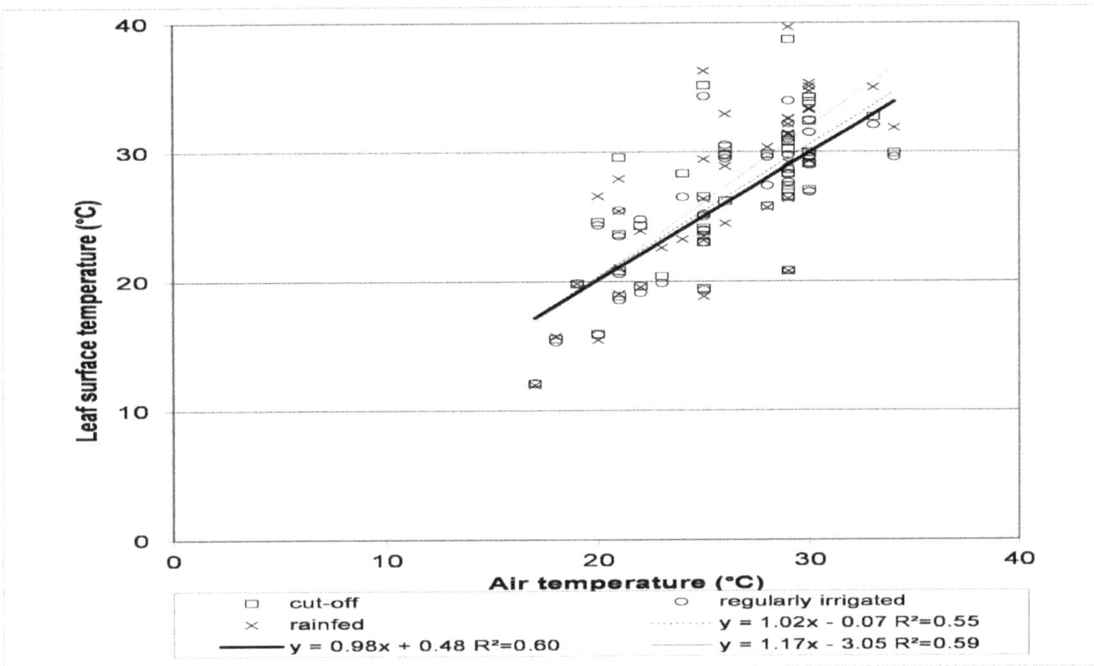

Figure 3. Relationship between the air and canopy temperature for the Kecskeméti jubileum tomato variety under rain-fed (thin line), cut-off (broken line) and regularly irrigated (thick line) conditions Source: Helyes et al. [31].

4.1.3. Stomatal Conductance

Stomatal conductance indicates the speed of water vapour evaporation that depends on more plant-specific characteristics such as stomatal density, leaf age and size, guard cell and cell turgor [97]. Stomatal conductance is related to the photosynthetic assimilation rates ensuring an appropriate balance between CO_2 uptake for photosynthesis and water loss through transpiration [98]. Variability in photosynthesis capacity can be explained by the CO_2 diffusion through stomata and leaf mesophyll which was influenced by the mesophyll thickness and porosity and size of stomata. In drought-acclimated tomato plants the decrease in mesophyll CO_2 conductance was due to an increased cell wall thickness [76]. Water stress significantly decreased the transpiration rate (37%) and stomatal conductance (26%) of maize [99]. Nevertheless, the extent of decrease in stomatal conductance depends on the growing period when the water deficiency occurred; at 7 days after anthesis of maize cultivars stomatal conductance decreased by 35% on average but at 21 days after anthesis this decrease was significantly larger (74%) under water deficiency than in well-watered cultivars [100]. In the case of tomatoes grown under non-irrigated conditions, stomatal conductance decreases from 14 to 73% depending on the weather and variety in comparison with the well-watered plants [3,32,87,101] (Table 3).

Table 3. Physiological traits related to water use and photosynthesis for vegetable crops under optimal water supply (OW) and water stress (WS) conditions.

Traits	Crops	Units	OW	WS	Difference %	References
Stomatal resistance	green pea	s cm^{-1}	2.87	3.22	12.2	Nemeskéri et al. [14]
	snap bean	s cm^{-1}	1.33	2.54	90.9	Nemeskéri et al. [15]
	sweet corn	s cm^{-1}	2.13	2.85	33.8	Nemeskéri et al. [88]
Stomatal conductance	green pea	mmol m^{-2} s^{-1}	0.57	0.32	−43.9	Gurumurthy et al. [35]
	tomato	mmol m^{-2} s^{-1}	1200	125	−89.6	Nankishore and Farrell [32]
		μmol m^{-2} s^{-1}	457.26	394.95	−13.6	Nemeskéri et al. [3]
		mol m^{-2} s^{-1}	20.2–37.9	6.3–10.2	−68.8 −73.1	Helyes et al. [101]
Chlorophyll fluorescence	snap bean	Fv/Fm	0.80	0.78	−2.5	Tari et al. [102]
	maize	Fv/Fm	0.810	0.695	−14.2	Yan et al. [103]
	tomato	Fv/Fm	0.785	0.745	−5.1	Nankishore and Farrell [32]
		Fv/Fm	0.748	0.696	−7.0	Nemeskéri et al. [3]
		Fq'/Fm'	0.4	0.25	−37.5	Zhou et al. [66]
Chlorophyll content	green pea	SPAD *	48.16	49.02	1.8	Nemeskéri et al. [14]
	snap bean	SPAD	34.57	38.94	12.6	Nemeskéri et al. [16]
	sweet corn	SPAD	47.48	44.67	−5.9	Nemeskéri et al. [2]
	tomato	SPAD	50.97	52.63	3.3	Nemeskéri et al. [3]
Vegetation index	green pea	NDVI	0.679	0.693	2.3	Nemeskéri et al. [14]
	snap bean	NDVI	0.778	0.681	−12.5	Nemeskéri et al. [16]
	sweet corn	NDVI	0.743	0.711	−4.3	Nemeskéri et al. [2]

* SPAD = relative chlorophyll content of leaves; NDVI = normalized differential vegetation index.

Under water scarcity, stomatal conductance for both water and CO_2 flow decreased by closing the stomata [104], thus it can be said that stomatal resistance increased. The extent of stomatal resistance mainly gives information about the speed of water vapour. Under severe water deficit conditions, stomatal resistance increased by 91% for snap beans, 34% for sweet corn and 12% for green peas in comparison with the well-watered plants (Table 3). The studies shown in Table 3 proved that snap beans responded more intensively to severe water stress than sweet corn and green peas. Flowering and pod development periods of legume crops are the most sensitive to water stress when stomatal resistance changes depending on the varieties and the degree of water stress. Under moderate water deficiency, the late ripening green pea varieties had high stomatal resistance (>3.0 s cm^{-1}), while that of green-podded snap bean varieties was relatively low (0.8–1.2 s cm^{-1}) and yellow-podded snap beans showed different values depending on the varieties (1.0–1.43 s cm^{-1}) [14,15]. During tasselling, the late ripening sweet corn hybrids responded with higher stomatal resistance (3 s cm^{-1}) to medium water deficiency than during the silking period [88].

4.2. Photosynthesis in Drought

The aspects of photosynthesis of selected vegetable crops which can be measured by remote sensing methods and used for the evaluation for drought tolerance of genotypes have to be taken into consideration. In the photosynthesis process the light capture and conversion of light energy to chemical energy is made by photosynthetic pigments in the photochemistry photosystems (PSI, PSII) of leaves. The light energy in the leaf that is not used for photosynthesis is either emitted as fluorescence or released as heat [105]. The efficiency of photosynthesis can be measured by the efficiency of PSII photochemistry or by the amount of photosynthetic pigments [106].

4.2.1. Chlorophyll Fluorescence

Intense dry conditions of soil cause stomatal closure, reduced CO_2 mesophyll conductance [107] and decreasing activity of PSII [108], which contributes to the decrease in photosynthesis. Photosystem II (PSII) is highly sensitive to light and drought [60] and the maximum quantum yield of PSII photochemistry (Fv/Fm) indicates an undisturbed or deficient operation of photosynthesis. Chlorophyll *a* fluorescence is considered to be suitable for the measurement of activity of photosynthesis because environmental stresses significantly affect the emission of chlorophyll fluorescence [109]. For example, UV-B radiation decreased the chlorophyll fluorescence of green peas [110] and ozone stress decreases the Fv/Fm and chlorophyll *a* concentration of leaves [111]. In higher plants, Fv/Fm fluorescence ranged from 0.78 to 0.84 [112], however this change depended on the variety and intensity of water stress.

In snap beans, the Fv/Fm ratio was relatively high (0.82–0.83) under optimal water supply conditions and it only decreased to 0.80 in the drought sensitive genotype under water stress conditions [102], which proved that chlorophyll fluorescence was not highly sensitive to water deficit.

In dry years, tomatoes grown under non-irrigated conditions had low photosynthetic activity (Fv/Fm = 0.662) and under moderate and optimal water supply conditions the Fv/Fm value ranged between 0.753 and 0.758 [3]. Likewise, the above-mentioned results from Nankishore and Farrell [32] showed a small decrease in Fv/Fm (5.1%) in tomatoes under drought (Table 3).

The maximum efficiency of PSII (Fv/Fm) of well-watered maize plants stayed constant while that of drought stressed plants stayed at control level during the first 2 days then decreased sharply as the soil became drier [103].

Use of Fv/Fm to evaluate the drought tolerance of crops is contradictory. Under controlled conditions, Fv/Fm for pot-grown grapevines decreased when water potential dropped but it seemed to be a good indicator to distinguish the moderate and severe drought stress in the field [113]. Drought stress affected the Fv/Fm parameter of the asparagus bean (*Vigna unguiculata* L.) [114]. Contrary to these results, no change was detected in the Fv/Fm for strawberries [115] and soybeans [116] grown under drought. Others [117,118] stated that PSII activity expressed by the Fv/Fm of drought tolerant tomato genotypes was less decreased under intensive water stress than sensitive ones. Likewise, Li et al. [119] found that Fv/Fm in drought tolerant barley varieties was higher than those of the drought sensitive group under drought stress. Under 4 day waterlogging conditions, the chlorophyll fluorescence (Fv/Fm) of flooding stress tolerant wax maize hybrids did not change significantly, while the photosynthesis efficiency of sensitive hybrids was relatively low and the Fv/Fm value decreased by 5.2% in comparison with the control [6]. The measurement of chlorophyll fluorescence as a rapid non-destructive method can be easily carried out in the field, thus it can be recommended for screening for drought tolerance [120].

4.2.2. Photosynthetic Pigments

Environmental stresses change not only the activity of the photochemistry apparatus but the chlorophyll concentration in the leaf due to metabolic disturbance [121], whereupon the light absorption decreases. Water also absorbs the radiation in the infrared wavelength of the spectrum and as the water content of leaf decreases, the light absorption decreases and reflectance increases due to the radiative attributes of water [122,123]. Therefore, the water content of leaves and the amount of photosynthetic pigments in leaves both influence the light absorption by leaves. The light absorption of the leaf can be indirectly measured by portable chlorophyll meter. In this way the calculated SPAD values correlated with the chlorophyll content of leaves [124] expresses the efficiency of photosynthesis by the intercepted photosynthetic active radiation. The high SPAD value indicates the low water and chlorophyll concentration simultaneously in the leaf, resulting in a decrease in light absorption and increase in reflectance that is larger in extent in snap beans and smaller in green peas and tomatoes (Table 3). Iturbe-Ormaetxe et al. [125] stated that the decrease in chlorophyll *a* concentration of leaves was larger (−30%) than that of chlorophyll *b* (−20.8%) for green peas exposed to water stress than in well-watered plants.

5. Relationship between Drought Stress Markers and Yield

During reproductive periods of plants that are most sensitive to water deficiency, the changes in physiological responses can be used to screen the water stress tolerance of genotypes. During this time the water supply determines the yield production. Stomatal resistance and the relative chlorophyll content of leaves (SPAD) of the individual plants indicate the disturbance of water circulation and photosynthesis. Spectral vegetation indices indicate the absorption of solar energy of the canopy in the visible light spectrum. Health status and water deficit of vegetation can be monitored by different vegetation indices and it can also determine the need for irrigation [126–129]. The normalized differential vegetation index (NDVI) expresses the ratio of spectral reflectance on the canopy in the infrared and red region and it is used to monitor the effect of water stress on plant growth and forecast biomass [130,131].

The question is how closely the physiological traits are related to water circulation and photosynthesis and can be used to predict the expected yield. Nevertheless, the physiological traits measured during the generative stages of plant species are different (Table 4). On the basis of long-term experiments, stomatal resistance measured during flowering of snap beans and tomatoes correlated with the pod yield of individual plants and weight of tomato fruits under severe drought. A close correlation between the relative chlorophyll content of leaves (SPAD) and weight of tomato fruits and final yield was detected under both mild and severe water deficiency which can be used for selection of genotypes with water stress tolerance. During tasselling of sweet corn, the expected yield of plants can be less predicted by stomatal resistance (47%) and to a higher extent by spectral traits (58–68%) under moderate water deficiency. During flowering of green peas, stomatal resistance and chlorophyll content of leaves showed a close correlation with the expected yield only under severe drought (Table 4).

Table 4. Correlation coefficients between physiological traits measured during flowering and yield under drought.

Crops	Water Supply	NI		DI	
	Traits	Yield g plant^{-1}	Yield t ha^{-1}	Yield g plant^{-1}	Yield t ha^{-1}
Green peas	SR	0.3885	0.7648	0.3541	0.4371
	SPAD	0.4685	0.7027	0.6378	0.5301
	NDVI	0.5550	0.7192	0.6200	0.2891
Snap beans	SR	0.6075	0.4687	0.5249	0.7163
	SPAD	0.4326	0.4671	0.6567	0.4385
	NDVI	0.4251	0.7300	0.3356	0.7665
Sweet corns *	SR	0.6184	0.5756	0.6866	0.6214
	SPAD	0.5346	0.4614	0.8221	0.6250
	NDVI	0.6804	0.4619	0.7648	0.4907
Tomato	gs	0.6851 ^{y}	0.7153	0.3026 ^{y}	0.3018
	SPAD	0.8655 ^{y}	0.8405	0.9256 ^{y}	0.8482
	Fv/Fm	0.4505 ^{y}	0.3669	0.1103 ^{y}	0.0961

* during tasselling ^{y} = fruit weight (g), gs =stomatal conductance, SR = stomatal resistance, NI = non-irrigation, DI = deficit irrigation Source: [2,3,14–16].

Other researchers used the normalized differential vegetation index (NDVI) for yield prediction; it was successful for castor beans [132], soybeans [133] and beans [134]. According to Spitkó et al. [38], a medium correlation (r = 0.5–0.6) was detected between NDVI and final yield at 15 days after flowering of maize but not during the flowering period. Different stress indices such as stress degree days (SDD) or crop water stress index (CWSI) can be used to evaluate the water stress tolerance of genotypes [25] for scheduling of irrigation [93] and maybe for prediction of yield. In the case of sweet corn, significant negative correlation was detected between the CWSI and chlorophyll content of leaves ($r = 0.802$) but for the CWSI, a significant positive ($r = 0.478$) correlation was observed with the yield [25].

Helyes et al. [31] found a close correlation between the stress degree days (SDD) and yield of tomatoes. If the canopy temperature exceeded the air temperature (at noon), transpiration was reduced, which indicated water stress and resulted in yield reduction and quantity. Figure 4 shows the interrelation between the canopy and air temperature difference values and the yield. In our experiments the correlation was significant at $p = 0.01$ level with $r^2 = 0.57$ correlation coefficient. High yield per hectare can be achieved if the difference between the cumulative canopy and the air temperature is negative during the growing season.

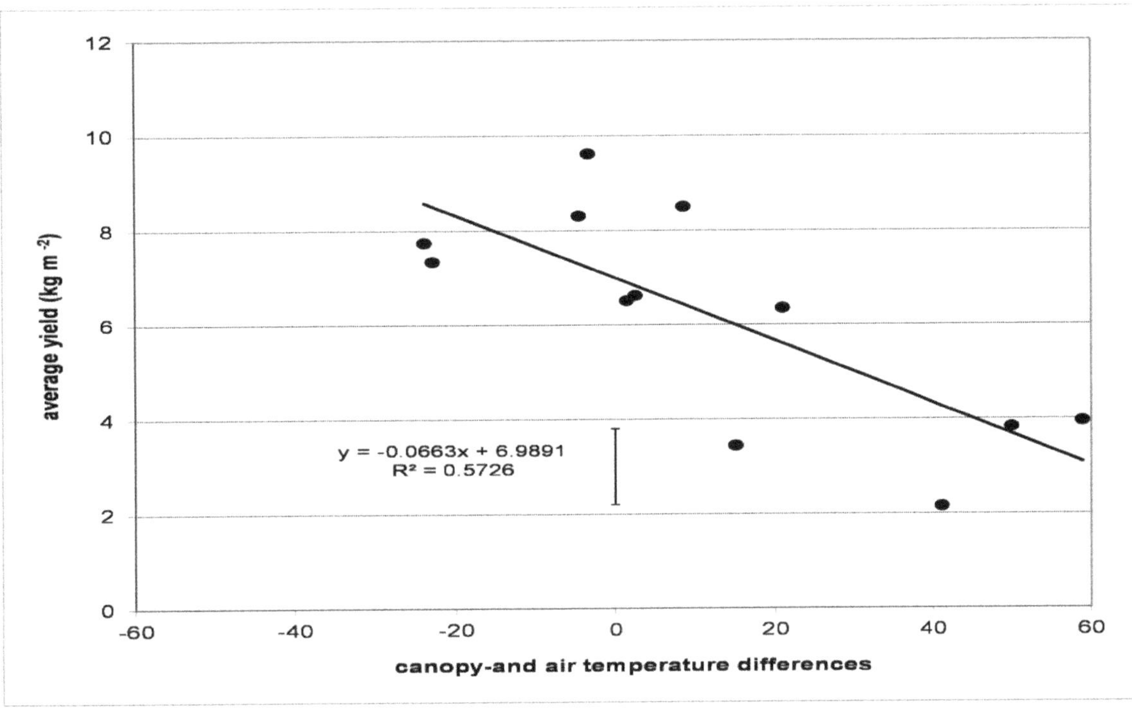

Figure 4. Correlation between canopy and air temperature differences and yield Source: [31].

6. Use of Physiological Characteristics

The use of physiological traits in a breeding program, either directly by selection or stress markers, depends on their genetic correlation with the yield, heritability and genotype × environment interaction [11,135]. Under water stress, high heritability of stomatal resistance, photosynthesis rate and transpiration rate ($h^2 = 0.91$–0.99) was found for *Vigna mungo* that gives a possibility for successful selection of genotypes [35]. Under severe drought, stomatal conductance and relative chlorophyll content of leaves (SPAD) measured during flowering correlated with the expected yield therefore they are suitable for the selection of individual genotypes for green peas and tomatoes, while the use of these traits for the selection of sweet corn can be efficient only under moderate water stress (Table 4). In the case of snap beans, because the water deficiency has a significant effect on leaf area, the normalized differential vegetation index (NDVI) measured during flowering can predict the expected yield more efficiently than the SPAD value of the leaves of individual plants.

Application of remote sensing techniques makes monitoring the water status of plants and real time irrigation easy [39,136]. The trend in the canopy temperature and the difference between the leaf temperature and air temperature (SDD) can be considered to be the water stress markers of plants [92]. The relationships between the canopy temperature, air temperature and transpiration involving the atmospheric and soil conditions and plant characteristics [40], was used to develop the crop water stress index (CWSI), indicating the need for irrigation. During drought, the decrease in NDVI occurred to a larger extent in snap beans, while it was less in sweet corn and hardly changed in green peas in comparison with optimal water supply conditions (Table 3). This explained why the NDVI was used as spectral indicator for irrigation scheduling mainly in snap beans [136].

In summary, some of the physiological traits influencing the decrease of water loss and biomass production of plants can be used to evaluate the water status of vegetable crops and the water stress tolerance of genotypes. During the generative period, under water deficit conditions, the changes in the stomatal conductance and chlorophyll content of leaves for individual plants is suitable for the estimation the productivity of genotypes. Nevertheless, leaf area of crops should be taken into consideration as they determine the transpiration and their chlorophyll density influences the intensity of photosynthesis and finally the yield. Water stress indices and spectral vegetation indices seemed to be more appropriate in the detection of perceived water deficiency than for the prediction of final yield.

Author Contributions: E.N. planned and wrote the first draft of the review. L.H. contributed to the writing and reviewed the final draft.

References

1. Ray, P. Hi-Tech Horticulture and Climate Change. In *Climate Dinamics in Horticultural Science, Principles and Applications*; Choudhary, M.L., Patel, V.B., Siddiqui, M.W., Mahdi, S.S., Eds.; Apple Academic Press: Oakville, ON, Canada; CRC Press Taylor & Francis Group: Boca Raton, FL, USA, 2015; Volume 1, pp. 1–22.
2. Nemeskéri, E.; Molnár, K.; Rácz, C.; Dobos, A.C.; Helyes, L. Effect of Water Supply on Spectral Traits and Their Relationship with the Productivity of Sweet Corns. *Agronomy* **2019**, *9*, 63. [CrossRef]
3. Nemeskéri, E.; Neményi, A.; Bőcs, A.; Pék, Z.; Helyes, L. Physiological Factors and their Relationship with the Productivity of Processing Tomato under Different Water Supplies. *Water* **2019**, *11*, 586. [CrossRef]
4. Aroca, R.; Porcel, R.; Ruiz-Lozano, J.M. Regulation of root water uptake under abiotic stress conditions. *J. Exp. Bot.* **2012**, *63*, 43–57. [CrossRef] [PubMed]
5. Limami, A.; Diab, H.; Lothier, J. Nitrogen metabolism in plants under low oxygen stress. *Planta* **2014**, *239*, 531–541. [CrossRef] [PubMed]
6. Zhu, M.; Li, F.H.; Shi, Z.S. Morphological and photosynthetic response of waxy corn inbred line to waterlogging. *Photosynthetica* **2016**, *54*, 636–640. [CrossRef]
7. Tian, L.; Bi, W.; Liu, X.; Sun, L.; Li, J. Effects of waterlogging stress on the physiological response and grain-filling characteristics of spring maize (*Zea mays* L.) under field conditions. *Acta Physiol. Plant.* **2019**, *41*, 63. [CrossRef]
8. Food and Agriculture Organization. FAOSTAT Crops Production. 2017. Available online: http://www.fao.org/faostat/en/#data/QC (accessed on 9 July 2018).
9. Heszky, L. Szárazság és a növény kapcsolata. *Agrofórum* **2007**, *18*, 37–41. (In Hungarian)
10. Patanè, C.; Tringali, S.; Sortino, O. Effects of deficit irrigation on biomass, yield, water productivity and fruit quality of processing tomato under semi-arid Mediterranean climate conditions. *Sci. Hortic.* **2011**, *129*, 590–596. [CrossRef]
11. Mir, R.R.; Zaman-Allah, M.; Sreenivasulu, N.; Trethowan, R.; Varshney, R.K. Integrated genomics, physiology and breeding approaches for improving drought tolerance in crops. *Theor. Appl. Genet.* **2012**, *125*, 625–645. [CrossRef] [PubMed]
12. Kang, S.; Shi, W.; Zhang, J. An improved water-use efficiency for maize grown under regulated deficit irrigation. *Field Crop. Res.* **2000**, *67*, 207–214. [CrossRef]
13. Kirda, C. *Deficit Irrigation Scheduling Based on Plant Growth Stages Showing Water Stress Tolerance*; Deficit Irrigation Practice Water Report 22; FAO: Rome, Italy, 2002; pp. 3–10.
14. Nemeskéri, E.; Molnár, K.; Vígh, R.; Nagy, J.; Dobos, A. Relationships between stomatal behaviour, spectral traits and water use and productivity of green peas (Pisum sativum L.) in dry seasons. *Acta Physiol. Plant.* **2015**, *37*, 1–16. [CrossRef]

15. Nemeskéri, E.; Molnár, K.; Pék, Z.; Helyes, L. Effect of water supply on the water use-related physiological traits and yield of snap beans in dry seasons. *Irrig. Sci.* **2018**, *36*, 143–158. [CrossRef]
16. Nemeskéri, E.; Molnár, K.; Helyes, L. Relationships of spectral traits with yield and nutritional quality of snap beans (Phaseolus vulgaris L.) in dry seasons. *Arch. Agron. Soil Sci.* **2018**, *64*, 1222–1239. [CrossRef]
17. Fang, X.; Turner, N.C.; Yan, G.; Li, F.; Siddique, K.H.M. Flower numbers, pod production, pollen viability, and pistil function are reduced and flower and pod abortion increased in chickpea (Cicer arietinum L.) under terminal drought. *J. Exp. Bot.* **2010**, *61*, 335–345. [CrossRef] [PubMed]
18. Behboudian, M.H.; Ma, Q.; Turner, N.C.; Palta, J.A. Reactions of chickpea to water stress: Yield and seed composition. *J. Sci. Food Agric.* **2001**, *81*, 1288–1291. [CrossRef]
19. Boutraa, T.; Sanders, F.E. Influence of Water Stress on Grain Yield and Vegetative Growth of Two Cultivars of Bean (Phaseolus vulgaris L.). *J. Agron. Crop. Sci.* **2001**, *187*, 251–257. [CrossRef]
20. Beshir, H.; Bueckert, R.; Tar'An, B. Effect of temporary drought at different growth stages on snap bean pod quality and yield. *Afr. Crop. Sci. J.* **2016**, *24*, 317–330. [CrossRef]
21. Durigon, A.; Evers, J.; Metselaar, K.; Lier, Q.D.J.V. Water Stress Permanently Alters Shoot Architecture in Common Bean Plants. *Agronomy* **2019**, *9*, 160. [CrossRef]
22. Baigorri, H.; Antolín, M.C.; Sánchez-Díaz, M. Reproductive response of two morphologically different pea cultivars to drought. *Eur. J. Agron.* **1999**, *10*, 119–128. [CrossRef]
23. Moser, S.B.; Feil, B.; Jampatong, S.; Stamp, P. Effects of pre-anthesis drought, nitrogen fertilizer rate, and variety on grain yield, yield components, and harvest index of tropical maize. *Agric. Water Manag.* **2006**, *81*, 41–58. [CrossRef]
24. Öktem, A. Effect of water shortage on yield, and protein and mineral compositions of drip-irrigated sweet corn in sustainable agricultural systems. *Agric. Water Manag.* **2008**, *95*, 1003–1010. [CrossRef]
25. Uçak, A.B.; Öktemb, A.; Sezerc, C.; Cengizc, R.; İnald, B. Determination of arid and temperature resistant sweet corn (Zea mays saccharata Sturt) lines. *Int. J. Environ. Agric. Res.* **2016**, *2*, 79–88.
26. Helyes, L.; Varga, G. Irrigation demand of tomato according to the results of three decades. *Acta Hortic.* **1994**, *376*, 323–328. [CrossRef]
27. Helyes, L.; Varga, G.; Dimény, J.; Pék, Z. The simultaneous effect of variety, irrigation and weather on tomato yield. *Acta Hortic.* **1999**, *487*, 499–506. [CrossRef]
28. Patane, C.; Cosentino, S.; Cosentino, S. Effects of soil water deficit on yield and quality of processing tomato under a Mediterranean climate. *Agric. Water Manag.* **2010**, *97*, 131–138. [CrossRef]
29. Pires, R.C.D.M.; Furlani, P.R.; Ribeiro, R.V.; Junior, D.B.; Sakai, E.; Lourenção, A.L.; Neto, A.T. Irrigation frequency and substrate volume effects in the growth and yield of tomato plants under greenhouse conditions. *Sci. Agric.* **2011**, *68*, 400–405. [CrossRef]
30. Bahadur, A.; Chatterjee, A.; Kumar, R.; Singh, M.; Naik, P.S. Physiological and biochemical basis of drought tolerance in vegetables. *Veg. Sci.* **2011**, *38*, 1–16.
31. Helyes, L.; Bőcs, A.; Pék, Z. Effect of water supply on canopy temperature, stomatal conductance and yield quantity of processing tomato (Lycopersicon esculentum Mill.). *Int. J. Hortic. Sci.* **2010**, *16*, 13–15. [CrossRef]
32. Nankishore, A.; Farrell, A.D. The response of contrasting tomato genotypes to combined heat and drought stress. *J. Plant Physiol.* **2016**, *202*, 75–82. [CrossRef]
33. Agbna, G.H.; Dongli, S.; Zhipeng, L.; Elshaikh, N.A.; Guangcheng, S.; Timm, L.C. Effects of deficit irrigation and biochar addition on the growth, yield, and quality of tomato. *Sci. Hortic.* **2017**, *222*, 90–101. [CrossRef]
34. Rodriguez-Ortega, W.; Martinez, V.; Rivero, R.; Cámara-Zapata, J.-M.; Mestre, T.; Garcia-Sanchez, F. Use of a smart irrigation system to study the effects of irrigation management on the agronomic and physiological responses of tomato plants grown under different temperatures regimes. *Agric. Water Manag.* **2017**, *183*, 158–168. [CrossRef]
35. Gurumurthy, S.; Sarkar, B.; Vanaja, M.; Lakshmi, J.; Yadav, S.K.; Maheswari, M. Morpho-physiological and biochemical changes in black gram (Vigna mungo L. Hepper) genotypes under drought stress at flowering stage. *Acta Physiol. Plant.* **2019**, *41*, 42. [CrossRef]
36. Stone, P.J.; Wilson, D.R.; Jamieson, P.D.; Gillespie, R.N. Water deficit effects on sweet corn. Part II. Canopy development. *Aust. J. Agric. Res.* **2001**, *54*, 115–126. [CrossRef]
37. Genc, L.; Inalpulat, M.; Kizil, U.; Mirik, M.; Smith, S.E.; Mendes, M. Determination of water stress with spectral reflectance on sweet corn (Zea mays L.) using classification tree (CT) analysis. *Zemdirbyste-Agriculture* **2013**, *100*, 81–90. [CrossRef]

38. Spitkó, T.; Nagy, Z.; Zsubori, Z.T.; Szőke, C.; Berzy, T.; Pintér, J.; Márton, L. Connection between normalized difference vegetation index and yield in maize. *Plant Soil Environ.* **2016**, *62*, 293–298. [CrossRef]
39. Zhou, J.; Khot, L.R.; Boydston, R.A.; Miklas, P.N.; Porter, L. Low altitude remote sensing technologies for crop stress monitoring: A case study on spatial and temporal monitoring of irrigated pinto bean. *Precis. Agric.* **2018**, *19*, 555–569. [CrossRef]
40. Chatterjee, A.; Solankey, S. Functional Physiology in Drought Tolerance of Vegetable Crops: An Approach to Mitigate Climate Change Impact. In *Climate Dynamics in Horticultural Science, Principles and Applications*; Choudhary, M.L., Patel, V.B., Siddiqui, M.W., Mahdi, S.S., Eds.; Apple Academic Press: Oakville, ON, Canada; CRC Press Taylor & Francis Group: Boca Raton, FL, USA, 2015; Volume 1, pp. 149–171.
41. Cattivelli, L.; Baldi, P.; Crosatti, C.; Di Fonzo, N.; Faccioli, P.; Grossi, M.; Mastrangelo, A.M.; Pecchioni, N.; Stanca, A.M. Chromosome regions and stress-related sequences involved in resistance to abiotic stress in Triticeae. *Plant Mol. Biol.* **2002**, *48*, 649–665. [CrossRef]
42. Ramanjulu, S.; Bartels, D. Drought- and desiccation-induced modulation of gene expression in plants. *Plant Cell Environ.* **2002**, *25*, 141–151. [CrossRef]
43. Zhang, J.; Zheng, H.G.; Aarti, A.; Pantuwan, G.; Nguyen, T.T.; Tripathy, J.N.; Sarial, A.K.; Robin, S.; Babu, R.C.; Nguyen, B.D.; et al. Locating genomic regions associated with components of drought resistance in rice: Comparative mapping within and across species. *Theor. Appl. Genet.* **2001**, *103*, 19–29. [CrossRef]
44. Lanceras, J.C.; Pantuwan, G.; Jongdee, B.; Toojinda, T. Quantitative Trait Loci Associated with Drought Tolerance at Reproductive Stage in Rice. *Plant Physiol.* **2004**, *135*, 384–399. [CrossRef]
45. Cattivelli, L.; Rizza, F.; Badeck, F.-W.; Mazzucotelli, E.; Mastrangelo, A.M.; Francia, E.; Maré, C.; Tondelli, A.; Stanca, A.M. Drought tolerance improvement in crop plants: An integrated view from breeding to genomics. *Field Crop. Res.* **2008**, *105*, 1–14. [CrossRef]
46. De Leonardis, A.M.; Petrarulo, M.; Vita, P.D.; Mastrangelo, A.M. Genetic and Molecular Aspects of Plant Response to Drought in Annual Crop Species. In *Advances in Selected Plant Physiology Aspects*; Giuseppe, M., Dichio, B., Eds.; InTech Publisher: Rijeka, Croatia, 2012; pp. 45–74.
47. Kumar, R.; Solankey, S.S.; Singh, M. Breeding for drought tolerance in vegetables. *Veg. Sci.* **2012**, *39*, 1–15.
48. Augé, R.M.; Sylvia, D.M.; Park, S.; Buttery, B.R.; Saxton, A.M.; Moore, J.L.; Cho, K. Partitioning mycorrhizal influence on water relations of Phaseolus vulgaris into soil and plant components. *Can. J. Bot.* **2004**, *82*, 503–514. [CrossRef]
49. Mnasri, B.; Aouani, M.E.; Mhamdi, R. Nodulation and growth of common bean (Phaseolus vulgaris) under water deficiency. *Soil Biol. Biochem.* **2007**, *39*, 1744–1750. [CrossRef]
50. Dalton, D.A.; Moran, J.F.; Iturbe-Ormaetxe, I.; Matamoros, M.A.; Rubio, M.C.; Iturbe-Ormaetxe, I. Reactive oxygen species and antioxidants in legume nodules. *Physiol. Plant.* **2000**, *109*, 372–381.
51. Collados, C.; Barea, J.M.; Ruiz-Lozano, J.M.; Azcón, R. Arbuscular mycorrhizal symbiosis can alleviate drought-induced nodule senescence in soybean plants. *New Phytol.* **2001**, *151*, 493–502.
52. Esfahani, M.N.; Mostajeran, A. Rhizobial strain involvement in symbiosis efficiency of chickpea–rhizobia under drought stress: Plant growth, nitrogen fixation and antioxidant enzyme activities. *Acta Physiol. Plant.* **2011**, *33*, 1075–1083. [CrossRef]
53. Nicholas, S. Plant resistance to environmental stress. *Curr. Opin. Biotechnol.* **1998**, *9*, 214–219.
54. Nemeskéri, E. Water deficiency resistance study on soya and bean cultivars. *Acta Agron. Hung.* **2001**, *49*, 83–93. [CrossRef]
55. Marouelli, W.A.; Silva, W.L.C. Water tension thresholds for processing tomatoes under drip irrigation in Central Brazil. *Irrig. Sci.* **2007**, *25*, 411–418. [CrossRef]
56. Nemeskéri, E.; Horváth, K.; Pék, P.; Helyes, L. Effect of mycorrhizal and bacterial products on the traits related to photosynthesis and fruit quality of tomato under water deficiency conditions. *Acta Hortic.* **2019**, 61–66. [CrossRef]
57. Nilsen, E.T.; Forseth, E.N., Jr. The role of leaf movements for optimizing photosynthesis in relation to environmental variation. In *Advance in Photosynthesis and Respiration, The Leaf: A Platform for Performing Photosynthesis*; Adams, W.W., Terashima, I., Eds.; Springer: Basel, Switzerland, 2018; Volume 4, pp. 401–423.
58. Fernandez, D.; Castrillo, M. Maize leaf rolling initiation. *Photosynthetica* **1999**, *37*, 493–497. [CrossRef]
59. Sağlam, A.; Kadioglu, A.; Demiralay, M.; Terzi, R. Leaf Rolling Reduces Photosynthetic Loss in Maize Under Severe Drought. *Acta Bot. Croat.* **2014**, *73*, 315–332. [CrossRef]

60. Pastenes, C.; Pimentel, P.; Lillo, J. Leaf movements and photoinhibition in relation to water stress in field-grown beans. *J. Exp. Bot.* **2005**, *56*, 425–433. [CrossRef]
61. Gomes, A.M.M.; Lagoa, A.M.M.A.; Medina, C.L.; Machado, E.C.; Machado, M.A. Interactions between leaf water potential, stomatal conductance and abscisic acid content of orange trees submitted to drought stress. *Braz. J. Plant Physiol.* **2004**, *16*, 155–161. [CrossRef]
62. Parry, A.D.; Horgan, R. Abscisic acid biosynthesis in roots. *Planta* **1992**, *187*, 185–191. [CrossRef]
63. Reynolds-Henne, C.E.; Langenegger, A.; Mani, J.; Schenk, N.; Zumsteg, A.; Feller, U. Interactions between temperature, drought and stomatal opening in legumes. *Environ. Exp. Bot.* **2010**, *68*, 37–43. [CrossRef]
64. De Carvalho, M.H.C.; Laffray, D.; Louguet, P. Comparison of the physiological responses of Phaseolus vulgaris and Vigna unguiculata cultivars when submitted to drought conditions. *Environ. Exp. Bot.* **1998**, *40*, 197–207. [CrossRef]
65. Garcia, A.L.; Marcelis, L.; Garcia-Sanchez, F.; Nicolas, N.; Martinez, V. Moderate water stress affects tomato leaf water relations in dependence on the nitrogen supply. *Biol. Plant.* **2007**, *51*, 707–712. [CrossRef]
66. Zhou, R.; Kong, L.; Wu, Z.; Rosenqvist, E.; Wang, Y.; Zhao, L.; Zhao, T.; Ottosen, C.O. Physiological response of tomatoes at drought, heat and their combination followed by recovery. *Physiol. Plant.* **2019**, *165*, 144–154. [CrossRef]
67. Du, W.-J.; Yu, D.-Y.; Fu, S.-X. Analysis of QTLs for the trichome density on the upper and downer surface of leaf blade in soybean [Glycine max (L.) Merr.]. *Agric. Sci. China* **2009**, *8*, 529–537. [CrossRef]
68. Nobel, P.S. *Physicochemical and Environmental Plant Physiology*; Academic Press: Cambridge, MA, USA, 1991; p. 635.
69. Sam, O.; Jeréz, E.; Dell'Amico, J.; Ruiz-Sanchez, M.C. Water stress induced changes in anatomy of tomato leaf epidermis. *Biol. Plant.* **2000**, *43*, 275–277. [CrossRef]
70. Sánchez, F.J.; Manzanares, M.; De Andres, E.F.; Tenorio, J.L.; Ayerbe, L. Residual transpiration rate, epicuticular wax load and leaf colour of pea plants in drought conditions. Influence on harvest index and canopy temperature. *Eur. J. Agron.* **2001**, *15*, 57–70. [CrossRef]
71. Gomes, F.P.; Oliva, M.A.; Mielke, M.S.; Almeida, A.-A.F.; Aquino, L.A. Osmotic adjustment, proline accumulation and cell membrane stability in leaves of Cocos nucifera submitted to drought stress. *Sci. Hortic.* **2010**, *126*, 379–384. [CrossRef]
72. Sánchez, F.J.; Manzanares, M.; De Andres, E.F.; Tenorio, J.L.; Ayerbe, L. Turgor maintenance, osmotic adjustment and soluble sugar and proline accumulation in 49 pea cultivars in response to water stress. *Field Crop. Res.* **1998**, *59*, 225–235. [CrossRef]
73. Babita, M.; Maheswari, M.; Rao, L.; Shanker, A.K.; Rao, D.G. Osmotic adjustment, drought tolerance and yield in castor (Ricinus communis L.) hybrids. *Environ. Exp. Bot.* **2010**, *69*, 243–249. [CrossRef]
74. Nemeskéri, E.; Sárdi, E.; Remenyik, J.; Kőszegi, B.; Nagy, P. Study of defensive mechanisms against drought of French bean (Phaseolus vulgaris L.) varieties. *Acta Physiol. Plant.* **2010**, *32*, 1125–1134. [CrossRef]
75. Chen, Z.; Gallie, D.R. The Ascorbic Acid Redox State Controls Guard Cell Signaling and Stomatal Movement. *Plant Cell* **2004**, *16*, 1143–1162. [CrossRef]
76. Galmés, J.; Ochogavía, J.M.; Gago, J.; Roldán, E.J.; Cifre, J.; Conesa, M.A. Leaf responses to drought stress in Mediterranean accessions of Solanum lycopersicum: Anatomical adaptations in relation to gas exchange parameters. *Plant Cell Environ.* **1913**, *36*, 920–935. [CrossRef]
77. Muir, C.D.; Conesa, M.À.; Galmés, J. Independent evolution of ab- and adaxial stomatal density enables adaptation. *bioRxiv* **2015**, 1–25. [CrossRef]
78. Galmés, J.; Conesa, M.A.; Manuel Ochogavía, J.; Alejandro Perdomo, J.; Francis, D.M.; Ribas-Carbo, M.; Save, R.; Flexas, J.; Medrano, H.; Cifre, J. Physiological and morphological adaptations in relation to water use efficiency in Mediterranean accessions of *Solanum lycopersicum*. *Plant Cell Environ.* **2011**, *34*, 245–260. [CrossRef]
79. Hardy, J.P.; Anderson, V.J.; Gardner, J.S. Stomatal characteristics, conductance ratios, and drought-induced leaf modifications of semiarid grassland species. *Am. J. Bot.* **1995**, *82*, 1–7. [CrossRef]
80. Dejonge, K.C.; Taghvaeian, S.; Trout, T.J.; Comas, L.H. Comparison of canopy temperature-based water stress indices for maize. *Agric. Water Manag.* **2015**, *156*, 51–62. [CrossRef]
81. Jones, H.G.; Jones, H. Use of thermography for quantitative studies of spatial and temporal variation of stomatal conductance over leaf surfaces. *Plant Cell Environ.* **1999**, *22*, 1043–1055. [CrossRef]

82. Cornic, G.; Lawlor, D.W. Photosynthetic carbon assimilation and associated metabolism in relation to water deficits in higher plants. *Plant Cell Environ.* **2002**, *25*, 275–294.
83. Flagella, Z.; Campanile, R.G.; Stoppelli, M.C.; De Caro, A.; Di Fonzo, N. Drought tolerance of photosynthetic electron transport under CO_2-enriched and normal air in cereal species. *Physiol. Plant.* **1998**, *104*, 753–759. [CrossRef]
84. Pol, M.; Gołębiowska, D.; Miklewska, J. Influence of enhanced concentration of carbon dioxide and moderate drought on fluorescence induction in white clover (Trifolium repens L.). *Photosynthetica* **1999**, *37*, 537–542. [CrossRef]
85. Yordanov, I.; Velikova, V.; Tsonev, T. Plant Responses to Drought, Acclimation, and Stress Tolerance. *Photosynthetica* **2000**, *38*, 171–186. [CrossRef]
86. Ghanbari, A.A.; Shakiba, M.R.; Toorchi, M.; Choukan, R. Morpho-physiological responses of common bean leaf to water deficit stress. *Eur. J. Exp. Biol.* **2013**, *3*, 487–492.
87. Guida, G.; Sellami, M.H.; Mistretta, C.; Oliva, M.; Buonomo, R.; De Mascellis, R.; Patanè, C.; Rouphael, Y.; Albrizio, R.; Giorio, P. Agronomical, physiological and fruit quality responses of two Italian long-storage tomato landraces under rain-fed and full irrigation conditions. *Agric. Water Manag.* **2017**, *180*, 126–135. [CrossRef]
88. Nemeskéri, E.; Molnár, K.; Dobos, A.C. Csemegekukorica (Zea mays L. convar. saccharata) sztóma működése, és hatása a növekedésre és terméskomponensekre eltérő vízellátás alatt (Stomatal behaviour and its influence on the growing and yield components of sweet corn (Zea mays L. convar. saccharata). *Növénytermelés* **2017**, *66*, 75–95.
89. Yang, H.M.; Zhang, X.Y.; Wang, G.X. Relationships between stomatal character, photosynthetic character and seed chemical composition in grass pea at different water availabilities. *J. Agric. Sci.* **2004**, *142*, 675–681. [CrossRef]
90. Gonzalez-Dugo, M.P.; Moran, M.S.; Mateos, L.; Bryant, R. Canopy temperature variability as an indicator of crop water stress severity. *Irrig. Sci.* **2006**, *24*, 233–240. [CrossRef]
91. Bonanno, A.R.; Mack, H.J. Use of canopy—Air temperature differentials as a method for scheduling irrigation in snap beans. *J. Am. Hortic. Sci.* **1983**, *108*, 826–831.
92. Helyes, L. Relations among the water supply, foliage temperature and the yield of tomato. *Acta Hortic.* **1990**, *277*, 115–122. [CrossRef]
93. Helyes, L.; Dimény, J.; Varga, G. Az öntözés tervezése a lombfelszín-hőmérséklet alapján (Scheduling of irrigation with canopy temperature). *Növénytermelés* **2005**, *54*, 341–350.
94. Dufková, R. Difference in canopy and air temperature as an indicator of grassland water stress. *Soil Water Res.* **2006**, *1*, 127–138.
95. Taghvaeian, S.; Chávez, J.L.; Hansen, N.C. Infrared Thermometry to Estimate Crop Water Stress Index and Water Use of Irrigated Maize in Northeastern Colorado. *Remote Sens.* **2012**, *4*, 3619–3637. [CrossRef]
96. Helyes, L. A zöldségnövények vízellátottságának és öntözési igényének meghatározása a lombhőmérséklettel. Ph.D. Thesis, Agricultural Unversity of Gödöllő, Gödöllő, Hungary, 1991; p. 123. (In Hungarian).
97. Jones, H.G. *Plants and Microclimate*, 2nd ed.; Cambridge University Press: Cambridge, UK, 1992; p. 428.
98. Lawson, T.; Terashima, I.; Fujita, T.; Wang, Y. Coordination between Photosynthesis and Stomatal Behavior. In *The Leaf: A Platform for Performing Photosynthesis*; Adams, W.W., Terashima, I., Eds.; Springer: Basel, Switzerland, 2018; pp. 142–156.
99. Anjum, S.A.; Wang, L.C.; Farooq, M.; Hussain, M.; Xue, L.L.; Zou, C.M. Brassinolide Application Improves the Drought Tolerance in Maize through Modulation of Enzymatic Antioxidants and Leaf Gas Exchange. *J. Agron. Crop. Sci.* **2011**, *197*, 177–185. [CrossRef]
100. Sabagh, A.E.; Barutçular, C.; Islam, M.S. Relationships between stomatal conductance and yield under deficit irrigation in maize (Zea mays L.). *J. Exp. Biol. Agric. Sci.* **2017**, *5*, 15–21. [CrossRef]
101. Helyes, L.; Szuvandzsiev, P.; Neményi, A.; Pék, Z.; Lugasi, A. Different water supply and stomatal conductance correlates with yield quantity and quality parameters. *Acta Hortic.* **2013**, *971*, 119–125. [CrossRef]
102. Tari, I.; Camen, D.; Coradini, G.; Csiszár, J.; Fediuc, E.; Gémes, K.; Lazar, A.; Madosa, E.; Mihacea, S.; Poór, P.; et al. Changes in chlorophyll fluorescence parameters and oxidative stress responses of bush bean genotypes for selecting contrasting acclimation strategies under water stress. *Acta Biol. Hung.* **2008**, *59*, 335–345. [CrossRef]

103. Yan, H.; Wu, L.; Filardo, F.; Yang, X.; Zhao, X.; Fu, D. Chemical and hydraulic signals regulate stomatal behavior and photosynthetic activity in maize during progressive drought. *Acta Physiol. Plant.* **2017**, *39*, 125. [CrossRef]
104. Sing, S.K.; Reddy, K.R. Regulation of photosynthesis, fluorescence, stomatal conductance and water-use efficiency of cowpea [Vigna unguiculata (L.) Walp.] under drought. *J. Photochem. Photobiol. B* **2011**, *105*, 40–50. [CrossRef]
105. Lambrev, P.H.; Miloslavina, Y.; Jahns, P.; Holzwarth, A.R. On the relationship between non-photochemical quenching and photoprotection of Photosystem II. *Biochim. Biophys. Acta* **2012**, *1817*, 760–769. [CrossRef]
106. Bauerle, W.L.; Weston, D.J.; Bowden, J.D.; Dudley, J.B.; Toler, J.E. Leaf absorptance of photosynthetically active radiation in relation to chlorophyll meter estimates among woody plant species. *Sci. Hortic.* **2004**, *101*, 169–178. [CrossRef]
107. Flexas, J.; Bota, J.; Escalona, J.M.; Sampol, B.; Medrano, H. Effects of drought on photosynthesis in grapevines under field conditions: An evaluation of stomatal and mesophyll limitations. *Funct. Plant Biol.* **2002**, *29*, 461–471. [CrossRef]
108. Filek, M.; Łabanowska, M.; Kościelniak, J.; Biesaga-Kościelniak, J.; Kurdziel, M.; Szarejko, I.; Hartikainen, H. Characterization of barley leaf tolerance to drought stress by chlorophyll fluorescence and electron paramagnetic resonance studies. *J. Agron. Crop. Sci.* **2015**, *201*, 228–240. [CrossRef]
109. Guo, Y.; Tan, J. Recent advances in the application of chlorophyll a fluorescence from photosystem II. *Photochem. Photobiol.* **2015**, *91*, 1–14. [CrossRef]
110. Nogués, S.; Allen, D.J.; Morison, J.I.L.; Baker, N.R. Ultraviolet-B radiation effects on water relations, leaf development, and photosynthesis in droughted pea plants. *Plant Physiol.* **1998**, *117*, 173–181. [CrossRef]
111. Ismail, I.; Basahi, J.; Hassan, I. Gas exchange and chlorophyll fluorescence of pea (Pisum sativum L.) plants in response to ambient ozone at a rural site in Egypt. *Sci. Total Environ.* **2014**, *497*, 585–593. [CrossRef]
112. Demmig, B.; Björkman, O. Photon yield of O2 evolution and chlorophyll fluorescence characteristics at 77 K among vascular plants of diverse origins. *Planta* **1987**, *170*, 489–504.
113. Zulini, L.; Rubinigg, M.; Zorer, R.; Bertamini, M. Effects of drought stress on chlorophyll fluorescence and photosynthetic pigments in grapevine leaves (vitis vinifera cv. 'white riesling'). *Acta Hortic.* **2007**, *754*, 289–294. [CrossRef]
114. Wang, B.; Liu, Y.; Wu, X.; Lu, Z.; Li, G. The relationship of chlorophyll fluorescence parameters and drought tolerance in asparagus bean seedlings under drought stress. *Acta Agric. Zhejiang* **2009**, *21*, 246–249.
115. Razavi, F.; Pollet, B.; Steppe, K.; Van Labeke, M.C.; Labeke, M.C. Chlorophyll fluorescence as a tool for evaluation of drought stress in strawberry. *Photosynthetica* **2008**, *46*, 631–633. [CrossRef]
116. Ohashi, Y.; Nakayama, N.; Saneoka, H.; Fujita, K. Effects of drought stress on photosynthetic gas exchange, chlorophyll fluorescence and stem diameter of soybean plants. *Biol. Plant.* **2006**, *50*, 138–141. [CrossRef]
117. Srinivasa Rao, N.K.; Bhatt, R.M.; Mascarenhas, J.B.D.; Naren, A. Influence of moisture stress on leaf water status, osmotic potential, chlorophyll fluorescence and solute accumulation in field grown tomato cultivars. *Veg. Sci.* **1999**, *26*, 129–132.
118. Bahadur, A.; Kumar, R.; Mishra, U.; Rai, A.; Singh, M. Physiological approaches for screening of tomato genotypes for moisture stress tolerance. In Proceedings of the National Conference of Plant Physiology (NCPP-2010) BHU, Varanasi, India, 25–27 November 2010; p. 142.
119. Li, R.-H.; Guo, P.-G.; Michael, B.; Stefania, G.; Salvatore, C. Evaluation of Chlorophyll Content and Fluorescence Parameters as Indicators of Drought Tolerance in Barley. *Agric. Sci. China* **2006**, *5*, 751–757. [CrossRef]
120. Woo, N.S.; Badger, M.R.; Pogson, B.J. A rapid, non-invasive procedure for quantitative assessment of drought survival using chlorophyll fluorescence. *Plant Methods* **2008**, *4*, 27. [CrossRef]
121. Carter, G.A.; Knapp, A.K. Leaf optical properties in higher plants: Linking spectral characteristics to stress and chlorophyll concentration. *Am. J. Bot.* **2001**, *88*, 677–684. [CrossRef]
122. Bowman, W.D. The relationship between leaf water status, gas exchange, and spectral reflectance in cotton leaves. *Remote Sens. Environ.* **1989**, *30*, 249–255. [CrossRef]
123. Huntjr, E.; Rock, B. Detection of changes in leaf water content using Near- and Middle-Infrared reflectances. *Remote Sens. Environ.* **1989**, *30*, 43–54. [CrossRef]
124. Yadava, U.L. A rapid and nondestructive method to determine chlorophyll in intact leaves. *Hortic. Sci.* **1986**, *1*, 1449–1450.

125. Iturbe-Ormaetxe, I. Oxidative Damage in Pea Plants Exposed to Water Deficit or Paraquat. *Plant Physiol.* **1998**, *116*, 173–181. [CrossRef]
126. Anderson, M.C.; Zolin, C.A.; Sentelhas, P.C.; Hain, C.R.; Semmens, K.; Yilmaz, M.T.; Gao, F.; Otkin, J.A.; Tetrault, R. The Evaporative Stress Index as an indicator of agricultural drought in Brazil: An assessment based on crop yield impacts. *Remote Sens. Environ.* **2016**, *174*, 82–99. [CrossRef]
127. Xu, H.; Zhu, S.; Ying, Y.; Jiang, H. Application of multispectral reflectance for early detection of tomato disease. In Proceedings of the Optics for Natural Resources, Agriculture, and Foods, Boston, MA, USA, 1–4 October 2006; Volume 6381. [CrossRef]
128. Zwart, S.J.; Leclert, L.M.C. A remote sensing-based irrigation performance assessment: A case study of the Office du Niger in Mali. *Irrig. Sci.* **2010**, *28*, 371–385. [CrossRef]
129. Chawade, A.; Van Ham, J.; Blomquist, H.; Bagge, O.; Alexandersson, E.; Ortiz, R. High-Throughput Field-Phenotyping Tools for Plant Breeding and Precision Agriculture. *Agronomy* **2019**, *9*, 258. [CrossRef]
130. Romano, G.; Zia, S.; Spreer, W.; Sanchez, C.; Cairns, J.; Araus, J.L.; Müller, J. Use of thermography for high throughtput phenotyping of tropical maize adaptation in water stress. *Comput. Electron. Agric.* **2011**, *79*, 67–74. [CrossRef]
131. Thenkabail, P.S.; Smith, R.B.; De Pauw, E. Hyperspectral Vegetation Indices and Their Relationships with Agricultural Crop Characteristics. *Remote Sens. Environ.* **2000**, *71*, 158–182. [CrossRef]
132. Li, G.; Zhang, H.; Wu, X.; Shi, C.; Huang, X.; Qin, P. Canopy reflectance in two castor bean varieties (Ricinus communis L.) for growth assessment and yield prediction on coastal saline land of Yancheng District, China. *Ind. Crop. Prod.* **2011**, *33*, 395–402. [CrossRef]
133. Ma, B.L.; Dwyer, L.M.; Costa, C.; Cober, E.R.; Morrison, M.J. Early Prediction of Soybean Yield from Canopy Reflectance Measurements. *Agron. J.* **2001**, *93*, 1227–1234. [CrossRef]
134. Gutiérrez-Rodríguez, M.; Escalante-Estrada, J.A.; Gonzalez, M.T.R.; Reynolds, M.P. Canopy reflectance indices and its relationship with yield in common bean plants (Phaseolus vulgaris L.) with phosphorous supply. *Int. J. Agric. Biol.* **2006**, *8*, 203–207.
135. Scully, B.; Wallace, D.; Viands, D. Heritability and Correlation of Biomass, Growth Rates, Harvest Index, and Phenology to the Yield of Common Beans. *J. Am. Soc. Hortic. Sci.* **1991**, *116*, 127–130. [CrossRef]
136. Köksal, E.S. Hyperspectral reflectance data processing through cluster and principal component analysis for estimating irrigation and yield related indicators. *Agric. Water Manag.* **2011**, *98*, 1317–1328. [CrossRef]

11

Screening of Provitamin-A Maize Inbred Lines for Drought Tolerance: Beta-Carotene Content and Secondary Traits

Aleck Kondwakwenda *, Julia Sibiya, Rebecca Zengeni, Cousin Musvosvi and Samson Tesfay

School of Agriculture, Earth and Environmental Sciences, University of KwaZulu-Natal, Private Bag X01, Scottsville 3209, Pietermaritzburg, South Africa; sibiyaj@ukzn.ac.za (J.S.); zengeni@ukzn.ac.za (R.Z.); cousinmusvosvi@gmail.com (C.M.); tesfay@ukzn.ac.za (S.T.)

* Correspondence: alekondwa@gmail.com

Abstract: Provitamin A maize (*Zea mays* L.) biofortification is an ideal complementary means of combating vitamin A deficiency (VAD) in sub-Saharan Africa where maize consumption is high coupled by high VAD incidences. However, drought remains a major abiotic constraint to maize productivity in this region. Comprehensive drought screening of initial breeding materials before advancing them is important to achieve genetic gain. In this study, 46 provitamin-A inbred lines were screened for drought tolerance in the greenhouse and field under drought and optimum conditions using β-carotene content (BCC), grain yield (GY), and selected morphophysiological and biochemical traits. The results revealed that BCC, morphophysiological and biochemical traits were effective in discriminating among genotypes. Number of ears per plant (EPP), stomatal conductance (Gs), delayed leaf senescence (SEN), leaf rolling (RL), chlorophyll content (CC) and free proline content (PC) proved to be ideal traits to use when indirectly selecting for GY by virtue of having relative efficiency of indirect selection values that are greater than unity and considerable genetic variances under either or both conditions. The findings of this study form the basis of initial germplasm selection when improving provitamin A maize for drought tolerance.

Keywords: Provitamin A; maize; drought; morphological; physiological; biochemical; β-carotene

---

1. Introduction

Provitamin A maize has orange or yellow endosperm, which contains precursors of vitamin A in the form of carotenoids, hence the name "provitamin A maize". Provitamin A carotenoids include β-carotene, α-carotene and β-cryptoxanthin. Beta-carotene is the most important of the three carotenoids because it has higher provitamin A activity owing to its unique double ring molecular structure [1]. However, the ordinary (not improved) yellow maize grown and consumed throughout the world has β-carotene content of less than 1.5 $\mu g\ g^{-1}$ [2], which is too low given that the required target of total provitamin A carotenoid is 15 $\mu g\ g^{-1}$ [3]. Developing provitamin A maize cultivars with higher levels of provitamin A carotenoids through biofortification is a sustainable, cheap and effective complementary solution to VAD challenges faced by many developing nations [4].

An online Biofortification Priority Index (BPI) tool, developed and managed by HarvestPlus, shows that maize provitamin A biofortification as VAD intervention is most suitable for maize consuming developing countries, particularly the Southern Africa region (www.harvestplus.org/knowledge-market/BPI). However, maize production in this region is vulnerable to drought due to recurring low annual precipitation coupled by poor coping capacity of most farmers. For instance, in 1992 and 2002, most of the southern African countries experienced the worst droughts resulting in over 60% maize yield loss in the whole region [5]. In 2013, 770 million people in the Southern African Development

Community (SADC) were at risk of food insecurity due to severe mid season dry spells [6]. In 2016, eight of South Africa's nine provinces were declared food-insecure due to drought [7].

Drought stress affects maize at almost all growth stages, but the flowering and grain filling stages are the most susceptible, with yield losses of over 90% reported when drought coincides with these growth stages [8]. Genetic improvement of maize for drought tolerance through breeding is a sustainable solution to reduce the impacts of drought. However, breeding for drought tolerance is a complex task because the trait is controlled by many genes, and is highly affected by genotype and environment interaction. Furthermore, grain yield, which is the trait of interest, has very low variation and heritability under drought conditions, which makes selection difficult [9]. Comprehensive screening forms the foundation of any successful drought tolerance breeding program [10]. Additionally, screening materials for drought tolerance at the initial stages of breeding often imparts tolerance to other related stresses such as low nitrogen stress [11].

To increase the chances of selecting the appropriate genotypes, breeders should meticulously consider all the available information at screening phase. Screening maize for drought tolerance entails the selection of high yielding genotypes under water deficit stress and/or optimum conditions. Indirect selection for grain yield via related secondary traits helps to circumvent the challenge of poor grain yield variation and low heritability under drought conditions [12]. Stress-tolerant indices and multivariate statistics are often useful when selecting best genotypes.

Indirect selection for drought tolerance involves selecting for multi-secondary traits that are highly correlated to grain yield and have high heritability values. Furthermore, an ideal secondary trait should have efficiency of indirect selection relative to direct selection of greater than a unity [13]. Morphological and physiological (morphophysiological) traits that are associated with maize drought tolerance include anthesis-silking interval (ASI), leaf rolling (RL), chlorophyll content (CC), leaf senescence (SEN), and number of ears per plant (EPP) [12,14]. Stomatal conductance (Gs) analysis as a physiological response to drought stress has not been widely applied as a screening criterion for drought tolerance in maize and therefore, information about its correlation with grain yield and heritability under drought stress and optimum conditions is not well established.

Biochemical changes that are drought-induced in plants include increase in stress signaling hormones and proteins regulators such as proline and abscisic acid (ABA) among others [15,16]. Proline is an amino acid, which plays an osmoregulatory role in plants under drought conditions [17]. Despite the presence of genetic variation of proline content in plants under drought stress and wide application of proline analysis in understanding drought tolerance of other crops such as wheat [18], cowpea [19], and peanut [20], it has not been applied in large-scale maize drought tolerance screening.

Given the importance of maize provitamin A biofortification in maize-consuming developing countries and the prevailing devastating impacts of drought to maize productivity, it is important to investigate the effectiveness of integrated application of morphophysiological and biochemical traits in screening provitamin A maize inbred lines for drought tolerance. The objectives of the study were to: (i) determine the level of genotypic variation for drought tolerance among tropical provitamin A maize inbred lines in regards to secondary traits, (ii) screen candidate lines for drought tolerance based on grain yield and β-carotene content, and (iii) identify secondary traits that can be effectively used for indirect selection of grain yield under drought stressed and non-stressed conditions. The study forms the basis of germplasm selection for use in provitamin A maize drought tolerance breeding programs.

2. Materials and Methods

2.1. Plant Materials and Study Sites

Fifty inbred lines were screened for drought stress tolerance under managed drought conditions. Inbred lines consisted of 46 provitamin A (orange endosperm) and four drought tolerant white endosperm (non-provitamin A) checks. The 46 provitamin-A inbred lines were sourced from the provitamin A biofortification nurseries of International Maize and Wheat Improvement Center

(CIMMYT) (33) and International Institute of Tropical Agriculture (IITA) (13). The four drought tolerant checks were obtained from the Agricultural Research Council (ARC), Grain Crops Institute, Potchefstroom, South Africa. Names, codes and other information about the inbred lines are given in supplementary Table S1. The respective institutions could not provide pedigree information. The study was carried out across four environments (Env), which were two greenhouse and two field in KwaZulu-Natal Province of South Africa. Greenhouse trials were carried out at the University of KwaZulu-Natal (UKZN), Agriculture Pietermaritzburg campus (29°46′ S, 30°58′ E) from January to April 2017 (Env1) and May to September 2017 (Env2). The field trials were carried out at Ukulinga Research Farm in Pietermaritzburg (29°40′ S, 30°24′ E) from April to August 2017 (Env3) and at Makhathini Research station, Jozini, South Africa (27°39′ S, 32°17′ E) from May to September 2018 (Env4). Supplementary Table S2 shows monthly weather data for the four environments during the respective growing periods.

2.2. *Experimental Design and Crop Establishment*

An alpha lattice design was used to screen the 50 genotypes with two replications containing five incomplete blocks with ten genotypes each and two water regimes (water stress, S and optimum conditions, W) across all the four environments. In the field, the plot size was two rows of 5 m with 0.75 m between the rows and intra row spacing of 0.30 m. Plots were planted with two seeds per station and thinned to one plant 2 weeks after crop emergence. In the greenhouse, a plot was made of four 5 L perforated plastic pots with two plants in each pot, which were thinned to one plant per pot 2 weeks after crop emergence. Pine-bark growing media mixed with loam soil at a ratio of 3:1 was used in the greenhouse. In the field, the soil was predominantly black clay loam soil at both sites. The water stress treatment (S) for all the experiments was implemented in accordance with CIMMYT protocols of withholding irrigation at two weeks prior to expected anthesis date [21]. The water stress condition was maintained until 5 weeks after the flowering of 50% of the genotypes then a single irrigation was applied at grain filling stage. In the field, the optimum treatment (W) involved a 10-day interval sprinkler irrigation throughout the growing period. In the greenhouse, W consisted of drip irrigation for 3 min, four times per day. Across all the environments, compound fertilizer was applied at the rate of 150 kg N, 65 kg P and 65 kg K ha^{-1} at the time of planting and top-dressing fertilizer was applied at five weeks after emergence at a rate of 60 kg N ha^{-1}. In the field, weeds were controlled using Gramoxone® (Syngenta, Greensboro, NC, USA) at a rate of 5 L ha^{-1} and manual weeding whilst hand weeding was done in the greenhouse. Coragen® (Dupont, Washington, DC, USA) and Karate® (Syngenta, Greensboro, NC, USA) insecticides were used to control insects at a rate of 1 L ha^{-1} across all environments when it was necessary.

2.3. *Plant Characteristics*

2.3.1. Morphophysiological Traits

Data for the following morphophysiological and biochemical traits were collected under both water regimes across all the four environments. Most traits were measured following a CIMMYT protocol described by [21]. Both field and greenhouse grain yield (GY) was estimated per plot area bases and converted to tonnes per hectare. The plot area in the greenhouse was determined by multiplying the cylinder-shaped pot area by 4 since a plot was made up of four pots. Similarly, the plot area for the field experiments was determined by calculating the area of a rectangular shaped plot. Number of ears per plant (EPP) was computed as the number of ears with at least one fully developed grain divided by the number of harvested plants. Days to anthesis (DA) is the number of days after planting to when 50% of the plants in a plot start to shed pollen. Days to silking (DS) is the number of days after planting when 50% of plant in a plot produces silks. Anthesis-silking interval (ASI) was calculated as DS minus DA. Leaf rolling (RL) was scored using a scale of 0 to 10 where 0 = unrolled leaf and 10 = leaf rolled like an onion and scores were converted to percentage (%) with measurements taken twice after imposing drought treatment. Leaf senescence (SEN) was scored using a scale from 1 to 10 (1 = 10%; 2 = 20%; 3 = 30%; 4 = 40%; 5 = 50%; 6 = 60%; 7 = 70%; 8 = 80%; 9 = 90%; and 10 = 100% dead leaf area) at 3, 5 and 7 weeks after 50% of the plants reached anthesis. Chlorophyll content (CC) was measured from the adaxial surface of the second top fully expanded leaf of five plants per plot at 3, 5 and 7 weeks after 50% of the plants reached anthesis using SPAD-502-Plus chlorophyll meter (Konica Minolta, Osaka, Japan). Stomatal conductance (Gs) was measured from the abaxial surface of the second top fully expanded leaf using a SC–1 leaf porometer (Decagon Devices®, Pullman, WA, USA) at 3, 5 and 7 weeks after 50% of the plants reached anthesis. Chlorophyll content (CC) and stomatal conductance were measured at midday periods (1200–1400 h). Beta carotene content (BCC) and proline (PC) content were determined and measured as described in the following section.

2.3.2. Biochemical Traits

Beta-carotene content was measured from kernels harvested from the self pollinated plants of each of the 46 provitamin inbred lines. A sample of 20 g, which contained about 30 to 50 kernels, was randomly collected from each of the 46 provitamin-A inbred lines and dispatched to Agricultural Research Council (ARC), Science Analytical Laboratory, Pretoria, South Africa (http://www.arc.agric.za) for β-carotene analysis. High-performance liquid chromatography (HPLC) was used for analysis following a protocol for dried maize kernels as described by [22]. The β-carotene analysis was done three times per sample, giving three data points per each genotype.

Proline analysis was performed at UKZN, Crop Science laboratory following a protocol by [23]. Fresh leaf samples were collected from the second top fully expanded leaves for the S and W treatments of both field and greenhouse experiments at 3 weeks after imposing the S treatment. The leaf samples were freeze-dried at very low temperature (−74 °C) using liquid nitrogen before grinding them into fine powder. A 0.5 g ground leaf sample was homogenized in 10 mL of 3% aqueous sulfosalicylic acid and the homogenate was filtered. Two ml of filtrate was mixed with 2 mL acid-ninhydrin and 2 mL of glacial acetic acid for 1 h in a water bath at 100 °C. After cooling, 4 mL of toluene were added and then mixed vigorously using a rotor. The top mixture containing proline within toluene was decanted from the aqueous phase then taken for UV visible spectrophotometer analysis for the absorbance measurement at a wavelength of 520 nm using a model UV-1800 spectrophotometer (Shimadzu Corporation, Kyoto, Japan). The proline concentration was calculated using the formula shown in Equation (1) [23].

$$\begin{aligned}\text{Proline Content} \quad & (\mu\text{g per gram of dry leaf tissue}) \\ & = [(\mu\text{gproline/mL}) \times \text{mL toluene})/115.5\ \mu\text{g} \\ & /\mu\text{mole}]/[(\text{g sample})/5]\end{aligned} \tag{1}$$

where 115.5 is the molecular weight of proline.

2.4. Data Analysis

2.4.1. Analysis of Variance, Mean Performance and Stress-Tolerant Index

Analysis of variance (ANOVA) for all morphophysiological and biochemical traits was done after carrying out a test of homogeneity of variances. A lattice procedure of R software version 3.5.1 [24] was used to carry out the ANOVA following a mixed model (Equation (2)). Genotypes were considered as fixed effects and environments as random:

$$Y_{ijklm} = \mu + Re_i + B(Re)_{ij} + G_k + E_l + W_m + GE_{kl} + GW_{km} + EW_{lm} + GEW_{klm} + \mathcal{E}_{ijklm} \tag{2}$$

where Y_{ijklm} is the trait of interest, μ is the mean effect, Re_i is the effect of the ith replicate, $B(Re)_{ij}$ is the effect of the jth incomplete block within the ith replicate, G_k is the effect of the kth genotype, E_l is the effect of the lth environment, W_m is effect of the mth water regime while GE_{kl}, GW_{km}, EW_{lm} and GEW_{klm} are the respective interactions and $\mathcal{E}_{ijklm}$ is the residual error term.

Beta-carotene was analysed separately using the general linear model given in Equation (3).

$$Yij = \mu + Gj + S(G)ij \tag{3}$$

where Yij—the performance of ith sample of the jth genotype; Gj—the effect of jth genotype; $S(G)ij$—sample within genotype, which is the error term.

Least significant difference (LSD) test was carried out at 0.5 α level to separate the means.

Stress tolerance index (STI) described by [25] was used to select high yielding genotypes under both conditions as shown in Equation (4):

$$STI = \frac{Ypi \times Ysi}{Yp^2} \tag{4}$$

where Ys—grain yield of a genotype i under drought-stressed condition; Yp—grain yield of i genotype under non-stressed condition, and Xp—mean yield of genotypes under non-stressed condition. Genotypes with high STI value and β-carotene content greater than 1.5 μg g^{-1} were selected.

2.4.2. Variance Components and Heritability

Variance components and broad sense heritability (H) were computed for morphophysiological and biochemical traits in R software following a procedure described by [26] as shown in Equations (5) and (6).

$$\delta^2 p = \delta^2 g + \frac{\delta^2 ge}{e} + \frac{\delta^2}{re} \tag{5}$$

$$H = \frac{\delta^2 g}{\delta^2 p} \tag{6}$$

where $\delta^2 p$—phenotypic variance, $\delta^2 g$—genotypic variances, $\delta^2 ge$—genotype by environment interaction variance, δ^2—error variance, r—number of replications, e—number of environments and H—broad sense heritability.

Heritability classifications guidelines described by [27] were used to describe the heritability levels exhibited by the measured traits in this study in which values from 0 to 0.3 was low, 0.3 to 0.6 was moderate and >0.6 was high.

2.4.3. Principal Component Biplot, Trait Correlations and Relative Selection Efficiency

Phenotypic and genotypic correlations among morphophysiological and biochemical traits were computed separately for each water regime using the META-R software version 6.04 [28]. Correlation classifications guidelines described by [29] were used to explain the correlations among the traits in this study in which correlation coefficients values from ±0.9 to ±1.00 were considered very high correlations, ±0.7 to ±0.9 were high, ±0.5 to ±0.7 were moderate, ±0.3 to ±0.5 were low and ±0.00 to ±0.3 were negligible. A combined principal component PCA biplot was computed to graphically show the traits associated with each water regime. Equation (7) was used to test the efficiency of indirect selection of grain yield via secondary traits relative to direct selection as outlined by [13].

$$\text{Relative efficiency of indirect selection} = \frac{|r_g| h_X}{h_{GY}} \tag{7}$$

where $|r_g|$ is the value of the genotypic correlation between GY and a secondary trait, h_X is the square root of the broad sense heritability of trait, and h_{GY} is the square root of the broad sense heritability of grain yield. According to [13] the most desirable secondary traits should have an efficiency of indirect selection relative to direct selection of greater than unity.

3. Results

3.1. Analysis of Variance, Mean Performances and Stress-Tolerant Index

A combined ANOVA was carried out after separate ANOVA had shown significant ($p < 0.05$) effects of genotype, water regime and their interaction for the studied traits (Table 1). The combined ANOVA revealed that the genotype, water regime, environment and their respective interactions had significant ($p < 0.001$ and $p < 0.05$) effects on GY and other traits except DA, EPP and PC, which were not significantly affected by the interaction of the environment and water regime. Genotypes exhibited significantly ($p < 0.001$) different mean BCC (Table 1).

Mean performance of the top ten and bottom five genotypes with >1.5 μg g^{-1} BCC and ranked in descending order of STI values are presented in Table 2. Thirty-one genotypes had >1.5 μg g^{-1} BCC. Mean BCC was 2.05 μg g^{-1} with genotype 24 (CLHP0022) ranked first in the STI ranking of genotypes with >1.5 μg g^{-1} BCC. Genotypes 1 (CLHP00306) and 11 (CLHP00378) had the highest BCC of 4.22 μg g^{-1} whilst 14 (CLHP0343) had the lowest value of 0.64 μg g^{-1}. The mean performance of all the 50 genotypes under the two water regimes are given in supplementary Table S3. Mean STI was 0.53. Fifty percent of genotypes surpassed the mean STI value. The highest STI value was 0.94 exhibited by entry 27 (CLHP0005) whilst entry 20 (CLHP0364) had the least STI value of 0.23. Genotype 50 (CML569) had the highest STI value of 0.71 among the four drought tolerant checks. The proportion of test genotypes that were ranked higher than the best check on the STI ranking was 10.86%.

The mean GY under optimum and stress conditions were 1.72 t ha^{-1} and 0.88 t ha^{-1}, respectively. This resulted in 51.2% mean yield loss due to drought stress with the highest and lowest percentage yield losses of 64.17% and 29.21%, respectively. Genotype 50 (CML569) was the best yielding check under both drought-stressed and non-stressed conditions with 0.94 t ha^{-1} and 2.26 t ha^{-1}, respectively. The proportion of test genotypes that yielded higher than the best check was 45.57% and 8.70% under drought stressed and non-stressed conditions, respectively. Mean DA was reduced from 75.03 days under optimum conditions to 69.40 days under water stress conditions. Mean ASI increased from a mean of 1.97 days under optimum conditions to a mean of 8.56 days under drought stress conditions. Mean EPP was reduced by drought stress from 2.24 under optimum to 1.68 under drought stress conditions.

Mean RL increased from 3.04% under optimum conditions to 49.78% under water stress. Stomatal conductance was severely reduced from a mean of 368.94 mmol m^{-2} s^{-1} under optimum conditions to a mean of 49.78 mmol m^{-2} s^{-1} under drought stress conditions. Leaf senescence increased due to drought stress from a mean of 11.60% under optimum conditions to 50.18% under water stress conditions. Proline content increased from a mean of 31.83 μg g^{-1} under optimum conditions to a mean of 149.23 μg g^{-1} under water stress conditions with genotype 31 (CML486) having the highest PC of 230.63 μg g^{-1} under water stress conditions.

Table 1. Mean squares and significant tests after combined analysis of variance for nine morphophysiological traits of 50 inbred lines.

SOV	DF	GY	ASI	DA	EPP	RL	Gs	SEN	CC	PC	SOV	DF	BCC
Rep	1	0.33 ***	0.02 *	1.62	0.24	5.28	3370.51	28.13	11.05	570.12	Gen	45	2.758 ***
Rep.Bloc	8	0.37 ***	9.20 ns	156.74 ***	0.19	346.60 ***	10,845.54 ***	239.00 ***	35.05 **	2076.57	Error	92	0.006
Gen	49	1.70 ***	124.35 ***	487.55 ***	2.23 ***	1737.56 ***	17,231.90 ***	1993.14 ***	131.57 ***	6212.13 ***			
Env	3	24.22 ***	10.71 ***	54.82 ***	5.04 ***	516.00 ***	13,750.30 *	1472.46 ***	495.93 ***	889.32			
WR	1	142.85 ***	8685.62 ***	6339.38 ***	63.85 ***	287,471.53 ***	20,372,674.06 ***	297,606.12 ***	88,914.06 ***	2,756,772.93 ***			
Gen.Env	147	0.53 ***	23.21 ***	117.80 ***	0.48 ***	357.14 ***	10,469.86 ***	231.03 ***	31.12 ***	1517.61 ***			
Gen.WR	49	0.28 ***	102.51 ***	281.83 ***	0.92 ***	1715.20 ***	21,625.12 ***	1847.35 ***	94.69 ***	6224.34 ***			
Env.WR	3	13.07 ***	97.87 ***	89.21	0.47	670.90 ***	11,839.37 ***	1964.46 ***	284.66 ***	106.78			
Gen.Env.WR	147	0.22 ***	18.13 ***	119.03 ***	0.43 ***	341.42 ***	10,995.21 ***	215.09 ***	33.01 ***	1506.32 ***			
Error	391	0.01	1.99	9.36	0.20	10.23	1789.44	29.28	6.92	468.59			

SOV—Source of Variation, Rep—Replication, Bloc—incomplete block, Gen—genotype, Env—environment, WR—water regime, DF—degrees of freedom, GY—grain yield, ASI—anthesis-silking interval, DA—days to anthesis, EPP—ears per plant, RL—leaf rolling, SEN—leaf senescence, Gs—stomatal conductance, CCI—chlorophyll content index, PC—proline content, BCC—β-carotene content, * $p < 0.05$, ** $p < 0.01$, *** $p < 0.001$ and ^{ns}, not significant.

Table 2. Mean performance of the top 15 and bottom 5 provitamin A maize inbred lines with >1.5 $\mu g\ g^{-1}$ BCC and ranked in descending order of STI values.

			GY		ASI		DA		EPP		RL		Gs		SG		CC		PC	
Gen	STI	BCC	S	W	S	W	S	W	S	W	S	W	S	W	S	W	S	W	S	W
Top 10																				
24	0.77	2.73	1.00	2.27	6.88	0.75	77.63	79.00	1.88	2.75	27.50	2.00	26.03	420.63	31.25	11.25	19.18	44.45	148.93	33.20
1	0.75	4.22	1.00	2.23	12.75	2.38	74.88	71.25	1.63	2.50	27.50	1.63	26.60	330.11	37.50	10.00	22.21	41.93	175.82	24.88
8	0.70	2.26	1.00	2.08	4.75	2.13	73.63	78.38	2.00	2.13	12.50	1.88	35.53	353.66	33.75	13.75	20.99	37.17	193.53	25.57
17	0.69	1.59	0.99	2.07	4.00	2.25	80.13	74.50	1.88	2.25	26.25	3.25	40.64	413.58	26.25	10.00	19.23	44.53	178.11	40.96
42	0.68	2.64	0.95	2.12	5.25	0.25	53.88	71.75	2.50	2.63	13.75	3.50	24.12	406.85	27.50	13.75	24.33	39.06	165.51	34.52
38	0.64	2.76	0.91	2.08	0.50	1.38	63.50	71.38	2.38	2.75	30.00	2.50	20.14	347.01	47.50	12.50	16.73	43.74	140.98	28.30
44	0.60	2.53	0.94	1.88	5.50	1.38	58.00	64.88	2.13	2.00	16.25	3.00	37.85	361.02	30.00	10.00	25.19	35.24	213.15	25.96
7	0.58	2.61	1.00	1.73	4.63	2.25	79.25	80.75	2.00	2.38	18.75	3.88	34.79	376.43	25.00	10.00	19.79	40.66	205.98	35.33
39	0.58	1.69	0.91	1.90	6.88	1.75	62.00	64.63	1.75	2.25	21.25	4.00	17.52	292.60	43.75	10.00	21.73	39.95	175.97	23.09
25	0.58	1.67	0.85	2.01	4.75	2.63	79.50	75.13	1.75	2.13	22.50	2.88	44.85	273.27	27.50	11.25	23.01	38.75	184.89	28.27
Bottom 5																				
21	0.34	1.97	0.75	1.32	11.38	1.13	78.63	81.25	1.25	2.13	78.75	4.00	66.66	384.13	71.25	12.50	16.32	41.94	103.21	31.60
32	0.31	2.92	0.81	1.15	12.13	2.88	77.50	74.13	1.00	2.50	66.25	3.88	92.69	376.40	85.00	12.50	12.19	36.62	89.56	35.81
19	0.31	1.81	0.75	1.21	12.38	3.50	74.88	78.00	1.13	2.13	58.75	2.50	73.76	349.29	81.25	11.25	15.14	38.61	100.77	27.59
12	0.27	2.88	0.70	1.16	10.63	1.63	75.63	77.88	1.38	1.63	58.75	2.13	66.85	314.55	66.25	13.75	16.04	38.04	139.47	28.92
20	0.23	3.39	0.69	0.97	-9.88	0.88	80.88	72.00	1.13	1.75	87.50	3.50	68.86	338.51	88.75	12.50	10.07	34.20	95.98	28.24
Mean		2.05	0.88	1.72	8.56	1.97	69.40	75.03	1.68	2.24	40.95	3.04	49.78	368.94	50.18	11.60	18.80	39.88	149.23	31.83
CV		3.65	7.87	5.33	17.96	63.75	1.93	5.51	20.69	23.40	10.85	27.36	1.38	16.11	13.26	26.78	4.28	9.09	20.24	13.45
LSD		0.12	0.07	0.09	1.52	1.24	1.32	4.08	0.34	0.52	4.38	0.82	0.68	58.60	6.56	3.06	0.79	3.58	29.79	4.22

Gen—Genotype/inbred line, STI—Stress-Tolerant Index, GY—grain yield, BCC—β-carotene content; ASI—anthesis-silking interval, DA—days to anthesis, EPP—number of ears per plant, SEN—leaf senescence, RL—leaf rolling, Gs—stomatal conductance, CC—chlorophyll content, PC—proline content, CV—coefficient of variation, LSD—least significant difference, S—Water stress conditions and W—well-watered condition.

3.2. Heritability and Variance Components

Grain yield, DA, EPP, Gs and CC exhibited high heritability and genotypic variance values under non-stressed conditions whilst ASI, RL, SEN and PC had high values under drought conditions (Table 3). Heritability estimates ranged from 0.250 to 0.998 under non-drought conditions and 0.127 to 0.987 under drought conditions. Most traits, except Gs, were characterized by sharp differences in heritability values between the two water regimes. For instance, heritability estimates for GY yield dropped from 0.621 under non-stressed conditions to 0.398 under drought conditions whilst PC had heritability values of 0.263 and 0.777 under non-stressed and drought conditions, respectively. Generally, variance due to genotype by environment interaction was higher under drought conditions than under non-drought conditions.

Table 3. Variance components and heritability of measured morphophysiological traits.

Variance Components	Traits								
	GY	ASI	DA	EPP	RL	Gs	SEN	CC	PC
Non-stressed conditions (W)									
Gen ($\delta^2 g$)	0.098	0.189	73.122	0.189	0.074	506.778	0.656	18.235	7.753
Env ($\delta^2 e$)	0.238	0.025	0.000	0.025	0.214	0.385	0.323	1.471	0.000
Gen.Env ($\delta^2 ge$)	0.232	1.052	51.808	0.138	0.910	3.887	3.094	16.008	78.695
Error (δ^2)	0.012	0.683	2.580	0.123	1.618	0.525	9.500	0.714	16.873
Phenotypic ($\delta^2 p$)	0.157	0.537	86.396	0.239	0.504	507.815	2.617	22.326	29.536
Heritability (H)	0.621	0.352	0.846	0.792	0.147	0.998	0.251	0.817	0.263
Stressed conditions (S)									
Gen ($\delta^2 g$)	0.091	23.487	12.852	0.055	345.544	405.282	424.332	2.265	1200.237
Env ($\delta^2 e$)	0.021	0.463	2.244	0.127	4.748	0.036	29.510	5.685	0.000
Gen.Env ($\delta^2 ge$)	0.417	17.578	69.455	0.143	343.666	7.164	194.567	9.653	977.106
Error (δ^2)	0.273	2.328	30.552	0.005	19.500	26.971	48.751	12.963	917.053
Phenotypic ($\delta^2 p$)	0.193	28.173	34.034	0.091	433.898	410.444	479.067	6.298	1559.145
Heritability (H)	0.398	0.834	0.378	0.602	0.796	0.987	0.886	0.360	0.770

3.3. Principal Component Biplot Analysis

A combined principal components biplot was constructed to show traits that were more outstanding in discriminating among genotypes per each water regime (Figure 1). First and second principal components explained 80.4% of the total variation. Performance of genotypes with respect to the measured nine traits under water-stressed conditions (prefixed with SG) were located on the negative side of the biplot whilst majority of genotypes under optimum conditions (prefixed with WG) were on the positive side of the biplot. Traits ASI, RL, SEN and PC were more discriminating among genotypes under drought conditions whilst GY, Gs, CC, DA and EPP were more discriminating among genotypes under non-stressed conditions. Vector length is relative to the discriminating power of the respective trait. Hence, the trait discriminating power within the non-stressed condition in descending order was EPP, DA, GS, CC and GY. On the other hand, PC, RL, SEN and ASI, was the respective order of discriminating under drought stress conditions. Genotypes positioned at or close to a vector of a traits are more associated with that trait. Also genotypes that are located at the tip end of trait vector excelled in the respective trait. For instance, genotype 39 (TZM1224) excelled in proline content (PC) under drought conditions whilst genotypes 2 (CLHP00306) and 25 (CLHP0113) were further on the GY vector under well-watered conditions. Very few genotypes were associated and excelled in EPP and DA under non-stress conditions.

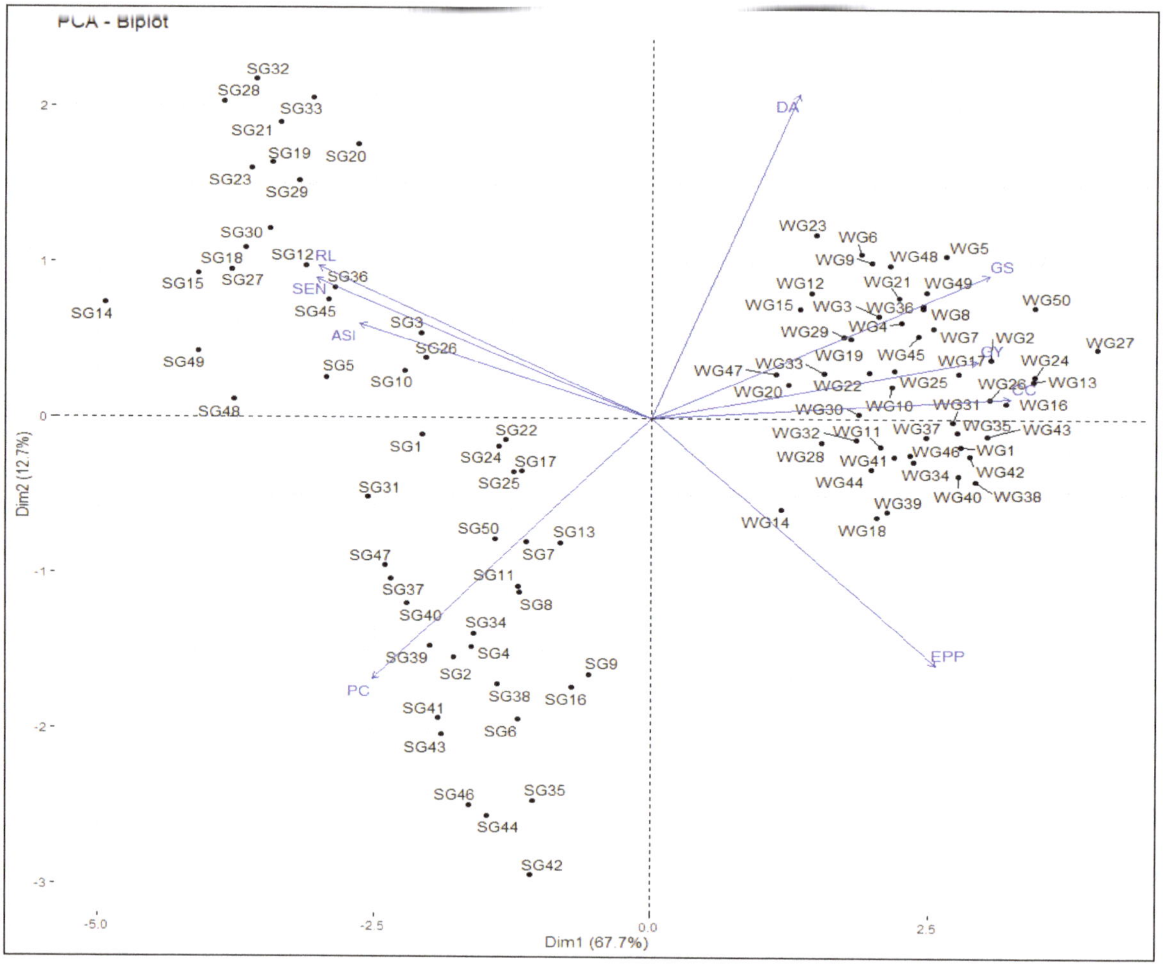

Figure 1. A combined principal component biplot showing genotypes clustering under stress conditions and well-watered conditions. ASI—anthesis-silking interval, CC—chlorophyll content, DA—days to anthesis, EPP—ears per plant, GY—grain yield, Gs—stomatal conductance, RL—leaf rolling, SEN—leaf senescence, PC—proline content, SG—genotypes under water stress conditions, WG -genotypes grown under optimum conditions.

3.4. Phenotypic Correlation Analysis

Phenotypic correlation coefficients (r) among measured morphophysiological and biochemical traits under non-stressed conditions (upper diagonal) and drought conditions (lower diagonal) (Table 4). Under drought stress, GY had significant ($p < 0.001$) and positive correlations that were high with EPP, and low with RL, CC and PC. It also had a significant and negative correlation that was moderate with ASI and Gs, and low with DA, and SEN. Number of ears per plant had significant ($p < 0.001$) and negative correlations which were high with ASI, moderate with DA and Gs, and low with SEN, CC and PC. Stomatal conductance had significant ($p < 0.001$) and moderate positive correlations with RL and SEN. In addition, PC had a significant ($p < 0.001$,) low positive correlation with RL and SEN. It also had a significant ($p < 0.05$), low negative correlation with CC. Under non-stressed conditions, GY had significant ($p < 0.001$) and positive correlation, which was moderate with CC and EPP, low with Gs and DA, and negligible with RL. It also had significant and negative correlation, which was moderate with ASI, low with SEN and negligible with PC. Number of ears per plant had a significant ($p < 0.001$) negative correlation which was moderate with ASI and low with SEN. On the other hand, EPP had a significant ($p < 0.001$) positive correlation, which was moderate with Gs and low with CC. Proline content had negligible correlation with SEN and CC under non-stressed conditions.

Table 4. Phenotypic correlation coefficients describing association of traits under S (lower diagonal) and W (upper diagonal) conditions.

		Non-Stressed Condition (W)								
		GY	**ASI**	**DA**	**EPP**	**RL**	**Gs**	**SEN**	**CC**	**PC**
Water Stressed (S)	GY	1	−0.593 ***	0.462 ***	0.502 ***	0.187 ***	0.496 ***	−0.205 **	0.612 ***	0.109 *
	ASI	−0.694 ***	1	0.446 ***	−0.520 ***	−0.013	−0.307	0.166	−0.004	−0.067
	DA	−0.444 **	0.510 **	1	0.363 ***	−0.308	0.046	0.106	0.267	0.068
	EPP	0.774 ***	−0.711 ***	−0.563 ***	1	0.096 ***	0.626 ***	−0.338 ***	0.324 ***	0.088
	RL	0.446 *	0.458	0.207	0.513 **	1	0.152	0.128	−0.061	0.262
	Gs	−0.566 *	0.333 *	0.232	−0.573 ***	0.538***	1	−0.120 ***	0.246 *	0.492
	SEN	−0.423 ***	0.365	0.041	−0.486 **	0.334 **	0.584 **	1	−0.250 **	0.038 *
	CC	0.406 ***	−0.209	−0.24	0.455 ***	−0.466 *	−0.390 *	−0.396 ***	1	−0.188 *
	PC	0.317 ***	0.374	−0.076	0.415 ***	0.472 ***	−0.235	0.489 ***	−0.356 *	1

GY—grain yield, ASI—anthesis-silking interval, CC—chlorophyll content, DA—days to anthesis, EPP—ears per plant, Gs—stomatal conductance, RL—leaf rolling, PC -proline content, SEN—leaf senescence, *,**,*** indicate level of significance of the correlation at $p < 0.05$, $p < 0.01$ and $p < 0.001$, respectively.

Across environments phenotypic correlation coefficients for grain yield only were computed in order to compare greenhouse and field environments (supplementary Table S4). Environments were further subdivided into greenhouse non-stressed, greenhouse-stressed, field-non-stressed and field-stressed. Correlations among all the four environmental subdivisions ranged from moderate to high ($r = 0.456$ to $r = 0.751$). Significant ($p < 0.001$) and high correlation ($r = 0.751$) was observed between greenhouse and field-non-stressed whilst greenhouse and field-stressed had moderate correlations ($r = 0.643$). Significant ($p < 0.001$) moderate correlations were also observed between greenhouse non-stressed and stressed ($r = 0.553$), and field-stressed and non-stressed ($r = 0.585$). Greenhouse non-stressed and field-stressed had a significant ($p < 0.05$) moderate correlation ($r = 0.502$) whilst greenhouse stressed and field-non-stressed had a significant ($p < 0.05$) moderate correlation ($r = 0.456$).

3.5. *Relative Efficiency of Indirect Selection*

Relative efficiency of indirect selection through secondary traits for grain yield ranged from 0.142 for DA to 1.370 for Gs under drought conditions and from 0.012 for PC under to 1.235 for Gs under non-drought conditions (Table 5). Traits that exhibited relative selection efficiency of greater than unity are EPP, RL, Gs, SEN and PC under drought conditions, and EPP, Gs and CC under optimum conditions. All the genetic correlations coefficients used in calculating the relative efficiency of indirect selection are given in supplementary Table S5.

Table 5. Genetic correlations and the relative efficiency of indirect selection through secondary traits for grain yield improvement under drought stress and optimum conditions.

Secondary Traits	Genetic Correlation (rg) with GY		Relative Efficiency of Indirect Selection	
	Stressed	Non-Stressed	Stressed	Non-Stressed
Anthesis silking interval	0.352 *	0.19 ^{ns}	0.510	0.143
Days to anthesis	0.146 ***	0.163 ^{ns}	0.142	0.190
Ears per plant	0.909 ***	0.912 ***	1.118	1.030
Leaf Rolling	0.934 ***	0.334 ^{ns}	1.321	0.163
Stomatal conductance	0.715 ***	0.974 ***	1.126	1.235
Leaf senescence	0.918 ***	0.842 ***	1.370	0.535
Chlorophyll content	0.738 ***	0.996 ***	0.702	1.142
Proline content	0.829 ***	0.018 ^{ns}	1.153	0.012

* $p < 0.05$, *** $p < 0.001$, ^{ns}, not significant.

4. Discussion

Comprehensive screening of germplasm forms the basis of selecting the right materials, which in turn increases the chances of achieving genetic gain in plant breeding. In this study, the ANOVA showed highly significant genotypic differences with respect to all the studied traits. This indicates the feasibility of genetic improvement for drought tolerance and β-carotene content given that genetic variation is a prerequisite for genetic advance [30]. In a related study, [31] also reported a significant genetic variation among inbred lines obtained from CIMMYT and IITA stress-tolerant breeding programs. The significance of the water regime and its interaction with the genotype for all the traits indicate that the imposed drought stress was effective in discriminating among the genotypes. Furthermore, the highly significant genotype by environment interaction effects observed for most of the traits showed that there were performance differences by inbred lines across the environments. This confirms the complexity of drought tolerance breeding as this makes performance ranking difficult [12]. However, the significant high and moderate positive correlations observed between greenhouse and field environments after subdividing them along water regimes suggest that the two screening environments had almost similar discriminating abilities. This validates the ranking of genotypes within water regimes across the four environments.

Trait performance inconsistence was observed between the two water regimes. That is, combined principal component biplot revealed that RL, SEN, PC and ASI were more discriminating under drought conditions while EPP, GY, CC, Gs and DA were more discriminating under non-stressed conditions. Similarly, variables Gs, GY, EPP, DA and CC had higher heritability and genotypic variance estimates under optimum conditions than under drought conditions whilst the opposite pattern was observed for ASI, SEN, PC and RL (Table 2). In this regard, our findings concur with [32], who reported a decline in the genotypic variances for GY and EPP whilst that of ASI increased under drought conditions. This inconsistency justifies the importance of using a combination of multi-traits when screening germplasm for contrasting growing conditions [33]. This also suggests the need of separating the breeding programs to target different growing conditions in which different secondary traits would be utilized for selection. South Africa, like most SSA countries, has mixed growing conditions; therefore, separating the breeding target environments would be ideal.

The low heritability and genotypic variance estimates exhibited by GY under drought stress agree with findings by some of the previous researchers [34,35]. The observed yield loss (51.2%) due to drought stress was lower than the 81% reported by [36] who attributed greater variation under drought stress to kernel number. In this study, greater variation can be attributed to EPP, RL, Gs, SEN and PC under drought conditions, and EPP, Gs and CC under non-stressed conditions. Stomatal conductance was one of the traits that consistently accounted for greater variation under both conditions. This coupled with having a relative efficiency greater than unity suggests that Gs can be effectively used as selection proxy for GY [13]. This concurs with [37], who reported Gs as the major physiological trait that can effectively discriminate among genotypes between drought tolerant and susceptible plant genotypes. However, its low correlation with GY under non-stressed conditions reduces its effective utilization as an indirect selection criterion for production under optimum conditions.

The EPP and ASI are by far the most applied traits in maize drought tolerance studies [21,33,38]. In the current study, EPP was one of the largest contributors to the total genetic variation as demonstrated by high heritability and genotypic variance estimates under both drought stress and optimum conditions. Furthermore, by having higher discriminatory power as indicated by a longer PCA vector and high correlation with GY under both conditions confirms that EPP is an important trait in maize drought tolerance studies as reported by other researchers [9,36]. This, in addition to having a relative efficiency value greater than a unity justifies the use of EPP in indirect selection for GY. Despite having a moderate correlation with GY and moderate to high heritability estimates under both conditions, the effective use of ASI in indirect selection for GY would be limited by having a relative efficiency value of less than a unity. This is contrary to the findings by [32,39], who reported a relative efficiency of indirect selection for ASI, which was greater than a unity. However, it should be noted that although ASI did not demonstrate to be an ideal trait for indirect selection for GY, it could still be utilized in drought tolerant maize breeding by virtue of having moderate correlation with GY and moderate to high heritability estimates.

The observed significant correlations between GY and photosynthesis related traits such as SEN, RL, CC and Gs confirms the importance of photosynthesis for maize yield [40,41]. Leaf senescence and RL had relative efficiency values of greater than unity under drought and therefore can be used in indirect selection for GY. This concurs with [42] who found delayed leaf senescence to be useful in indirectly selecting for GY in maize under drought conditions. Moderate correlation between PC and GY coupled with high heritability estimate and relative efficiency of greater than unity under drought conditions, infers that genotypes that exhibited high PC under drought stress can be selected as drought tolerant. This supports the claim that under drought conditions proline is released to effect plant cell osmotic adjustments which helps to conserve cell turgor [17,43]. In a related study, [44] reported an increase in PC of 47% and 114% after exposing wheat genotypes to reproductive and grain filling drought stresses, respectively. However, lack of high correlation between PC and GY under non-stressed conditions hinders its effective use as a selection proxy for GY for optimum production.

Although this study did not investigate the effects of water regime, environment and the interaction of genotype and environment on BCC, there is enough evidence in literature that provitamin A content cannot be influenced by genotype and environment interaction but can be affected by the environment [45–47]. This makes selection for provitamin A an easy task. Furthermore, in their studies, [47] and [48] reported no significant correlation between GY and BCC, indicating that the two key traits can be improved simultaneously. The maximum BCC of 4.22 $\mu g\ g^{-1}$ observed in this study was lower than 13.22 $\mu g\ g^{-1}$ reported by [2]. This difference could be attributed to the natural superiority of temperate maize, which was studied by the later, over the its tropical counterpart [49].

5. Conclusions

This study showed that BCC, and the morphophysiological and biochemical traits applied in the screening of provitamin-A inbred lines for drought tolerance were effective in discriminating among the evaluated genotypes. There was considerably high genetic variation among the provitamin A genotypes under study that can be utilized when breeding for drought tolerance. The study also demonstrated that EPP, Gs, RL, SEN and PC can be effectively used in indirect selection of GY under drought-stressed conditions whilst EPP, Gs, and CC were ideal traits for GY indirect selection under non-stressed conditions. By having more secondary traits with a relative efficiency greater than unity under drought stress than under non-stressed, the study confirmed that indirect selection would be more useful than direct selection under drought conditions where GY exhibited low heritability and genotypic variance estimates. Consequently, based on BCC >1.5 $\mu g\ g^{-1}$, which was greater than that of ordinary maize and the ranking for STI values, 30 inbred lines were selected to be utilized in developing drought tolerant provitamin A maize varieties.

Supplementary Materials
Table S1: List of maize inbred lines used in the study, Table S2: Monthly weather data during the greenhouse trials (Env & Env2) at UKZN and field trials (Env3 and Env4) at Ukulinga and Makhatini research stations in South Africa, Table S3: Mean performance of all the genotypes evaluated under drought (S) and non-stressed (W) conditions across four environments ranked by the selection index (SI) values, Table S4: Pearson's correlation coefficients (r) describing association of four environments grouped according to water regimes, Table S5: Genetic correlation describing association of traits under S (lower diagonal) and W (upper diagonal) conditions.

Author Contributions: Conceptualization, A.K. and J.S.; Data curation, A.K. and C.M.; Formal analysis, A.K., C.M., and S.T.; Funding acquisition, J.S.; Investigation, A.K.; Methodology, A.K., J.S., C.M. and S.T.; Project administration, J.S.; Supervision, J.S. and R.Z.; Validation, A.K., J.S., R.Z., C.M. and S.T.; Writing—original draft, A.K. Writing—review & editing, A.K., J.S., R.Z., C.M. and S.T.

Acknowledgments: The authors would like to thank Andile Mshengu for administrative duties. CIMMYT and IITA are also acknowledged for providing the germplasm.

References

1. Thurnham, D. Vitamin A and carotenoids. In *Essentials of Human Nutrition*; Mann, J., Truswell, A.S., Eds.; Oxfrd University Press: Oxford, UK, 2012; pp. 191–216.
2. Harjes, C.E.; Rocheford, T.R.; Bai, L.; Brutnell, T.P.; Kandianis, C.B.; Sowinski, S.G.; Stapleton, A.E.; Vallabhaneni, R.; Williams, M.; Wurtzel, E.T.; et al. Natural Genetic Variation in Lycopene Epsilon Cyclase Tapped for Maize Biofortification. *Science* **2008**, *319*, 330–333. [CrossRef]
3. Andersson, M.; Saltzman, A.; Singh Virk, P.; Pfeiffer, W. Progress update: Crop development of biofortified staple food crops under HarvestPlus. *Afr. J. Food Agric. Nutr. Dev.* **2017**, *17*, 11905–11935. [CrossRef]
4. Bouis, H.E.; Saltzman, A. Improving nutrition through biofortification: A review of evidence from HarvestPlus, 2003 through 2016. *Glob. Food Secur.* **2017**, *12*, 49–58. [CrossRef]
5. Magorokosho, C.; Tongooona, P. Selection for Drought Tolerance in Two Tropical Maize Populations. *Afr. Crop Sci. J.* **2003**, *11*, 151–161. [CrossRef]
6. DAFF. *Trends in the Agricultural Sector*; Department of Agriculture, Forestry and Fisheries: Pretoria, South Africa, 2013.

7. FAO. *Southern Africa el Niño Response Plan (2016/2017)*; Food and Agriculture Organization of the United Nations: Rome, Italy, 2017.
8. Lu, Y.; Hao, Z.; Xie, C.; Crossa, J.; Araus, J.-L.; Gao, S.; Vivek, B.S.; Magorokosho, C.; Mugo, S.; Makumbi, D.; et al. Large-scale screening for maize drought resistance using multiple selection criteria evaluated under water-stressed and well-watered environments. *Field Crop. Res.* **2011**, *124*, 37–45. [CrossRef]
9. Almeida, G.D.; Nair, S.; Borém, A.; Cairns, J.; Trachsel, S.; Ribaut, J.-M.; Bänziger, M.; Prasanna, B.M.; Crossa, J.; Babu, R. Molecular mapping across three populations reveals a QTL hotspot region on chromosome 3 for secondary traits associated with drought tolerance in tropical maize. *Mol. Breed.* **2014**, *34*, 701–715. [CrossRef] [PubMed]
10. Blum, A. Phenotyping and Selection. In *Plant Breeding for Water-Limited Environments*; Springer: New York, NY, USA, 2011; pp. 153–216.
11. Bänziger, M.; Edmeades, G.; Lafitte, H. Selection for drought tolerance increases maize yields across a range of nitrogen levels. *Crop Sci.* **1999**, *39*, 1035–1040. [CrossRef]
12. Betrán, F.J.; Beck, D.; Bänziger, M.; Edmeades, G.O. Secondary traits in parental inbreds and hybrids under stress and non-stress environments in tropical maize. *Field Crop. Res.* **2003**, *83*, 51–65. [CrossRef]
13. Falconer, D.S. *Introduction to Quantitative Genetics, 2nd edn London*; Longman: London, UK, 1981.
14. Edmeades, G. *Progress in Achieving and Delivering Drought Tolerance in Maize—An Update*; ISAAA: Ithaca, NY, USA, 2013; p. 130.
15. Yang, S.; Vanderbeld, B.; Wan, J.; Huang, Y. Narrowing Down the Targets: Towards Successful Genetic Engineering of Drought-Tolerant Crops. *Mol. Plant* **2010**, *3*, 469–490. [CrossRef] [PubMed]
16. Helander, J.D.M.; Vaidya, A.S.; Cutler, S.R. Chemical manipulation of plant water use. *Bioorganic Med. Chem.* **2016**, *24*, 493–500. [CrossRef]
17. Shao, H.-B.; Chen, X.-Y.; Chu, L.-Y.; Zhao, X.-N.; Wu, G.; Yuan, Y.-B.; Zhao, C.-X.; Hu, Z.-M. Investigation on the relationship of proline with wheat anti-drought under soil water deficits. *Colloids Surf. B Biointerfaces* **2006**, *53*, 113–119. [CrossRef]
18. Vendruscolo, E.C.G.; Schuster, I.; Pileggi, M.; Scapim, C.A.; Molinari, H.B.C.; Marur, C.J.; Vieira, L.G.E. Stress-induced synthesis of proline confers tolerance to water deficit in transgenic wheat. *J. Plant Physiol.* **2007**, *164*, 1367–1376. [CrossRef] [PubMed]
19. Zegaoui, Z.; Planchais, S.; Cabassa, C.; Djebbar, R.; Belbachir, O.A.; Carol, P. Variation in relative water content, proline accumulation and stress gene expression in two cowpea landraces under drought. *J. Plant Physiol.* **2017**, *218*, 26–34. [CrossRef] [PubMed]
20. Zhang, M.; Wang, L.-F.; Zhang, K.; Liu, F.-Z.; Wan, Y.-S. Drought-induced responses of organic osmolytes and proline metabolism during pre-flowering stage in leaves of peanut (*Arachis hypogaea* L.). *J. Integr. Agric.* **2017**, *16*, 2197–2205. [CrossRef]
21. Bänzinger, M.; Edmeades, G.O.; Beck, D.; Bellon, M. *Breeding for Drought and Nitrogen Stress Tolerance in Maize: From Theory to Practice*; CIMMYT: Mexico City, Mexico, 2000.
22. Menkir, A.; Liu, W.; White, W.S.; Maziya-Dixon, B.; Rocheford, T. Carotenoid diversity in tropical-adapted yellow maize inbred lines. *Food Chem.* **2008**, *109*, 521–529. [CrossRef]
23. Bates, L.S.; Waldren, R.P.; Teare, I.D. Rapid determination of free proline for water-stress studies. *Plant Soil* **1973**, *39*, 205–207. [CrossRef]
24. R Development Core Team. *R: A Language and Environment for Statistical Computing*; R Foundation for Statistical Computing: Vienna, Austria, 2018.
25. Fernandez, G.C. Effective Selection Criteria for Assessing Plant Stress Tolerance. In Proceedings of the International Symposium on Adaptation of Vegetables and Other Food Crops in Temperature and Water Stress, Shanhua, Taiwan, 13–16 August 1992; pp. 257–270.
26. Allard, R. *Principles of Plant Breeding*; Wiley: New York, NY, USA, 1999.
27. Robinson, H.F.; Comstock, R.E.; Harvey, P.H. Estimates of heritability and the degree of dominance in corn. *Agron. J.* **1949**, *41*, 353–359. [CrossRef]
28. Alvarado, G.; López, M.; Vargas, M.; Pacheco, Á; Rodríguez, F.; Burgueño, J.; Crossa, J. META-R (Multi Environment Trail Analysis with R for Windows) Version 6.04 CIMMYT Research Data & Software Repository Network: 2019. Available online: https://data.cimmyt.org/dataset.xhtml?persistentId=hdl:11529/10201 (accessed on 20 August 2019).

29. Mukaka, M.M. Statistics Corner: A guide to appropriate use of Correlation coefficient in medical research. *Malawi Med. J.* **2012**, *24*, 69–71.
30. Pixley, K.; Rojas, N.P.; Babu, R.; Mutale, R.; Surles, R.; Simpungwe, E. Biofortification of maize with provitamin A carotenoids. In *Carotenoids in Human Nutrition and Health*; Tanumihardjo, S.A., Ed.; Springer Science and Business Media: New York, NY, USA, 2013.
31. Wen, W.; Araus, J.L.; Shah, T.; Cairns, J.; Mahuku, G.; Bänziger, M.; Torres, J.L.; Sánchez, C.; Yan, J. Molecular Characterization of a Diverse Maize Inbred Line Collection and its Potential Utilization for Stress Tolerance Improvement. *Crop Sci.* **2011**, *51*, 2569. [CrossRef]
32. Bolaños, J.; Edmeades, G.O. The importance of the anthesis-silking interval in breeding for drought tolerance in tropical maize. *Field Crop. Res.* **1996**, *48*, 65–80. [CrossRef]
33. Araus, J.L.; Serret, M.D.; Edmeades, G. Phenotyping maize for adaptation to drought. *Front. Physiol.* **2012**, 3. [CrossRef] [PubMed]
34. Bänziger, M.; Cooper, M. Breeding for low input conditions and consequences for participatory plant breeding examples from tropical maize and wheat. *Euphytica* **2001**, *122*, 503–519. [CrossRef]
35. Betrán, F.J.; Beck, D.; Bänziger, M.; Edmeades, G.O. Genetic Analysis of Inbred and Hybrid Grain Yield under Stress and Nonstress Environments in Tropical Maize. *Crop Sci.* **2003**, *43*, 807–817. [CrossRef]
36. Cairns, J.E.; Sanchez, C.; Vargas, M.; Ordoñez, R.; Araus, J.L. Dissecting Maize Productivity: Ideotypes Associated with Grain Yield under Drought Stress and Well-watered Conditions. *J. Integr. Plant Biol.* **2012**, *54*, 1007–1020. [CrossRef] [PubMed]
37. Grzesiak, M.T.; Grzesiak, S.; Skoczowski, A. Changes of leaf water potential and gas exchange during and after drought in triticale and maize genotypes differing in drought tolerance. *Photosynthetica* **2006**, *44*, 561–568. [CrossRef]
38. Campos, H.; Cooper, M.; Edmeades, G.O.; Löffler, C.; Schussler, J.R.; Ibañez, M. Changes in drought tolerance in maize associated with fifty years of breeding for yield in the U.S. corn belt. *Maydica* **2006**, *51*, 369–381.
39. Ziyomo, C.; Bernardo, R. Drought tolerance in maize: Indirect selection through secondary traits versus genomewide selection. *Crop Sci.* **2013**, *53*, 1269–1275. [CrossRef]
40. Cabrera-Bosquet, L.; Molero, G.; Nogués, S.; Araus, J.L. Water and nitrogen conditions affect the relationships of Δ13C and Δ18O to gas exchange and growth in durum wheat. *J. Exp. Bot.* **2009**, *60*, 1633–1644. [CrossRef]
41. Vadez, V. Root hydraulics: The forgotten side of roots in drought adaptation. *Field Crop. Res.* **2014**, *165*, 15–24. [CrossRef]
42. Zheng, H.J.; Wu, A.Z.; Zheng, C.C.; Wang, Y.F.; Cai, R.; Shen, X.F.; Xu, R.R.; Liu, P.; Kong, L.J.; Dong, S.T. QTL mapping of maize (*Zea mays*) stay-green traits and their relationship to yield. *Plant Breed.* **2009**, *128*, 54–62. [CrossRef]
43. Marcińska, I.; Czyczyło-Mysza, I.; Skrzypek, E.; Filek, M.; Grzesiak, S.; Grzesiak, M.T.; Janowiak, F.; Hura, T.; Dziurka, M.; Dziurka, K.; et al. Impact of osmotic stress on physiological and biochemical characteristics in drought-susceptible and drought-resistant wheat genotypes. *Acta Physiol. Plant.* **2013**, *35*, 451–461. [CrossRef]
44. Moayedi, A.A.; Nasrulhaq-Boyce, A.; Tavakoli, H. Application of physiological and biochemical indices for screening and assessment of drought tolerance in durum wheat genotypes. *Aust. J. Crop Sci.* **2011**, *5*, 1014–1018.
45. Menkir, A.; Maziya-Dixon, B. Influence of genotype and environment on beta-carotene content of tropical yellow-endosperm maize genotypes [*Zea mays* L.; Nigeria]. *Maydica* **2004**, *49*, 313–318.
46. Pfeiffer, W.H.; McClafferty, B. HarvestPlus: Breeding Crops for Better Nutrition. *Crop Sci.* **2007**, *47*, S-88–S-105. [CrossRef]
47. Suwarno, W.B.; Pixley, K.V.; Palacios-Rojas, N.; Kaeppler, S.M.; Babu, R. Formation of Heterotic Groups and Understanding Genetic Effects in a Provitamin A Biofortified Maize Breeding Program. *Crop Sci.* **2014**, *54*, 14. [CrossRef]
48. Egesel, C.O.; Wong, J.C.; Lambert, R.J.; Rocheford, T.R. Combining Ability of Maize Inbreds for Carotenoids and Tocopherols. *Crop Sci.* **2003**, *43*, 818–823. [CrossRef]

12

Resilience Capacity Assessment of the Traditional Lima Bean (*Phaseolus lunatus* L.) Landraces Facing Climate Change

María Isabel Martínez-Nieto [1], Elena Estrelles [1], Josefa Prieto-Mossi [1], Josep Roselló [2] and Pilar Soriano [1,*]

[1] Jardí Botànic-ICBiBE, Universitat de València, Quart 80, 46008 Valencia, Spain; maria.isabel.martinez@uv.es (M.I.M.-N.); elena.estrelles@uv.es (E.E.); josefa.prieto-mossi@uv.es (J.P.-M.)

[2] Estación Experimental Agraria de Carcaixent, Carretera CV-5950 (Camino Del Barranquet), Carcaixent, 46740 Valencia, Spain; joseprosello@gmail.com

* Correspondence: pilar.soriano@uv.es

Abstract: Agriculture is highly exposed to climate warming, and promoting traditional cultivars constitutes an adaptive farming mechanism from climate change impacts. This study compared seed traits and adaptability in the germinative process, through temperature and drought response, between a commercial cultivar and Mediterranean *Phaseolus lunatus* L. landraces. Genetic and phylogenetic analyses were conducted to characterize local cultivars. Optimal germination temperature, and water stress tolerance, with increasing polyethylene glycol (PEG) concentrations, were initially evaluated. Base temperature, thermal time, base potential and hydrotime were calculated to compare the thermal and hydric responses and competitiveness among cultivars. Eight molecular markers were analyzed to calculate polymorphism and divergence parameters, of which three, together with South American species accessions, were used to construct a Bayesian phylogeny. No major differences were found in seed traits, rather different bicolored patterns. A preference for high temperatures and fast germination were observed. The 'Pintat' landrace showed marked competitiveness compared to the commercial cultivar when faced with temperature and drought tolerance. No genetic differences were found among the Valencian landraces and the phylogeny confirmed their Andean origin. Promoting landraces for their greater resilience is a tool to help overcome the worldwide challenge deriving from climate change and loss of agrobiodiversity.

Keywords: *Phaseolus*; landrace; seed; germination; drought tolerance; genetic approach; sustainable agriculture; climate change

---

1. Introduction

Agriculture is highly exposed to environmental changes, such as climate warming and aridification, as farming activities depend directly on climate conditions. Indeed, the role of agriculture is fundamentally improving natural resources management, rural development, food production and preserving environmental heritage by the conservation of seminatural habitats, landscape and biodiversity [1,2].

Accordingly, the cultivation and conservation of traditional landraces and crop diversification can be effective adaptation strategies to respond to these changing conditions [3], mainly given the increase in aridity and rainfall unpredictability that derive from these changes in environmental conditions.

Loss of crop diversity is a worldwide challenge. Modern cultivars have replaced local landraces, which are now threatened in food production systems, including cultural heritage, local knowledge and traditional farmer skills. This decline, supported by worldwide globalization, leads to reduced agrobiodiversity on a massive scale, and mainly in developed countries where the industrial food system moves towards genetic uniformity. With the disappearance of traditional species and cultivars, wide ranges of unharvested species also disappear. Promoting local cultivars, which are theoretically more competitive, is one of the major adaptive mechanisms of agriculture to climate change impacts [4–6].

Legumes, specifically *Phaseolus lunatus*, are considered one of the most valuable sources of nutrients in developing countries [7,8]. *Phaseolus lunatus* (Fabaceae), commonly known as "lima bean", and locally termed "garrofó", is the second genus *Phaseolus* species to follow *P. vulgaris* in terms of its economic interest. Attention is paid mainly to its food use worldwide [9], even though other relevant aspects are under study, such as the role on plant protection of the cyanogenic glycosides present in the seeds of this species [10,11].

Cultivated varieties have a South American origin and initially concentrated in northern Peru, where an in-depth selection was developed by the Inca civilization for a long time [12]. According to current germplasm and herbarium records, the conspecific wild ancestor of lima bean is widely distributed from Mexico to Argentina [13]. These landraces are classified into two major groups, Mesoamerican and Andean, according to their geographic origin and seed characteristics [14].

Although it originally comes from Mesoamerica and the Andes, it is currently cultivated throughout Latin America, the southern United States, Canada, and many other world regions including Mediterranean countries, where it is associated with local gastronomy.

On the coasts of the Mediterranean Basin, it is cultivated in warm sunny places in deep well-drained soil. Its strong roots allow plants to thrive on lands where other legumes cannot. It is a highly demanding crop with special requirements. These plants have a type IV climbing growth habit [15] with considerable vegetative development, which means they need a structure that supports, ventilates and illuminates their branches. Nowadays, this cultivation is maintained only for the value of its tender pods and dried grains, and for its special link with traditional cuisine. Currently, the traditional cultivars of this species are being replaced with commercial varieties and represent a testimonial crop in small areas on the European continent.

Hence the present research intends to compare and assess the resistance and adaptability of local cultivars and a commercial variety to face the environmental alterations deriving from climate change. The commercial cultivar is imported from Peru and can be purchased in most retail stores. Primitive landraces, known as 'Pintat', 'Ull de Perdiu' and 'Cella Negra', are traditionally used in the western Mediterranean Basin and are especially cultivated in east Spain (Valencian Community). The use of this species in the eastern Iberian Peninsula in that traditional cuisine is very ancient. Today, we only have references to using these four cultivars in the last 100 years in this region. Our main aim in this work was to recover forgotten crops for the future. In fact, some of the studied cultivars, in particular 'Cella Negra', have practically disappeared today and it has been very difficult to find seeds of this plant.

Furthermore, barcoding is a method to identify taxonomic units using short DNA sequences that allow the determination of the genetic polymorphism and divergences between them. The aim is to identify a region or a combination of regions capable of discriminating taxonomic units, such as species, subspecies, cultivars, or even gene lineages within species [16] and references therein]. Although chloroplast DNA barcoding is utilized mainly to identify plant species, its application can be extended to the food industry, evolution studies and forensics [17]. Various regions of the plastid genome have been proposed to serve as DNA barcodes in plants, such as those put forward by Shaw et al. [18] or Taberlet et al. [19], internal transcribed spacers (ITS) [20] or other specific genes like FRO1 and Phs7 used in legumes phylogenetics by Diniz et al. [21]. This method has been useful in Leguminosae phylogenetics and wild gene pool identifications in *Phaseolus lunatus* [16,21–23]. Thus, it might be a useful tool for typifying local landraces.

This study focuses on seed characterization and providing new information about seed response to temperature and water stress tolerance, estimated during the germinative process, in the

Phaseolus lunatus traditional cultivars from Mediterranean Europe in line with the future global warming and water deficit scenario.

It also aims to characterize molecularly cultivars—by determining the genetic polymorphism and divergences among local, traditional and commercial, as well as American accessions—of *P. lunatus* in an attempt to genetically delimitate landraces, and to find the potential correlation of these genetic characteristics and germination responses. Moreover, the phylogenetic origin of the Valencian cultivars is studied as part of its molecular characterization.

2. Materials and Methods

Four lima bean (*Phaseolus lunatus*) cultivars were tested, three of which were from local Valencian traditional crops ('Pintat', 'Ull de Perdiu' and 'Cella Negra'), mainly provided by the Estación Experimental Agraria de Carcaixent (EEA-Carcaixent) (province of Valencia, Spain). A fourth commercial cultivar imported from Peru to Spain (hereinafter referred to as 'Peru') was bought for the study. The seeds provided by the EEA-Carcaixent were collected during the previous season, nearly one year before starting the germination tests. We did not collect data on the seeds of commercial origin.

2.1. Seed Features

Seed dimensions were measured on a digital image using the ImageJ software [24]. Seed weight was determined by an Orion Cahn C-33 microbalance. All the data were obtained from $n = 50$ seeds from each cultivar.

In order to detect differences in variance levels and to identify homogeneous groups, a one-way ANOVA and Tukey's test ($p < 0.05$) were applied, respectively, for each parameter among the different cultivars.

2.2. Germination Assays

Seed germination assays were performed with the 'Pintat' and the commercial cultivar, 'Peru', for the low seed availability of the rarest landraces, 'Ull de Perdiu' and 'Cella Negra'. Sporadic tests were conducted with them to provide the initial data for future studies. Data were included as Supplementary data.

Tests were carried out using four replications of 10–15 seeds (depending on seed availability) per treatment for each cultivar. Tests were conducted on 14-cm diameter Petri dishes with paper filters kept in climate-controlled cabinets. Illumination was provided by daylight fluorescent tubes with a 12-h photoperiod and a mean irradiance of 100 $\mu mol \cdot m^{-2} \cdot s^{-1}$. The germination process was evaluated for 15 days. Germinated seeds were counted daily.

Firstly, the optimum germination conditions for successive experiments were set. Temperature screening, using six constant temperatures (15 °C, 20 °C, 25 °C, 30 °C, 35 °C, 40 °C), was applied to determine the optimal germination temperature.

The water stress effect was evaluated by the controlled osmotic potential levels generated using polyethylene glycol (PEG 6000) solution at 30 °C according to Villela et al. [25] to obtain 0 (control), −1, −2, −3, −4 and −5 bar. In order to minimize the evaporation and concentration of the effect of solutions and to maintain the known osmotic potential stable, seeds were moistened every 24 h with fresh PEG solutions and plates were kept in double plastic zip lock bags. After 15 days, non-germinated seeds were transferred to distilled water. Thereby, germination capacity recovery was tested to check the potential influence of PEG exposure on seed germination behavior of *Phaseolus lunatus* cultivars.

Germination Percentage and Mean Germination Time (MGT) were considered to compare seed responses. The base temperature (Tb), by back extrapolation [26], and the thermal time requirement [27] were also calculated to compare thermal responses. Then, the base potential (Ψb) and hydrotime (θ) for each cultivar were calculated [28,29].

Variance levels and homogeneous groups were determined by the one-way ANOVA and Tukey's test ($p < 0.05$), respectively, for each parameter among cultivars.

2.3. Genetic Assays

2.3.1. Plant Material and DNA Extraction

The plant material used in the molecular analysis was obtained from the seeds germinated in Germplasm Bank (UV) or Estación Experimental Agraria (EEA-Carcaixent). Eight individuals of each cultivar were analyzed, except for 'Cella Negra', where only five individuals were available at the time of when the genetic assays were done. For the 'Ull de Perdiu' and 'Pintat', we studied two different samples; one seed accession was obtained from the EEA-Carcaixent, while the other was bought from a traditional market. All the 'Cella Negra' seeds came from EEA-Carcaixent and all the 'Peru' ones were obtained from a market as local farmers do not traditionally cultivate them. All the accessions were identified according to seeds' distinctive morphological features. DNA was extracted from young leaves using Doyle and Doyle [30] protocol, modified by Soltis laboratory (2002; https://www.floridamuseum.ufl.edu/wp-content/uploads/sites/95/2014/02/CTAB-DNA-Extraction.pdf). In order to phylogenetically locate the Valencian cultivars, all the accessions provided by Serrano-Serrano et al. [22] in NCBI were used.

2.3.2. Molecular Analyses

A pool of five chloroplast and three nuclear markers (see Table 1) was analyzed to characterize the Valencian landraces. These markers were variable in other studies related specifically to *Phaseolus lunatus* or *Phaseolus* spp. [21–23]. A standard PCR protocol following GoTaq® Polymerase (Promega, Madison, WI, USA) instructions was used for all the markers, except for *Phs7* and *FRO3*, which were amplified following Diniz et al. [21]. The PCR products were purified using the Real Clean PCR Kit (Durviz, Valencia, Spain) and sequenced in an ABI 3100 Genetic Analyzer with the ABI BigDye Terminator Cycle Sequencing Ready Reaction Kit (Applied Biosystems, Foster City, CA, USA).

Table 1. The markers analyzed for the *P. lunatus* Valencian cultivars, primer names, Tm (primer melting temperature) and original references in which they were described.

Marker	Primer Names	Tm	Reference
atpB-rbcL	atpB-f	55	[31]
	rbcL-r		
trnL-trnF	trnL (UAA) 3′ exon f	58	[19]
	trnF (GAA) r		
trnL intron + *trnL-trnF*	trnL (UAA) 5′ exon f	58	[19]
	trnF (GAA) r		
rpoB-trnC	rpoB-f	55	[18]
	trnC-r		
psbA-trnH	psba-f	56	[32]
	trnH-r		
ITS	ITS1-f	58	[20]
	ITS4-r		
Phs7	Phs7-f	62	[21]
	Phs7-r		
FRO3	FRO3-f	62	[21]
	FRO3-r		

2.3.3. Phylogenetic Analyses

As Serrano-Serrano et al. [22] used ITS, *Atpb-rbcL* and *trnL-trnF* fragments to construct a wide *P. lunatus* phylogeny, these fragments were employed to locate the origin of the Valencian cultivars. Two individuals of each Valencian landrace and all the accessions provided by Serrano-Serrano et al. [22] in NCBI were analyzed. MAFFT v. 7.402 [33,34] was utilized to generate a multiple sequence alignment. The preconfigured MAFFT strategy, which favors accuracy with the FFT-NS-I algorithm (an iterative refinement method that performs 1000 iterations), and default parameters were selected. The ambiguously aligned regions were automatically dealt with using GBlocks v. 091b [35] by implementing the least stringent parameters, but allowing for gaps in 50% of sequences. (NCBI accession numbers: MT072230–MT072258, ITS; MT080626–MT080654, *atpB-rbcL*; MT090972–MT091000, *trnL-trnF*; MT110491–MT110519, *rpoB-trnC*; MT124955–MT124983, *psbA-trnH*; MT154089– MT154117, phs7; MT154118–MT154146, FRO3).

A Bayesian phylogenetic MCMC analysis was run with MrBayes v. 3.2.2 [36]. Indels were coded with SeqState v. 1.4.1 [37] according to modified complex coding. The coded indels were considered to be a partition of standard data (states = 0, 1, 2, 3, ?), with the gamma rate and hyperprior fixed at 1.0 to allow different stationary state frequency proportions to be explored by the MCMC procedure. The optimal substitution models for the nucleotide section were inferred with PartitionFinder2 [38] by considering a model with linked branch lengths for the codificant and non-codificant regions of nrITS and chloroplast fragments, respectively, and using the Bayesian information criterion (BIC). Finally, three partitions were considered: two within the ITS (ITS1 + ITS2 and 5.8S), as well as *Atpb-rbcL* + *trnL-trnF*. This analysis favored the HKY + G model for the ITS1 + ITS2 partition, K80 + I for 5.8S, and also GTR + I for the chloroplast region. Then, a MrBayes analysis was conducted with two parallel and simultaneous four-chain runs, executed over 5×10^6 generations, starting with a random tree, and sampling after every 500th step. The first 25% of the data was discarded as burn-in. The 50% majority-rule consensus tree and the corresponding posterior probabilities were calculated from the remaining trees. Chain convergence was assessed by ensuring that the average standard deviation or split frequencies (ASDSF) values were below 0.01, and the potential scale reduction factor (PSRF) values approached 1.00. iTOL v. 4.4.2 [39,40] was used to construct the 50% majority rule consensus tree. The programs MAFFT, MrBayes and PartitionFinder2 were hosted at the CIPRES Science Gateway [41].

2.3.4. DNA Polymorphism and Divergence

The MAFFT original alignment without outgroups was employed to evaluate DNA polymorphism and divergence by taking in account the studied Valencian cultivars and all the accessions, including those used by Serrano-Serrano et al. [22], respectively. All the analyzed markers were utilized to study the Valencian landraces, as well as ITS, *Atpb-rbcL* and *trnL-trnF*, for the whole analysis. Five parameters were calculated by DnaSP v. 6 [42]: segregating sites (s), nucleotide diversity (π), number of haplotypes (h), haplotype diversity (H_d), and the nucleotide genetic differentiation estimate K_{st}.

3. Results

3.1. Seed Features

The seed dimensions of these four *Phaseolus lunatus* cultivars were similar (Table 2). It is noteworthy that the 'Pintat' seeds obtained higher values for the length and width parameters, and had a more rounded contour. The thickness analysis indicated significant differences among cultivars, with the lowest values for the traditional landraces. The 'Peru' and 'Pintat' seeds were the heaviest, while the 'Cella Negra' seeds were lightweight.

Seed coat color is an important consumer trait. In this group, it is a relevant distinctive character for these traditional cultivars (Figure 1; Table 2). The studied commercial cultivar, identified herein as 'Peru', has a completely white seed coat showing no type of pigmentation. The traditional 'Pintat' depicts an irregular spotted pigmentation over the whole external cover, from dark maroon to brown, depending on the maturation stage. The 'Ull de Perdiu' cultivar has a characteristic black eye surrounding the hilum seed zone. Finally, the cultivar known as locally 'Cella Negra' is identified by having a dark brown to black seed tip close to the embryo radicle lobe.

Table 2. Seed morphological features for the different studied cultivars. Length (L), width (W) and their relation (L/W), thickness, as well as weight and color trait of seed coat, are indicated. The same letters indicate homogeneous groups among temperatures ($p < 0.05$) for each cultivar.

	'Peru'	'Pintat'	'Ull de Perdiu'	'Cella Negra'
L (mm)	25.3 ± 0.21 b	26.4 ± 0.13 a	24.8 ± 0.17 b	25.0 ± 0.18 b
W (mm)	15.5 ± 0.14 b	17.3 ± 0.09 a	15.7 ± 0.12 b	15.9 ± 0.14 b
L/W	1.63 ± 0.16 a	1.53 ± 0.10 b	1.58 ± 0.14 ab	1.58 ± 0.21 ab
Thickness (mm)	6.72 ± 0.45 a	5.60 ± 0.42 c	5.96 ± 0.81 bc	6.37 ± 0.72 ab
Weight (g)	1.82 ± 0.07 a	1.81 ± 0.10 a	1.75 ± 0.39 ab	1.54 ± 0.15 b
Pigmentation	No	Yes	Yes	Yes

Figure 1. Seed morphological traits and pigmentation for the studied *Phaseolus lunatus* cultivars; 'Peru'; 'Pintat'; 'Ull de Perdiu'; 'Cella Negra'.

3.2. Germination Assays

3.2.1. Germination Response to Temperature

High germination percentages were achieved at almost all the tested temperatures. The lowest values were for 35 °C in the two studied cultivars, while no germination was observed in any of them above this temperature.

After taking into account the values obtained for the germination percentage and mean germination time, the optimal germination temperature for the studied group of *Phaseolus lunatus* cultivars was set at 30 °C (Figure 2; Table 3). Good results for germination percentages were also obtained at 15 °C and 25 °C, mainly for the 'Pintat' cultivar, but germination was slower in both cases. The values with the same letters did not significantly differ at the 5% level. No significant differences were found when comparing germination velocities among the cultivars at each specific temperature.

The regression lines, indicating the response of germination velocity to increasing temperature (Figure 3), showed a steeper slope for the local 'Pintat' cultivar than for the commercial one, labelled as 'Peru', given the shorter mean germination time; i.e., faster germination. This effect became evident at the temperatures exceeding 19 °C. When the thermal time, S and Tb parameters were calculated from the regression line equations, the 'Pintat' seeds gave values of 131.6 °C·day^{-1} and 5.2 °C respectively, with 185 °C·day^{-1} and −15.0 °C for 'Peru'.

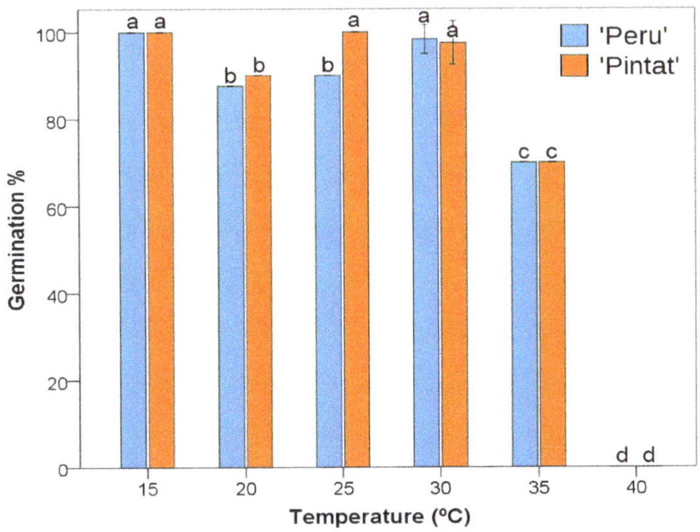

Figure 2. The germination percentage values obtained at different temperatures for the studied *Phaseolus lunatus* cultivars. The same letters indicate homogeneous groups ($p < 0.05$).

Table 3. Mean germination time (days) at different temperatures (°C) for 'Pintat' and 'Peru' cultivars. The same letters indicate homogeneous groups among temperatures ($p < 0.05$) for each cultivar.

	15 °C	20 °C	25 °C	30 °C	35 °C	40 °C
'Peru'	5.6 ± 0.2 b	5.9 ± 0.5 b	5.0 ± 0.3 b	3.9 ± 0.6 a	5.1 ± 0.6 b	-
'Pintat'	6.1 ± 0.2 cd	5.6 ± 0.2 c	4.6 ± 0.2 b	3.6 ± 0.4 a	6.3 ± 0.5 d	-

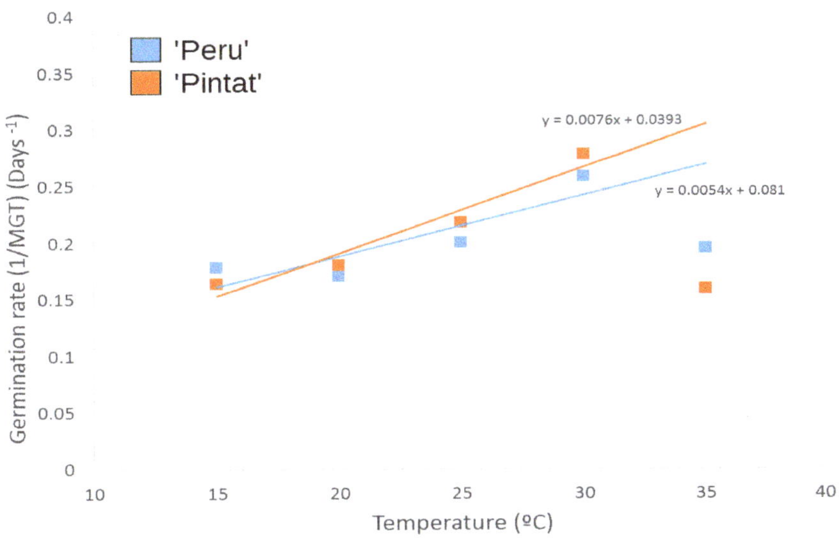

Figure 3. The linear regression of the germination rates (MGT) related to the tested temperatures for two cultivars: 'Pintat' and 'Peru'.

3.2.2. Germination Response to Drought Stress

Characteristically, germination was affected by rising PEG concentrations. In both cases, a drastic reduction in germination was recorded from −4 bars, and no germination took place at −5 bar. However, Figure 4 and Table 4 show better tolerance to induced water stress for the 'Pintat' cultivar, which obtained higher germination percentages and velocity under all the tested conditions. At −2 bar, no significant differences appeared in relation to the control for 'Pintat' landrace, while germination lowered by 28.8% for the cultivar 'Peru'.

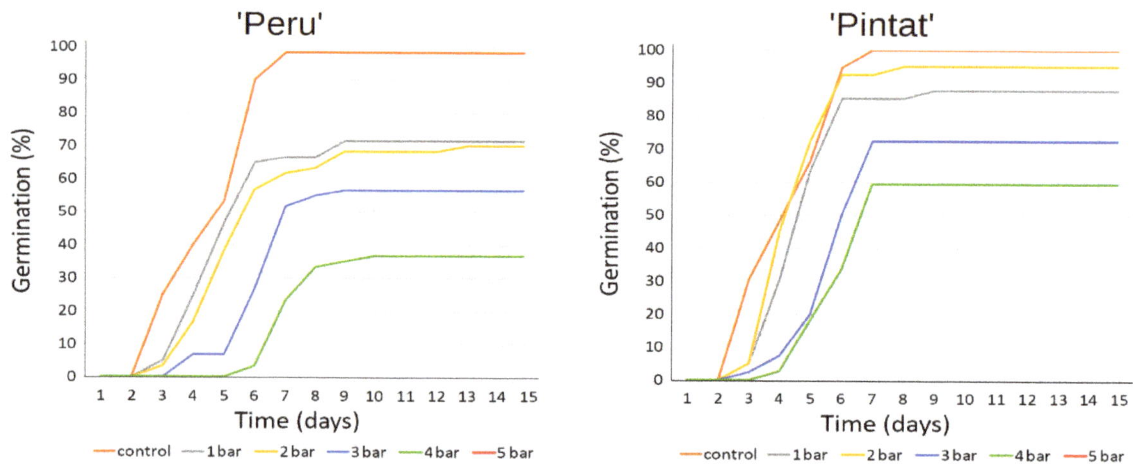

Figure 4. The accumulative germination percentages of the studied cultivars at increasing osmotic pressures obtained with PEG from 0, the control, to a maximum of −5 bar.

Table 4. Mean germination time (MGT), expressed as days, for the *Phaseolus lunatus* cultivars analyzed at increasing PEG 6000 concentrations. The same letters indicate homogeneous groups among cultivars and the tested concentrations ($p < 0.05$).

	Osmotic Potential (Bar)				
	0	**−1**	**−2**	**−3**	**−4**
'Peru'	3.9 ± 0.6 a	4.2 ± 0.4 ab	4.8 ± 1.2 ab	5.5 ± 0.6 abc	6.5 ± 0.6 bc
'Pintat'	3.6 ± 0.4 a	3.9 ± 0.8 a	3.8 ± 0.2 a	5.1 ± 0.2 ab	5.2 ± 0.6 ab

The germination test conducted at increasing water stress pointed out differences in the seeds of the studied cultivars for their physiological potential to face water deficit. A drastic drop in germination was recorded at −4 and −5 bars. The cultivar 'Pintat' demonstrated better tolerance to water stress, which obtained values above 50% for the germination percentage for all the tested osmotic potentials up to −4 bar.

The 'Pintat' cultivar displayed a faster response to germination velocity under all the conditions, and only showed a clear decrease from −2 bar (Table 4; Figure 5).

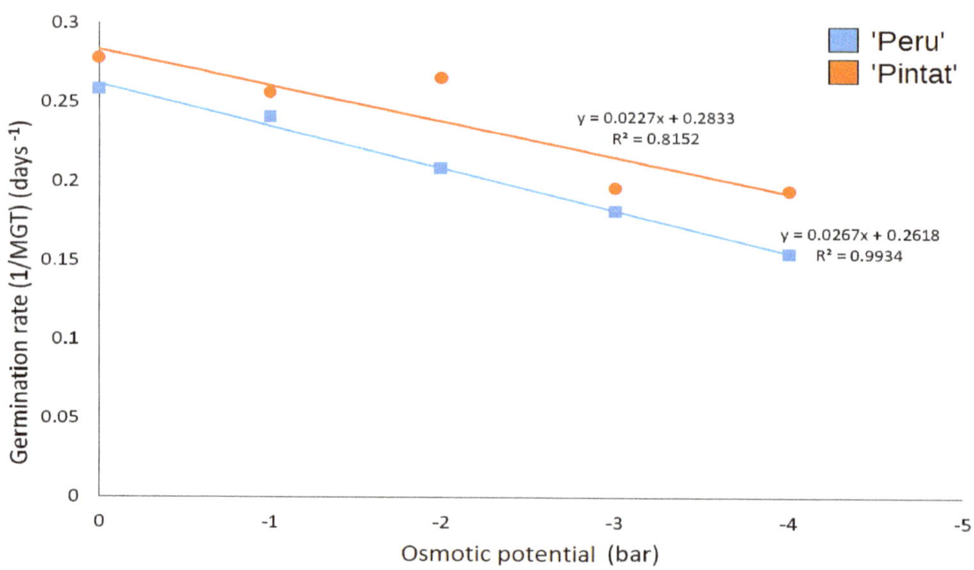

Figure 5. The relation between osmotic potential (bar) and germination rate (1/MGT) for the studied cultivars at 30 °C.

The hydrotime calculated from the linear regression slope was 37.5 and 44.1 bar·day for the cultivars 'Peru' and 'Pintat', respectively. The theoretical values calculated for the minimum osmotic potential (Ψb) at which radicle emergence was prevented were respectively −9.8 and −12.5 bar for these same cultivars. When PEG exposure ended, non-germinated seeds were transferred to the non-stressed medium. After 15 days of incubation in distilled water, no recovery was observed at any tested concentration.

3.3. Genetic Assays

The dataset herein considered comprised new 29 sequences, including three nuclear and five chloroplastic concatenated fragments that belong to the four more common *P. lunatus* cultivars in Spain. The phylogenetic analyses included the ITS, *Atpb-rbcL* and *trnL-trnF* fragments, two individuals of each Valencian landrace and all the accessions provided by Serrano-Serrano et al. [22] in NCBI. Seventy-eight individuals were analyzed. The MAFFT algorithm produced an alignment of 1828 bp with outgroups and 1410 without them. After the automatic removal of ambiguously aligned positions in GBlocks v. 0.91b, 97% (1781 nucleotides) of the original length, 13 selected blocks were kept after taking the outgroups into account. This final alignment included 110 variable positions, of which 73 were parsimony informative and 37 were singletons. The MrBayes analysis reached an average standard deviation of split frequencies of 0.01 after 156 generations. The resulting topology is presented in Figure 6, where the Valencian cultivars were clustered in the AI gene pool, together with the Andean Cordillera accessions from Ecuador and Peru with high clade support (BI ≥ 0.9). These landraces also formed a high supported clade inside the AI gene pool (BI = 0.98). The main groups also displayed good clade support (BI ≥ 0.9), except for the MII gene pool (BI = 0.61), which was clustered in a wider and well-supported Mesoamerican group, split inside.

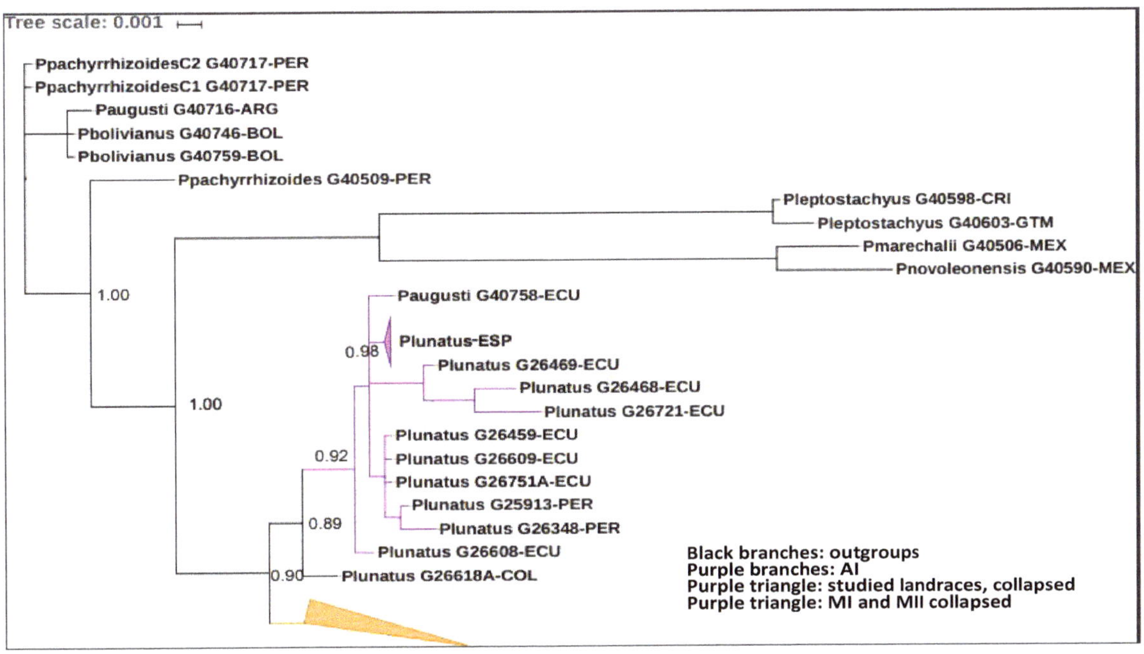

Figure 6. Phylogram depicting the phylogenetic relations among the *P. lunatus* accessions from Spain and South America obtained with MrBayes and based on nrITS and cpDNA data. Support values are given for the main nodes (BI). Colors correspond to the gene pools for wild *P. lunatus*: black branches belong to outgroups, purple branches to AI (Andean I) and the purple triangle inside represents collapsed clades of Valencian landraces (local and 'Peru') as they were almost genetically identical, the orange triangle represents the MI (Mesoamerican I) and MII (Mesoamerican II) collapsed clades as they did not provide any relevant information for our purposes. The whole tree is shown in the Supplementary Material (Figure S2). COL = Colombia, ECU = Ecuador, PER = Peru, ESP = Spain.

Polymorphism and divergence analyses were conducted throughout two groups: only the Valencian cultivars and Valencian and South American cultivars from Serrano-Serrano et al. [22], excluding outgroups. All the analyzed fragments were used in the 29 sequences of the Valencian group with very low genetic diversity estimates. There were no gaps and 4080 sites, of which only four were variable and none showed any pattern of change. This group presented nine haplotypes, 3.8×10^{-4} of nucleotide diversity and a non-significant genetic differentiation estimate K_{st} of 0.022. The group including the South American varieties comprised 67 sequences of concatenated ITS, *Atpb-rbcL* and *trnL-trnF* fragments, 1800 sites and 1413 sites excluding gaps, 44 of which were variable. The group showed 37 haplotypes, a nucleotide diversity of 3.89×10^{-3} and a significant genetic differentiation estimate K_{st} of 0.528 (Table 5).

Table 5. The polymorphism and divergence data of the two *P. lunatus* cultivars groups. The Valencian landraces included the most frequently used cultivars in Spain ('Peru', 'Pintat', 'Ull de Perdiu', 'Cella Negra'). The Valencian + South American group included the Valencian and South American accessions from Serrano-Serrano et al. [22]. N: number of individuals, n: number of sites, n′: number of sites excluding sites with gaps/missing data, S: number of variable sites, h: number of haplotypes, Π (s.d.): nucleotide diversity and standard deviation in brackets, H_d (s.d.): haplotype diversity and standard deviation in brackets, K_{st}: genetic differentiation estimate and its *p*-value (n.s.: non-significant, ***: $p < 0.001$).

	N	n	n′	S	H	Π (s.d.)	H_d (s.d.)	K_{st}
Valencian cultivars	29	4080	4080	4	9	$3.8 \cdot \times 10^{-4}$ (4×10^{-5})	0.862 (0.035)	0.022 n.s.
Valencian + S. American Accessions	67	1800	1413	44	37	$3.9 \cdot \times 10^{-3}$ (2.5×10^{-4})	0.954 ($2 \cdot 10^{-4}$)	0.528 ***

4. Discussion

A landrace differs from a variety that has been selectively modified to improve particular characteristics. These traditional landraces, cultivated continuously for years, are severely threatened by genetic extinction because they are replaced with modern varieties, selected mainly for their higher productivity instead of their resistance to climate change consequences [43].

Currently, the commercial white-seed bean ('Peru') is the cheapest and the most widely sold among lima beans in the Valencian Community, and probably the only one known to most people. Seeds of 'Pintat', and rarely of 'Ull de Perdiu' are sold only in a few local markets, while the cultivar 'Cella Negra' has practically disappeared. The EEA-Carcaixent conserves and multiplies a few accessions of the cultivar 'Cella Negra' for its preservation, from the few seeds that it has been able to find from some farmers who still cultivate it for their own use. We focused our research according to the assumption that the commercial predominance of the different cultivars is not a question based on consumer preferences, but on local farmers' low profitability.

Local landraces are associated with one specific geographical location and, therefore, present climatic adaptability. They are generally better adapted to abiotic stress than modern cultivars [44] and supporting the recovery of their cultivation can mean advantages to face the climate change threat, especially if consumer demand increases. Hence, this climatic adaptability reveals the need to conserve the landrace germplasm as a means to provide information about adaptations to drought and heat stress, and because it constitutes a tool to identify stress-tolerant alleles to improve productivity when faced with climate change [45,46].

Baudoin [47], after thoroughly reviewing the diversity of *Phaseolus lunatus*, already indicated this species as an underexploited crop with a very high cultivation potential given its ability to withstand several types of stress, including severe drought. Moreover, the necessity of carrying out preservation programs for germplasm banks of wild forms and landraces was highlighted.

Regarding seed morphology, clear variability that depends on cultivars is described in the literature. Additionally, variations in dimensions, test patterns and color in cultivars from different countries are known as Potato with small rounded seeds, along with Sieva, with medium-sized reniform seeds, while Andean ones are known as Big Lima and have large, but flat, seeds [14].

In the four studied landraces, no major differences in seed dimensions appeared. Seeds have the morphological characteristics of most of the individuals cultivated for commercial purposes, mainly with big, attractive and nutritional seeds, which indicate their Andean origin.

Regarding seed color and according to bibliographic references, the most frequent seed coat color of the cultivated plants differs depending on the considered geographic area. White seeds are one of the most frequently found among Cuban cultivars [48], while a predominant bicolored pattern is observed in the cultivars grown in Peru [12]. The studied landraces exhibit different bicolored patterns, which also agrees with their geographical provenance.

The seeds of the studied landraces underwent fast germination with no primary dormancy trait, even though dormancy was detected in some colored lima beans [49]. The marked preference for high temperatures in this species stood out, as clearly evidenced by our results with an optimal germination response at 30 °C. Indeed, Polock and Toole [50] and Polock [51] indicated that temperatures below 25 °C in the imbibition phase can be harmful. Other authors have indicated a good response of lima beans at high temperatures (25 °C and 30 °C), even when they were exposed to different salt concentration levels [52].

When we compared germination behavior of the commercial cultivar and the landrace 'Pintat' for the different tested temperature regimes, a stronger competitiveness of the local cultivar was observed from 19 °C to the optimum temperature, close to 30 °C.

Drought is one of the most important problems in agriculture as it leads to reduced yields and loss of crops. Water availability is essential for plants, as they need a good water supply throughout their life cycle. Therefore, water deficit in plants affects all phases of their development, physiological processes, growth and production which, under extreme conditions, can lead to plants dying [43,53,54]. Like the exposed response to temperatures, our results also support the hypothesis of the higher tolerance of those landraces cultivated for years that better adapt to changes in environmental conditions deriving from the Mediterranean climate. In fact, the 'Pintat' landrace was the most tolerant to water stress, simulated by lowering the osmotic potential of PEG solutions. In fact, the thermal time and hydrotime parameters have proven to be good discrimination tools to identify drought- and high temperature-tolerant common bean cultivars [55].

Conversely, DNA barcoding has not found any differences between the local and commercial cultivars used in the Valencian Community, not even when using different sorts: chloroplastic, nuclear, codificant and non-codificant markers. Nevertheless, their origin can be clearly situated. Recent phylogenetic studies have used genome-wide SNP markers polymorphisms [56] to indicate that the wild lima bean is structured into three gene pools, as previously proposed by Serrano-Serrano et al. [22]: the Mesoamerican one (MI); the Mesoamerican two (MII); the Andean one (AI). Their geographic ranges do not generally overlap. In addition, Chacón-Sánchez and Martínez-Castillo [56] also suggest the existence of another Andean gene pool (AII) in central Colombia. Our phylogenetic analyses, based on the data of Serrano-Serrano et al. [22], placed the Valencian cultivars in AI in relation to the 'Big Lima' morphology. These cultivars were phylogenetically grouped with the Andean Cordillera accessions from Ecuador, this being the domestication area of the Andean gene pool located between Ecuador and northern Peru [22,57].

For the Mesoamerican landraces, recent evidence indicates a scenario of a single domestication event in the gene pool MI for all the Mesoamerican landraces, perhaps in central-western Mexico, and the subsequent admixture among landraces and wild populations within the distribution range of gene pool MII, which gave rise to the MII landraces [56]. Therefore, and according to our results, these previous studies have shown that domestication was accompanied by strong founder effects that decreased the genetic diversity of the landraces in the Andes and MI of Mesoamerica. Thus, low polymorphism and divergence statistics have been found in the cultivars used in the Valencian Community (Spain), even between traditional ('Pintat', 'Ull de Perdiu', 'Cella Negra') and commercial ones ('Peru').

However, they all came from the same original gene pool in which a split occurred, as the earliest, when Europeans arrived in America 500 years ago, which is a negligible time in evolutionary terms.

Although other genome-wide barcoding techniques can be used [56,58], the different responses of these genetically close landraces can be explained by epigenetic mechanisms or by a few genes that play a relevant role in crop stress responses [59]. Indeed, rather than DNA barcoding, the search for relevant genes and local landrace alleles related to water stress tolerance could lead to new research works to help preserve these cultivars, by identifying the particular genetic features and their purity. When considering crop tolerance to overcome climate change-related stresses, natural variance among different cultivars can act as genetic reservoir for adaptation capability [60]. This idea, combined with an interest in providing added value to local landraces to defend their use recovery and agro-biodiversity conservation, could supply key future tools that promote local activities to face climate change effects on crops in order to contribute to the auto-sustainability of agronomy activities.

Author Contributions: Conceptualization, E.E., J.R. and P.S.; Investigation, M.I.M.-N., E.E. and P.S.; Methodology, M.I.M.-N., E.E., J.P.-M. and P.S.; Project administration, E.E. and P.S.; Visualization, M.I.M.-N., E.E. and P.S.; Writing—Original draft, M.I.M.-N., E.E. and P.S.; Writing—Review & Editing, M.I.M.-N., E.E., J.P.-M., J.R. and P.S. All authors have read and agreed to the published version of the manuscript.

Acknowledgments: The authors wish to thank the Estación Experimental Agraria de Carcaixent (Valencia, Spain) for its support, particularly Fernando Amorós Ortega for providing us with some of the plant materials (seeds and leaves) used in the experiments. The authors sincerely acknowledge the valuable comments, corrections and suggestions made by anonymous reviewers that have significantly improve the manuscript.

References

1. Crop Production Statistics at Regional Level, EUROSTAT. 2014. Available online: http://ec.europa.eu/eurostat/statistics-explained/index.php/Agricultural_production_-_crops (accessed on 20 December 2019).
2. Knox, J.; Daccache, A.; Hess, T.; Haro, D. Meta-analysis of climate impacts and uncertainty on crop yields in Europe. *Environ. Res. Lett.* **2016**, *11*, 113004. [CrossRef]
3. Climate Change and Agriculture. Houses of Parliament. POSTnote 600. Available online: http://researchbriefings.files.parliament.uk/documents/POST-PN-0600/POST-PN-0600.pdf (accessed on 18 December 2019).
4. Arteaga, S.; Al Hassan, M.; Chaminda Bandara, W.; Yabor, L.; Llinares, J.; Boscaiu, M.; Vicente, O. Screening for salt tolerance in four local varieties of *Phaseolus lunatus* from Spain. *Agriculture* **2018**, *8*, 201. [CrossRef]
5. van de Wouw, M.; Kik, C.; van Hintum, T.; van Treuren, R.; Visser, B. Genetic erosion in crops: Concept, research results and challenges. *Plant Genet. Resour.* **2010**, *8*, 1–15. [CrossRef]
6. What is agrobiodiversity? *Food and Agricultural Organization of the UN*; FAO: Rome, Italy, 2004. Available online: http://www.fao.org/3/a-y5609e.pdf (accessed on 18 December 2019).
7. Seidu, K.T.; Osundahunsi, O.F.; Osamudiamen, P.M. Nutrients assessment of some lima bean varieties grown in southwest Nigeria. *Int. Food Res. J.* **2018**, *25*, 848–853.
8. Seidu, K.T.; Osundahunsi, O.F.; Olaleye, M.T.; Oluwalana, I.B. Amino acid composition, mineral contents and protein solubility of some lima bean (*Phaseolus lunatus* L. Walp) seeds coat. *Food Res. Int.* **2015**, *73*, 130–134. [CrossRef]
9. López-Alcocer, J.J.; Lépiz-Ildefonso, R.; González-Eguiarte, D.R.; Rodríguez-Macías, R.; López-Alcocer, E. Morphological variability of wild *Phaseolus lunatus* from the western region of Mexico. *Rev. Fitotec. Mex.* **2016**, *39*, 49–58.
10. Cuny, M.A.; La Forgia, D.; Desurmont, G.A.; Glauser, G.; Benrey, B. Role of cyanogenic glycosides in the seeds of wild lima bean, *Phaseolus lunatus*: Defense, plant nutrition or both? *Planta* **2019**, *250*, 1281–1292. [CrossRef]
11. Gleadow, R.M.; Woodrow, I.E. Constraints on effectiveness of cyanogenic glycosides in herbivore defense. *J. Chem. Ecol.* **2002**, *28*, 1301–1313. [CrossRef]
12. Pesantes-Vera, M.F.; León-Alcántara, E.; De La Cruz-Araujo, E.; Rodríguez-Soto, J.C. Variabilidad morfo-agronómica en poblaciones de pallar, *Phaseolus lunatus*, cultivado en condiciones de Costa de la Provincia de Trujillo (Perú). *REBIOL* **2015**, *35*, 29–38.
13. Debouck, D.G. *Notes sur Les Différents Taxons de Phaseolus à Partir des Herbiers. Cahiers de Phaséologie—Section Paniculati*; International Center for Tropical Agriculture (CIAT): Cali, Colombia, 2008; p. 233. Available online: http://www.ciat.cgiar.org/urg (accessed on 11 February 2019).

14. Baudet, J.C. The taxonomic status of the cultivated types of Lima bean (*Phaseolus lunatus* L.). *Trop. Grain Legume Bull.* **1977**, *7*, 29–30.
15. Checa, O.; Ceballos, H.; Blair, M.W. Generation means analysis of climbing ability in common bean (*Phaseolus vulgaris* L.). *J. Hered.* **2006**, *97*, 456–465. [CrossRef] [PubMed]
16. Madesis, P.; Ganopoulos, I.; Ralli, P.; Tsaftaris, A. Barcoding the major Mediterranean leguminous crops by combining universal chloroplast and nuclear DNA sequence targets. *Genet. Mol. Res* **2012**, *11*, 2548–2558. [CrossRef] [PubMed]
17. Ferri, G.; Alù, M.; Corradini, B.; Beduschi, G. Forensic botany: Species identification of botanical trace evidence using a multigene barcoding approach. *Int. J. Legal Med.* **2009**, *123*, 395–401. [CrossRef] [PubMed]
18. Shaw, J.; Lickey, E.B.; Beck, J.T.; Farmer, S.B.; Liu, W.; Miller, J.; Siripun, K.C.; Winder, C.T.; Schilling, E.E.; Small, R.L. The tortoise and the hare II: Relative utility of 21 noncoding chloroplast DNA sequences for phylogenetic analysis. *Am. J. Bot.* **2005**, *92*, 142–166. [CrossRef]
19. Taberlet, P.; Gielly, L.; Pautou, G.; Bouvet, J. Universal primers for amplification of three non-coding regions of chloroplast DNA. *Plant Mol. Biol.* **1991**, *17*, 1105–1109. [CrossRef]
20. White, T.J.; Bruns, T.D.; Lee, S.; Taylor, J. Amplification and direct sequencing of fungal ribosomal RNA genes for phylogenetics. In *PCR Protocols*; Innis, M.A., Gelfand, D.H., Sninsky, J.J., White, T.J., Eds.; Academic Press: San Diego, CA, USA, 1990; pp. 315–322.
21. Diniz, A.L.; Zucchi, M.I.; Santini, L.; Benchimol-Reis, L.L.; Fungaro, M.H.P.; Vieira, M.L.C. Nucleotide diversity based on phaseolin and iron reductase genes in common bean accessions of different geographical origins. *Genome* **2014**, *57*, 69–77. [CrossRef]
22. Serrano-Serrano, M.L.; Hernández-Torres, J.; Castillo-Villamizar, G.; Debouck, D.G.; Sánchez, M.I.C. Gene pools in wild Lima bean (*Phaseolus lunatus* L.) from the Americas: Evidences for an Andean origin and past migrations. *Mol. Phylogenet. Evol.* **2010**, *54*, 76–87. [CrossRef]
23. Nicolè, S.; Erickson, D.L.; Ambrosi, D.; Bellucci, E.; Lucchin, M.; Papa, R.; Kress, W.J.; Barcaccia, G. Biodiversity studies in *Phaseolus* species by DNA barcoding. *Genome* **2011**, *54*, 529–545. [CrossRef] [PubMed]
24. Rasband, W.S. ImageJ, US National Institutes of Health, Bethesda, Maryland, USA, 1997–2018. Available online: https://imagej.nih.gov/ij/ (accessed on 16 September 2019).
25. Villela, F.A.; Doni Filho, L.; Sequeira, E.L. Tabela de potencial osmótico em função da concentração de polietileno glicol 6000 e da temperatura. *Pesqui. Agropecu. Bras.* **1991**, *26*, 1957–1968.
26. García-Huidobro, J.; Monteith, J.L.; Squire, G.R. Time, temperature and germination of pearl millet. *J. Exp. Bot.* **1982**, *33*, 288–296. [CrossRef]
27. Trudgill, D.L. Why do tropical poikilothermic organisms tend to have higher threshold temperatures for development than temperate ones? *Funct. Ecol.* **1995**, *9*, 136–137.
28. Kebreab, E.; Murdoch, A.J. Modelling the effects of water stress and temperature on germination rate of *Orobanche aegyptiaca* seeds. *J. Exp. Bot.* **1999**, *50*, 655–664. [CrossRef]
29. Bradford, K.J. A water relations analysis of seed germination rates. *Plant Physiol.* **1990**, *94*, 840–849. [CrossRef] [PubMed]
30. Doyle, J.J.; Doyle, J.L. A rapid procedure for DNA purification from small quantities of fresh leaf tissue. *Phytochem. Bull.* **1987**, *19*, 11–15.
31. Chiang, T.Y.; Schaal, B.A.; Peng, C.I. Universal primers for amplification and sequencing a noncoding spacer between the *atpB* and *rbcL* genes of chloroplast DNA. *Bot. Bull. Acad. Sin.* **1998**, *39*, 245–250.
32. Sang, T.; Crawford, D.J.; Stuessy, T.F. Chloroplast DNA phylogeny, reticulate evolution, and biogeography of *Paeonia* (Paeoniaceae). *Am. J. Bot.* **1997**, *84*, 1120–1136. [CrossRef]
33. Katoh, K.; Standley, D.M. MAFFT multiple sequence alignment software version 7: Improvements in performance and usability. *Mol. Biol. Evol.* **2013**, *30*, 772–780. [CrossRef]
34. Katoh, K.; Misawa, K.; Kuma, K.; Miyata, T. MAFFT: A novel method for rapid multiple sequence alignment based on fast Fourier transform. *Nucleic Acids Res.* **2002**, *30*, 3059–3066. [CrossRef]
35. Castresana, J. Selection of conserved blocks from multiple alignments for their use in phylogenetic analysis. *Mol. Biol. Evol.* **2000**, *17*, 540–552. [CrossRef]
36. Ronquist, F.; Teslenko, M.; Van der Mark, P.; Ayres, D.L.; Darling, A.; Höhna, S.; Larget, B.; Liu, L.; Suchard, M.A.; Huelsenbeck, J.P. MrBayes 3.2: Efficient Bayesian phylogenetic inference and model choice across a large model space. *Syst. Biol.* **2012**, *61*, 539–542. [CrossRef]
37. Müller, K. Seqstate: Primer design and sequences statistics for phylogenetic DNA data sets. *Appl. Bioinf.* **2005**, *4*, 65–69. [CrossRef] [PubMed]

38. Lanfear, R.; Frandsen, P.B.; Wright, A.M.; Senfeld, T.; Calcott, B. PartitionFinder 2: New methods for selecting partitioned models of evolution for molecular and morphological phylogenetic analyses. *Mol. Biol. Evol.* **2017**, *34*, 772–773. [CrossRef] [PubMed]
39. Letunic, I.; Bork, P. Interactive Tree Of Life (iTOL) v4: Recent updates and new developments. *Nucleic Acids Res.* **2019**, *47*, 256–259. [CrossRef] [PubMed]
40. Letunic, I.; Bork, P. Interactive Tree Of Life (iTOL): An online tool for phylogenetic tree display and annotation. *Bioinformatics* **2006**, *23*, 127–128. [CrossRef]
41. Miller, M.A.; Pfeiffer, W.; Schwartz, T. Creating the CIPRES science gateway for inference of large phylogenetic trees. In Proceedings of the Gateway Computing Environments Workshop (GCE), New Orleans, LA, USA, 14 November 2010; IEEE: New York, NY, USA, 2010; pp. 45–52.
42. Rozas, J.; Ferrer-Mata, A.; Sánchez-DelBarrio, J.C.; Guirao-Rico, S.; Librado, P.; Ramos-Onsins, S.E.; Sánchez-Gracia, A. DnaSP 6: DNA sequence polymorphism analysis of large data sets. *Mol. Biol. Evol.* **2017**, *34*, 3299–3302. [CrossRef]
43. Fita, A.; Rodríguez-Burruezo, A.; Boscaiu, M.; Prohens, J.; Vicente, O. Breeding and domesticating crops adapted to drought and salinity: A new paradigm for increasing food production. *Front. Plant Sci.* **2015**, *6*, 978. [CrossRef]
44. Mohammadi, R.; Haghparast, R.; Sadeghzadeh, B.; Ahmadi, H.; Solimani, K.; Amri, A. Adaptation patterns and yield stability of durum wheat landraces to highland cold rainfed areas of Iran. *Crop Sci.* **2014**, *54*, 944–954. [CrossRef]
45. Dwivedi, S.L.; Ceccarelli, S.; Blair, M.W.; Upadhyaya, H.D.; Are, A.K.; Ortiz, R. Landrace germplasm for improving yield and abiotic stress adaptation. *Trends Plant Sci.* **2016**, *21*, 31–42. [CrossRef]
46. Lopes, M.S.; El-Basyoni, I.; Baenziger, P.S.; Singh, S.; Royo, C.; Ozbek, K.; Aktas, H.; Ozer, E.; Ozdemir, F.; Manickavelu, A.; et al. Exploiting genetic diversity from landraces in wheat breeding for adaptation to climate change. *J. Exp. Bot.* **2015**, *66*, 3477–3486. [CrossRef]
47. Baudoin, J.P. Genetic Resources, Domestication and Evolution of Lima Bean, *Phaseolus lunatus*. In *Genetic Resources of Phaseolus Beans. Current Plant Science and Biotechnology in Agriculture*; Gepts, P., Ed.; Springer: Dordrecht, The Netherlands, 1988; Volume 6, pp. 393–407.
48. Castiñeiras, L.; Guzmán, F.A.; Duque, M.C.; Shagarodsky, T.; Cristóbal, R.; De Vicente, M.C. AFLPs and morphological diversity of *Phaseolus lunatus* L. in Cuban home gardens: Approaches to recovering the lost *ex situ* collection. *Biodivers. Conserv.* **2007**, *16*, 2847–2865. [CrossRef]
49. Wahua, T.A.T.; Tariah, N.M. Seed treatment to enhance germination of coloured lima bean (*Phaseolus lunatus* L.). *Field Crops Res.* **1984**, *8*, 361–369. [CrossRef]
50. Pollock, B.M.; Toole, V.K. Imbibition period as the critical temperature sensitive stage in germination of lima bean seeds. *Plant Physiol.* **1966**, *41*, 221–229. [CrossRef] [PubMed]
51. Pollock, B.M. Imbibition temperature sensitivity of lima bean seeds controlled by initial seed moisture. *Plant Physiol.* **1969**, *44*, 907–911. [CrossRef] [PubMed]
52. Rodrigues Do Nascimento, M.G.; Ursulino Alves, E.; Mauricio da Silva, M.L.; Marques Rodrigues, C. Lima bean (*Phaseolus lunatus* L.) seeds exposed to different salt concentrations and temperatures. *Rev. Caatinga* **2017**, *30*, 738–747. [CrossRef]
53. Sedlar, A.; Kidrič, M.; Šuštar-Vozlič, J.; Pipan, B.; Zadražnik, T.; Meglič, V. Drought Stress Response in Agricultural Plants: A Case Study of Common Bean (*Phaseolus vulgaris* L.). In *Drought-Detection and Solutions*; Ondrasek, G., Ed.; IntechOpen: London, UK, 2019. Available online: https://www.intechopen.com/books/drought-detection-and-solutions/drought-stress-response-in-agricultural-plants-a-case-study-of-common-bean-em-phaseolus-vulgaris-em- (accessed on 21 November 2019).
54. da Silva, E.C.; de Albuquerque, M.B.; de Azevedo Neto, A.D.; da Silva Junior, C.D. Drought and its consequences to plants—from individuals to ecosystem. In *Responses of Organisms to Water Stress*; Akinci, S., Ed.; Intechopen: London, UK, 2013. Available online: https://www.intechopen.com/books/responsesof-organisms-to-water-stress/drought-and-its-consequences-toplants-from-individual-to-ecosystem (accessed on 10 September 2019).
55. Cardoso, V.J.M.; Bianconi, A. Hydrotime model can describe the response of common bean (*Phaseolus vulgaris* L.) seeds to temperature and reduced water potential. *Acta Sci. Biol. Sci.* **2013**, *35*, 255–261. [CrossRef]
56. Chacón-Sánchez, M.I.; Martínez-Castillo, J. Testing domestication scenarios of lima bean (*Phaseolus lunatus* L.) in Mesoamerica: Insights from genome-wide genetic markers. *Front. Plant Sci.* **2017**, *8*, 1551. [CrossRef]

57. Motta-Aldana, J.R.; Serrano-Serrano, M.L.; Hernández-Torres, J.; Castillo-Villamizar, G.; Debouck, D.G. Multiple origins of Lima bean landraces in the Americas: Evidence from chloroplast and nuclear DNA polymorphisms. *Crop Sci.* **2010**, *50*, 1773–1787. [CrossRef]
58. Abu-Zaitoun, S.Y.; Chandrasekhar, K.; Assili, S.; Shtaya, M.J.; Jamous, R.M.; Mallah, O.B.; Nashef, K.; Sela, H.; Distelfeld, A.; Alhajaj, N.; et al. Unlocking the genetic diversity within a Middle-East panel of durum wheat landraces for adaptation to semi-arid climate. *Agronomy* **2018**, *8*, 233. [CrossRef]
59. Fortes, A.M.; Gallusci, P. Plant stress responses and phenotypic plasticity in the epigenomics era: Perspectives on the grapevine scenario, a model for perennial crop plants. *Front. Plant Sci.* **2017**, *8*, 82. [CrossRef]
60. Pereira, A. Plant abiotic stress challenges from the changing environment. *Front. Plant Sci.* **2016**, *7*, 1123. [CrossRef]

13

A *LEA* Gene from a Vietnamese Maize Landrace can Enhance the Drought Tolerance of Transgenic Maize and Tobacco

Bui Manh Minh [1], Nguyen Thuy Linh [1], Ha Hong Hanh [1], Le Thi Thu Hien [1,2], Nguyen Xuan Thang [3], Nong Van Hai [1,2] and Huynh Thi Thu Hue [1,2,*]

1 Institute of Genome Research, Vietnam Academy of Science and Technology, 18 Hoang Quoc Viet Street, Cau Giay District, Hanoi 100000, Vietnam; minhmb@igr.ac.vn (B.M.M.); linhnt@igr.ac.vn (N.T.L.); hahonghanh@igr.ac.vn (H.H.H.); hienlethu@igr.ac.vn (L.T.T.H.); vhnong@igr.ac.vn (N.V.H.)

2 Graduate University of Science and Technology, Vietnam Academy of Science and Technology, 18 Hoang Quoc Viet Street, Cau Giay District, Hanoi 100000, Vietnam

3 Maize Research Institute of Vietnam, Vietnam Academy of Agricultural Sciences, 229 Nguyen Thai Hoc Street, Dan Phuong District, Hanoi 100000, Vietnam; nxthangnmri@gmail.com

* Correspondence: hthue@igr.ac.vn

Abstract: Maize (*Zea mays*) is a major cereal crop worldwide, and there is increasing demand for maize cultivars with enhanced tolerance to desiccation. Late embryogenesis abundant (LEA) proteins group 5C is involved in plants' responses to various osmotic stresses such as drought and salt. A putative group 5C LEA gene from *Z. mays* cv. Tevang 1 was isolated, named *ZmLEA14tv*, and cloned into a T-DNA for expression in plants. The deduced amino acid of ZmLEA14tv showed a conserved Pfam LEA_2 domain and a high proportion of hydrophobic residues, characteristic of group 5C LEA proteins. Transgenic tobacco and maize plants expressing *ZmLEA14tv* were generated. During drought simulation conditions, the *ZmLEA14tv*-expressing plants of tobacco showed improved recovery ability, while those of maize enhanced the seed germination in comparison with the non-transgenic control plants. In addition, the survival rate of *ZmLEA14tv* transgenic maize seedlings was twice as high as the control. These results indicated that *ZmLEA14tv* might be involved in the drought tolerance of plants and could be a candidate gene for developing enhanced drought-tolerant crops.

Keywords: drought tolerance; *LEA*; Tevang 1 maize; tobacco

1. Introduction

Late embryogenesis abundant (LEA) proteins are mostly hydrophilic proteins, which can reduce the damage caused by severe environmental conditions. LEA proteins were reported to contribute to various developmental processes and to accumulate in response to drought, low temperature, salt stress, or treatment with the phytohormone ABA [1–4]. The first LEA was reported in cotton seeds [5,6]. LEA proteins accumulated during the late stages of embryogenesis and associated with the desiccation of seeds' embryos [7,8]. The members of the LEA protein family are also expressed during water deficit in bacteria (*Escherichia coli*) and yeast (*Saccharomyces cerevisiae*), suggesting a ubiquitous protective role of these proteins against osmotic stresses [9,10].

Following the Battaglia's classification, LEA proteins are categorized into seven different groups [11]. The LEA proteins of groups 1, 2, 3, 4, 6, and 7 are hydrophilic or typical LEA proteins, which have a low proportion of cysteine and tryptophan residues, and a high proportion of glycine, glutamic acid, lysine, and threonine residues. In contrast, the group 5 LEA protein has high content of hydrophobic residues. Based on amino acid sequences and conserved motifs, the group 5 LEA protein was classified into three subgroups, namely 5A, 5B, and 5C [11]. Subgroup 5C LEA proteins were characterized by a low instability index, low proportion of polar (hydrophilic) and small residues, a higher proportion of non-polar residues, and heat-unstable conformation [11–13]. Moreover, the 5C LEA proteins are folded intrinsically and have more β-sheets than α-helices, which is also different from group 5A and 5B [8,13]. These differences in residue proportion and physical characteristics of group 5C from other LEA protein groups may refer to alternative functions involving stress tolerance.

Recently, due to the development of new sequencing technologies, the whole genome sequences of valuable plants such as rice (*Oryza sativa* L.), maize (*Zea mays* L.), and cotton were published and made available to researchers [14–16]. Based on the conserved domains of LEA proteins, LEA protein families could be identified and characterized through whole-genome prediction approaches. In rice, 34 rice candidate *LEA* (*OsLEA*) genes were identified through a HMMER search (http://hmmer.janelia.org/) [17]. By using a similar method, 242, 136, and 142 candidate DNA regions that encode for LEA proteins were identified in three upland cotton namely *Gossypium hirsutum*, *G. arboreum*, and *G. raimondii*, respectively [18]. The LEA protein profile of maize was also reported with 32 LEA genes distributed non-randomly across chromosomes [19]. The accumulation of LEA profiles in various plants provided fundamental knowledge for functional analysis and *LEA* gene engineering in the future.

A small number of group 5C LEA proteins have been characterized, but their physical characteristics and biological functions are largely unknown. Some members of group 5C were identified in other plants such as cotton LEA14A, soybean D95-4, tomato ER5, hot pepper CaLEA6, *Arabidopsis* AtLEA14A, sweet potato IbLEA14A, rice OSLEA5, foxtail millet SiLEA14A, and wild peanut LEA [8,13,20–26]. The expression of LEA 5C proteins is upregulated by ABA and multiple abiotic stresses including salt and drought [13,23]. Functional studies of group 5C proteins showed that an overexpression of CaLEA6 protein, which originates from hot pepper (*Capsicum annuum*), could improve drought and salt tolerance significantly in tobacco [23]. In addition, the overexpression of other *LEA14A* genes such as *IbLEA14A* and *SiLEA14A* remarkably raised the level of lignification, free proline, and soluble sugar in transgenic sweet potato (*Ipomoea batatas*) calli, *Arabidopsis*, and foxtail millet (*Setaria italica*) [13,23,24]. Recombined OsLEA5 in *E. coli* could protect lactate dehydrogenase from misfolding under different abiotic stresses, resulting in stress tolerance [25].

Maize (*Zea mays* L.) is an important monocot crop worldwide; its production was more than 1.06 billion tons in 2016 [27]. Drought is the major factor that accounts for significant losses in maize productivity. A water reduction of 40% could decrease maize production by 39.3% [28]. Recently, a predicted profile of LEA family in maize has been reported through both bioinformatic and practical approaches [19]. A putative maize *LEA* gene located on chromosome 8 that contained a Pfam LEA_2 domain was described; however, any function of this gene in protecting plants against osmotic stress remains unknown [19]. In the current study, a putative *LEA* gene was isolated from the Tevang 1 landrace, designated as *ZmLEA14tv* and cloned into a T-DNA for expression in plants. The expression patterns of transgenic *ZmLEA14tv* in both tobacco and maize models were investigated to determine the importance of this gene in enhancing drought tolerance in selected plants.

2. Materials and Methods

2.1. Plant Materials and Drought-Stimulating Growth Conditions

Maize seeds (*Z. mays* cv. Tevang 1) were provided by the Maize Research Institute in Vietnam. Tevang 1 cultivar is a well-known landrace maize from the rocky mountain region in Northern Vietnam. The cultivar is well adapted to low annual rainfall and water-deficit cultivation. Maize seeds were germinated and grown under greenhouse conditions at 22/26 °C (night/day) and a photoperiod of 14/10 h (day/night) for two weeks. The genetic background line for transformation was K7, a selected maize that has a higher rate of regeneration and successful transformation through *A. tumefaciens*-mediated methods. The plant material was evaluated for several morphological and physiological traits. For analysis of drought tolerance at the germination stage, 30 seeds of both WT and homozygous *ZmLEA14tv* maize (T2 generation) were germinated on filter paper in a Petri dish wetted with water (as control) or 10% PEG, 20% PEG (*w*/*v*) solution for one day at 30 °C. Shoot and root lengths were measured after eight days of the treatment, followed by taking photographs. For the drought tolerance assay, the five-leaf-stage maize seedlings were assigned to a withholding water period for 14 days followed by a three-day re-watering. At least five seedlings were grown in each plot. The survival rate, fresh stem weight, and fresh root weight of drought-treated and control seedlings were measured. The experiment was replicated three times.

Seeds of WT tobacco (*Nicotiana tabacum*) cultivar K326 and transgenic tobacco were sterilized with 70% ethanol and 5% bleach, which follows a method described previously [29]. The tobacco explant was germinated in Murashige and Skoog (MS) medium containing 200 mg/μL kanamycin under light/dark cycle conditions of 16/8 h at 25 °C. The 20-day-old T3 tobacco plants were put under drought conditions for 15 days and then re-watered to observe the morphological modification.

2.2. Isolation of the Putative LEA Coding Region in Tevang 1 Landrace Maize

Based on the putative *ZmLEA14A* sequence published in GenBank (accession number EU976614.1) and the flanking sequence of this gene located on maize chromosome 8 in the maizeGBD database (http://www.maizegdb.org/), PCR-specific primers for amplification of this region were designed. The DNA regions containing the putative *ZmLEA14A* open reading frame (ORF) and 5′ UTR of this gene were amplified from genomic DNA of Tevang 1 cultivar using the *ZmLEA14tv_F* forward primer sequence (5′TCACTTCTCTTCCAGCGAGTAC3′) and *ZmLEA14tv_R* reverse primer sequence (5′ TCTCGTACTACTCAAGCAGCAC3′). The PCR product was denoted as *ZmLEA14tv* and purified by Thermo Scientific PCR product purification kit (Cat. number #K0702, Waltham, MA, USA) then cloned into a pJET1.2 cloning vector following the manual of the producer. The cloning vector pJET1.2, which contains the PCR fragment, was transformed into *E. coli* DH10β competent cell by heat shock at 42 °C for 1 min. The colonies harboring the pJET1.2–ZmLEA14tv plasmid were checked by colony PCR with isolation primers and the restriction enzyme *Bgl*II.

2.3. ZmLEA14tv Expression Vector Construction and Transgenic Plant Generation

The coding region of *ZmLEA14tv* was amplified with the *ZmLEA14tv_CloneF* forward primer sequence (5′ ATTACCATGGCGCAGTTGGTG3′) and *ZmLEA14tv_CloneR* reverse primer sequence (5′ ATTAGCGGCCGCGAAGATGCTGG3′) that generates the recognition sites of *Nco*I and *Not*I at the 5′ and 3′ ends of the PCR products, respectively. The thermocycler program (Eppendorf, Hamburg, Germany) starts at 95 °C for 3 min, followed by 30 cycles of amplification (95 °C for 1 min; 56 °C for 30 s; 72 °C for 1 min); the final extension step was 72 °C for 10 min. The 463-bp PCR product was then treated with *Nco*I and *Not*I, purified, and ligated into the pRTRA7/3 vector to generate a 35S promoter-ZmLEA14tv-35S terminator construct. The cassette was cut and combined into the T-DNA region of the pCAMBIA1300 (8958 bp) binary vector. The recombined binary vector pCAMBIA1300/ZmLEA14tv was transformed into *A. tumefaciens* strain EHA 105 after validation by sequencing. The transgenic tobacco and maize plants were regenerated by modified *Agrobacterium*-mediated transformation methods described in the studies of Topping (1998) and Frame et al. (2011), respectively [30,31].

2.4. Sequence Alignment and Gene Evolution Analysis

The sequence of PCR products was verified by 3500 Series Genetic Analyzers (Applied Biosystems, Foster City, CA, USA) followed the Sanger method. A deduced amino acid sequence of ZmLEA14tv protein was generated using the ExPaSy web tool (https://web.expasy.org). The isoelectric point molecular mass, the proportion of amino acid, and grand average of hydropathy (GRAVY) index of the putative ZmLEA14tv peptide were estimated using the ProtParam web tool (https://web.expasy.org/protparam/). Motif analysis was performed using the Pfam program (http://www.ebi.ac.uk/Tools/InterProScan/). The completed amino acid and deduced amino acid sequences of subgroup 5C

LEA proteins were used to construct a phylogenetic tree. Sequence alignment was carried out with ClustalW and adjusted manually. The phylogenetic tree was constructed with the neighbor-joining bootstrap method using the MEGA v6.0 program [32].

3. Results

3.1. Isolation and Vector Construction of *ZmLEA14tv* Gene

A putative LEA coding sequence from a native maize landrace Tevang 1 (*Z. mays* cv. Tevang 1) was isolated and named *ZmLEA14tv*. The isolated *ZmLEA14tv* had a size of 693 base pairs (bp) with an open reading frame of 459 bp in length encoding for a deduced 152 amino acid (aa) protein. Sequence analysis exhibited the highest similarity at 99% with a *Z. mays LEA14A* coding sequence (accession number NM_001159174), followed by *LEA14A* of *Shorgum bicolor* (XM_002454858.2) and *LEA14A*-like gene of *S. italica* (93% and 87%, respectively). The comparison with the reference sequence number NM_001159174 in GenBank showed two variations, G381A and C456T; however, the deduced amino acid sequence was not changed. The putative protein was predicted at 15.96 kDa in molecular weight with a pI of 6.08. The protein was rich in Val (12.6%), Leu (11.3%), and Gly (9.2%), but contained low quantities of Trp (0.7%), Asn (1.3%), Cys (0.7%), and Gln (1.3%). The GRAVY and instability index of the putative *ZmLEA14tv* were 0.047 and 16.01, respectively, suggesting stability and hydrophobic nature. A conserved "LEA_2" motif (PF03186), which was classified into subgroup 5C according to Battaglia's classification of LEA proteins, was found on the ZmLEA14tv through an InterProScan search [11]. Further analysis showed that ZmLEA14tv contained a low percentage of polar amino acids (23.02%) and a high level of hydrophobic residues (47.02%). The ZmLEA14tv protein displayed diverse homology with other group 5C LEA proteins and broadly matches similar segments in related LEA proteins, indicating a close evolutionary relationship among these proteins (Figure 1A,B). The phylogenetic analysis also showed that the ZmLEA14tv protein has the closest relationship with maize LEA14A protein (accession number NM_001159174), followed by an LEA-like protein from rice, namely Os01g0225600 (accession number NM_001048996), supported by high bootstrap values (99% and 95%, respectively).

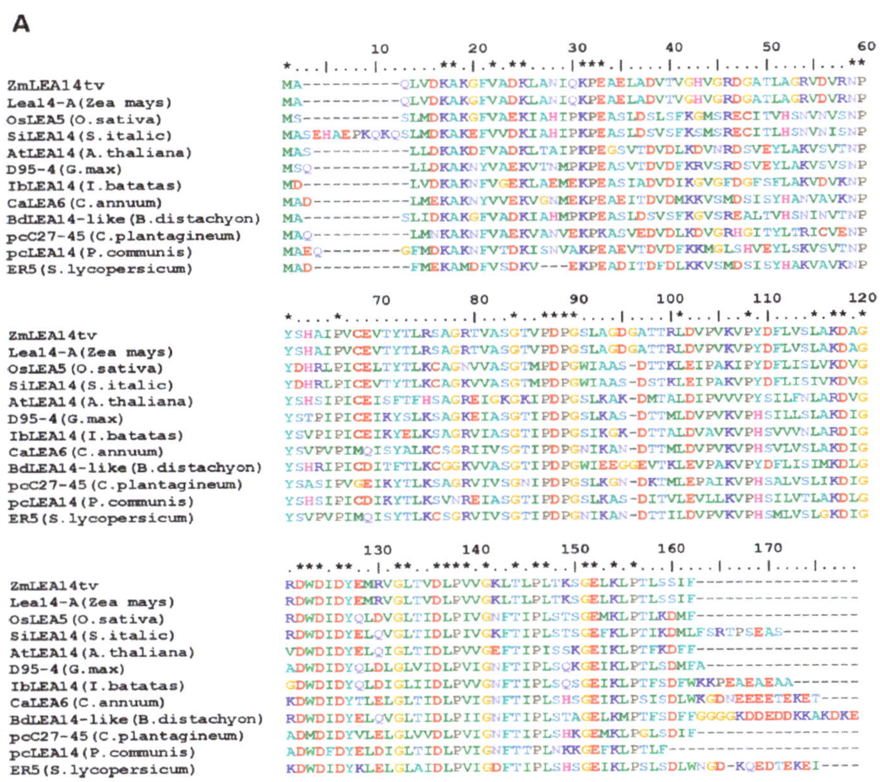

Figure 1. *Cont.*

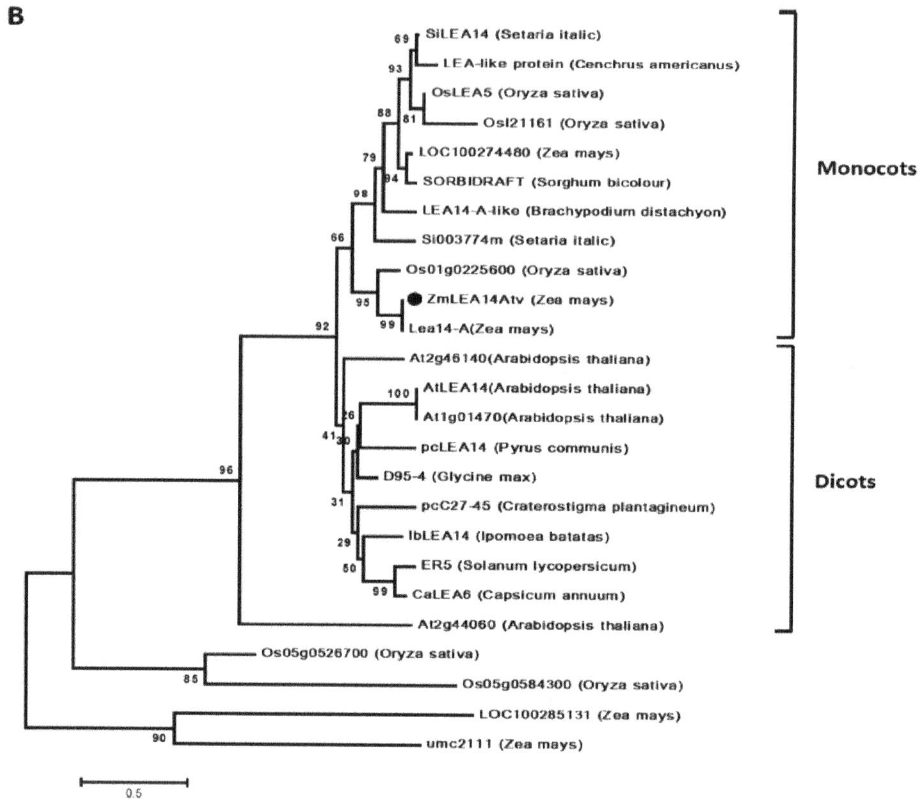

Figure 1. Sequence alignment and phylogenetic relationship for the putative ZmLEA14tv protein and its homologs. (**A**) Multiple sequence alignment of ZmLEA14tv with its homologs (LEA group 5 protein) from various plant species. The conserved positions were marked as stars. (**B**) Neighbor-Joining phylogenetic trees of ZmLEA14tv and its homologs. The clades of monocots and dicots are marked. ZmLEA14tv in *Z. mays* cv. Tevang 1 branch is highlighted by a solid dark circle. The GenBank accession numbers are as follows: SiLEA14 (*S. italic*, KJ767551), AtLEA14 (*Arabidopsis thaliana*, NM100029), Lea14-A (*Z. mays*, NM001159174), D95-4 (*Glycine max*, U08108), IbLEA14 (*Ipomoea batatas*, GU369820), ER5 (*Solanum lycopersicum*, U77719), Lemmi9 (*S. lycopersicum*, Z46654), CaLEA6 (*Capsicum annuum*, AF168168), OsLEA5 (*Oryza sativa*, JF776156), pcC27-45 (*Craterostigma plantagineum*, M62990), pcLEA14 (*Pyrus communis*, AF386513), At1g01470 (*A. thaliana*, BT015111), D95-4 (*G. max*, U08108), At2g46140 (*A. thaliana*, NM130176), Os01g0225600 (*O. sativa*, NM001048996), LEA14-A-like (*Brachypodium distachyon*, XM003567779), BdLEA14-like (*B. distachyon*, XM003567779), LOC100274480 (*Z. mays*, NM001148839), SORBIDRAFT (*Sorghum bicolour*, XM002441543), LEA-like protein (*Cenchrus americanus*, AY823547), OsI21161 (*O. sativa*, CM000130), Os05g0526700 (*O. sativa*, NM001062639), Os05g0584300 (*O. sativa*, NM001062985), At2g44060 (*A. thaliana*, BT024723), LOC100285131 (*Z. mays*, EU970969) and umc2111 (*Z. mays*, NM001155750).

3.2. *ZmLEA14tv Gene Expression in Drought Resistance for Transgenic Tobacco*

To evaluate the function of the *ZmLEA14tv* transgenic structure in plant osmotic tolerance, the transgenic tobacco plants that expressed ZmLEA14tv under the control of the CaMV 35S promoter were selected for further analysis (Figure 3A). Thirty transgenic plants were obtained, and three homozygous T3 transgenic lines (LEAtv-L1, LEAtv-L3, LEAtv-L7) with high expression levels of ZmLEA14tv (Figure 2B) were chosen for further investigation. To investigate the drought tolerance of the transgenic tobacco, the seedlings were treated with a shortage of water for 15 days. Subsequently, the plants were re-watered and grown for three days. Under normal and drought conditions, there were no significant differences in morphological features such as height, weight, and leaf surface between the transgenic and non-transgenic wild-type (WT) plants. Leaves of the transgenic and control plants became curled and wilted after 15 days of drought. However, 100% of the transgenic tobacco was restored after a three-day re-watering, unlike the control plant (Figure 2C). The fastest recovery was observed in LEAtv-L1, which also expressed the highest level of the *ZmLEA14tv* transgene. This result indicated a correlation between the *ZmLEA14tv* expression and the recovery ability of the plant after the drought conditions.

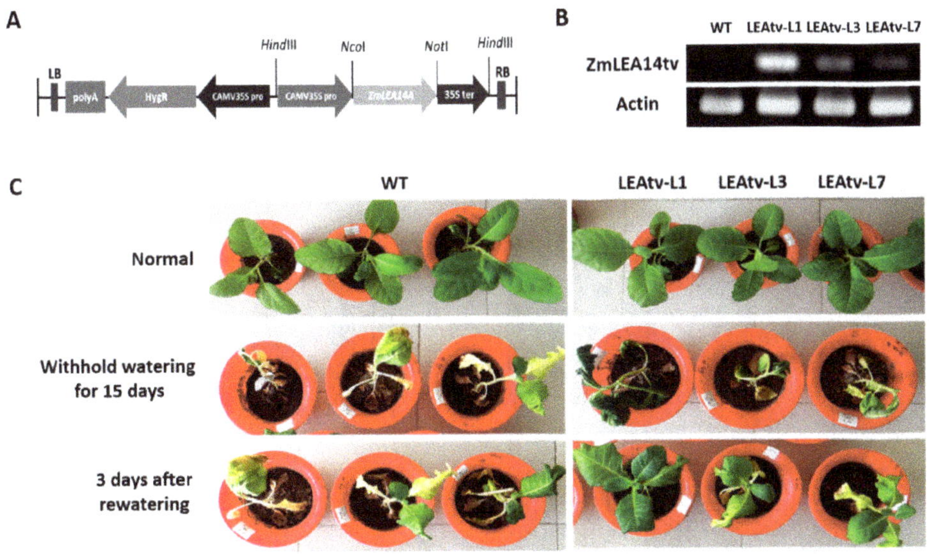

Figure 2. The expression of the *ZmLEA14tv* in tobacco (*Nicotiana tabacum*). (**A**) Schematic description of T-DNA involving in the pCAMBIA1300/ZmLEA14tv plasmid for the expression of *ZmLEA14tv* in plants. LB: left T-DNA border; RB: right T-DNA border; HygR: Hygromycin resistant gene; CaMV35S pro: Cauliflower mosaic virus 35S promoter; 35S ter: 35S terminator; *ZmLEA14A*: coding region of *ZmLEA14tv* gene; *Nco*I, *Not*I, and *Hin*dIII: restriction site of *Nco*I, *Not*I và *Hin*dIII, respectively. (**B**) RT-PCR analysis of ZmLEA14tv in transgenic tobacco lines (LEAtv-L1, LEAtv-L3, LEAtv-L7). WT: the wild type was used as a control. (**C**) The phenotype of transgenic and WT tobacco explants under normal and drought stress conditions.

3.3. Transgenic Maize with ZmLEA14tv Gene in Drought Tolerance

The ZmLEA14tv integration was confirmed by genomic PCR using pairs of primer sets specific to the Hpt and 35S promoter regions, respectively. RT-PCR and qRT-PCR of ZmLEA14tv were performed to validate the expression of the transgenic structure of T2 transgenic maize lines. Three T2 lines, namely L1453, L1482, and L1510, showed the highest expression level (2.7-, 9.2-, and 5.8-fold higher than the control, respectively) and the T2 seeds of these re-watered lines were used for further analysis (Figure 3A,B).

The drought tolerance of transgenic maize seeds during germination was examined. When germinated in water for eight days, the transgenic lines showed better growth than the WT line used as the regeneration material. Under normal conditions (H_2O), no significant differences in shoot height were observed between the transgenic lines and the WT ones. However, the root length of the L1510 and L1482 lines, which have higher ZmLEA14tv transcripts accumulation, was significantly higher than that of the wild type. In comparison with the control group in water, the germination of both the WT and transgenic lines was severely suppressed under 10% and 20% PEG stress (Figure 3C). None of the experimental seeds could germinate in 20% PEG. However, better germination and subsequent development were observed with the seeds of L1510 and L1482, which had higher expression of ZmLEA14tv transcripts in 10% PEG than the wild type (Figure 3D).

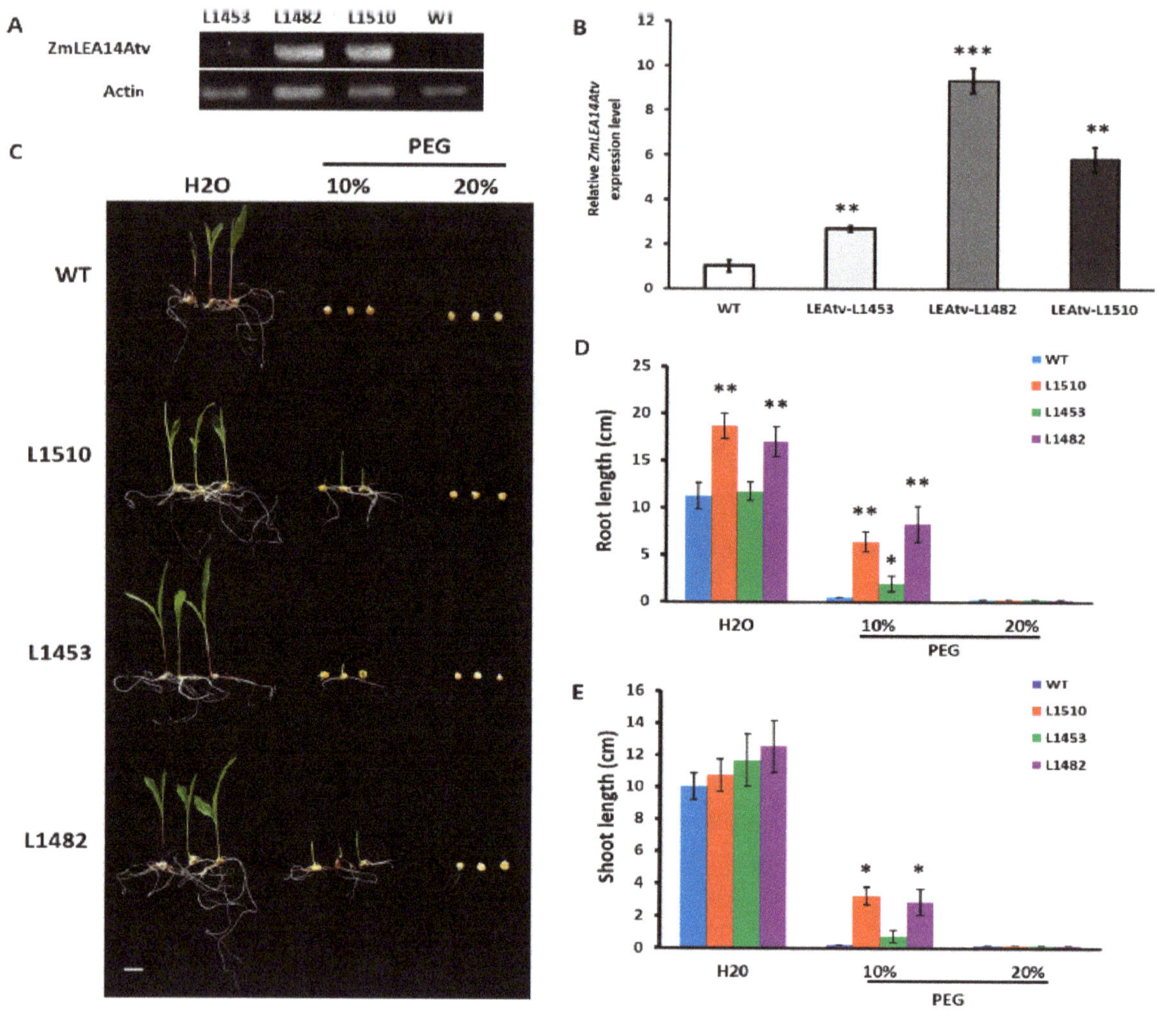

Figure 3. The expression of *ZmLEA14tv* in maize (*Z. mays*). (**A**,**B**) *ZmLEA14tv* expression in three lines of T2 transgenic maize (L1453, L1482, and L1510) determined by RT-PCR (**A**) and qRT-PCR (**B**). Data in (**B**) represent means and standard errors for three biological replicates. (**C**) The phenotype of transgenic and WT maize under various abiotic stress treatment during the germination stage. The T2 of transgenic seeds were soaked in water (as control) or in 10% PEG, 20% PEG solution for drought simulation for one day at 30 °C and then placed on filter paper in plastic boxes wetted with the same solutions mentioned above for eight days. Each experiment was replicated three times. (**D**,**E**) The root and shoot length of transgenic and wild-type maize germinated under control (H_2O) and drought-simulating conditions (10% PEG and 20% PEG). Statistical significance was determined by Student's *t*-test. * $p < 0.05$; ** $p < 0.01$.

Furthermore, the drought tolerance of transgenic maize seedlings in soil was examined (Figure 4B–E). No significant differences in survival rate and fresh weight were observed between the transgenic and WT plants under well-wateredconditions (H_2O). However, after drought stress for 14 days and re-watering, only 40% of WT seedlings could be restored, while this ratio in transgenic lines (LEAtv-L1 and LEAtv-L2) was almost doubled (87% and 80%, respectively) (Figure 4C). In addition, the fresh stem weight and fresh root weight of transgenic lines were significantly higher than the K7 wild type, suggesting a better growth rate of these lines in drought condition (Figure 4D,E). Taken together, these results indicate that the *ZmLEA14tv* gene showed improved drought resistance in transgenic maize.

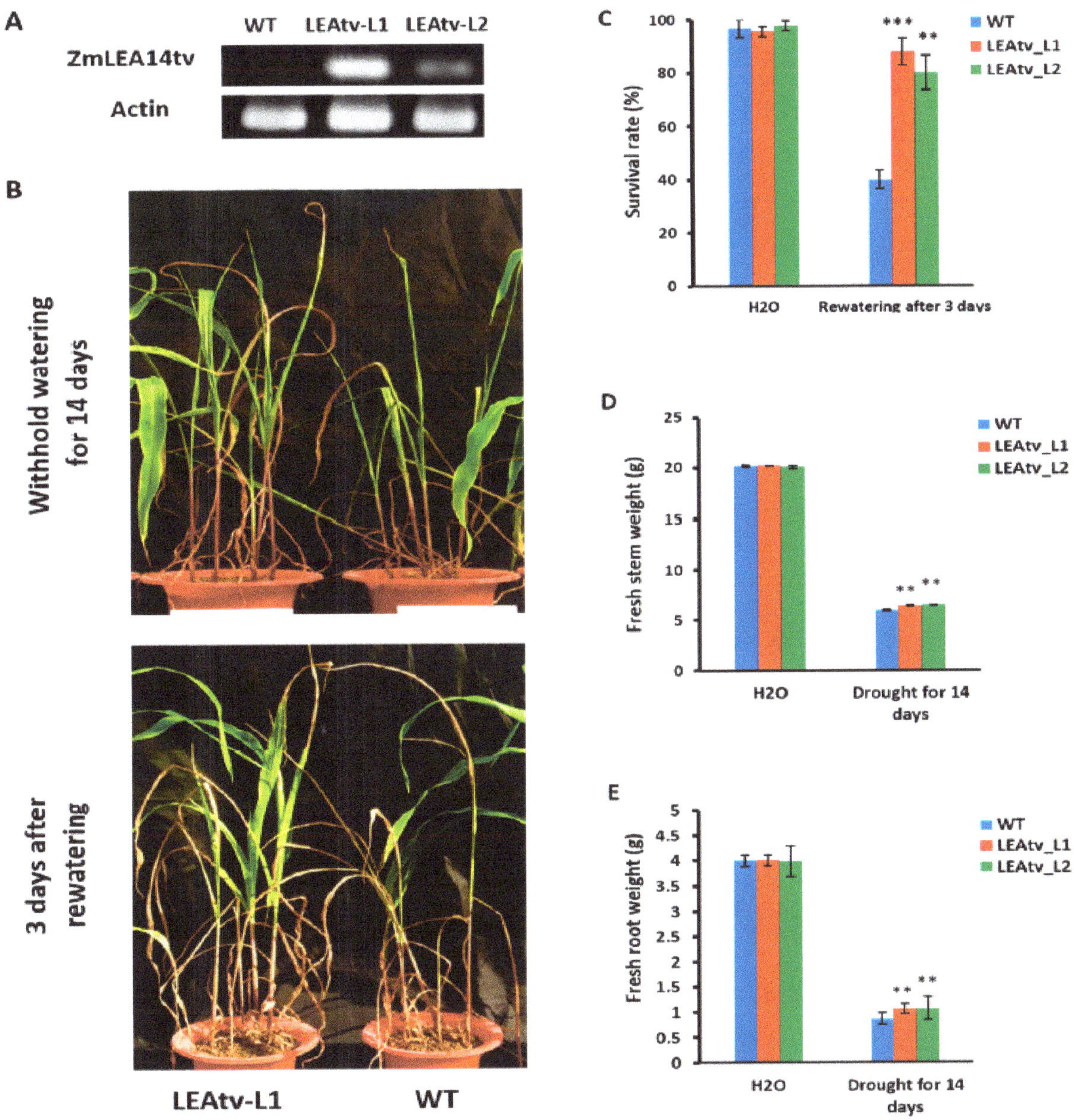

Figure 4. Drought tolerance of maize seedlings overexpressing *ZmLEA14tv*. (**A**) The RT-PCR analysis of *ZmLEA14tv* expression in transgenic maize lines (LEAtv-L1, LEAtv-L2). (**B**) The phenotype of transgenic and WT maize seedlings under drought stress treatment and re-watering conditions. The five-leaf-stage maize seedlings had watering withheld for 14 days, followed by a three-day re-watering. At least five seedlings were grown in each plot and the experiment was replicated three times. WT: wild type. (**C–E**) The survival rate, fresh stem weight, and fresh root weight of WT and transgenic maize seedling after the drought and re-watering treatments. Each experiment was replicated three times. Statistical significance was determined by Student's *t*-test. * $p < 0.05$, ** $p < 0.01$, *** $p < 0.001$.

4. Discussion

The identification and characterization of ZmLEA14tv, a putative atypical LEA group 5C member of the Tevang 1 maize cultivar, were reported in the present study. The deduced amino acid sequence of ZmLEA14tv possessed characteristics of a 5C LEA protein that contains a "LEA_2" domain (Pfam cluster PF03168). The LEA proteins are normally known as hydrophilins with a hydrophilicity index of more than 1 and a glycine (Gly) content more than 6% [11]. Typical LEA proteins can retain water and protect other soluble protein from the aggregation due to their highly hydrophilic properties [33]. Group 5C LEA proteins had a higher proportion of hydrophobic residues than typical LEA proteins; however, they were also involved in various kind of stress tolerance. The estimated GRAVY index of ZmLEA14tv was 0.047, much lower than typical LEA proteins. The ZmLEA14tv protein sequence deduced from the isolated DNA showed a "LEA_2" domain that was characteristic of 5C LEA proteins and a high level of homology with other 5C members.

Functional analysis of 5C LEA proteins showed that the molecular mechanisms of the protective ability against desiccation stress were diverse. The overexpression of AdLEA, a 5C LEA protein from wild peanut, could help maintain the photosynthetic efficiency, reduce the ROS level, and induce the expression of some drought-responsive genes in transgenic tobacco [26]. Meanwhile, overexpressing *SiLEA14* from foxtail millet enhanced a higher level of proline and sugar accumulation in transgenic *Arabidopsis* [13]. The present study showed that the expression of *ZmLEA14tv* in tobacco significantly improved the recoverability of transgenic plants suffering from a short desiccation (Figure 2). This result suggested that *ZmLEA14tv* could function properly in tobacco and enhance its drought tolerance.

One of the methods to generate drought-tolerant maize is enhancing the expression of superior drought-tolerant genes in commercial lines. The overexpression of the *OsSta2* gene, encoding for a AP2/ERF protein, under a maize ubiquitin promoter improved the salt tolerance and grain yield of transgenic rice [34]. Furthermore, enhanced expression of rice dehydrin, namely *OsDhn1*, could increase the tolerance to oxidative stress under salt and drought conditions [35]. Group 5C LEA is well known as an atypical group of proteins that are involved in various abiotic stress response in plants [8,13,24]. In this study, a putative LEA gene namely *ZmLEA14tv* was isolated from the genomic DNA of *Z. mays* cv. Tevang 1, which is well adapted to drought stress in the northern mountains of Vietnam. Our data revealed the potential application of *ZmLEA14tv* in genetic engineering for improving crop performance in the context of climate change. Interestingly, transgenic maize seeds expressing *ZmLEA14tv* showed an improved germination ability in drought-simulated condition in comparison with the WT and did not cause any delay in shoot and root development in transgenic plants.

5. Conclusions

In summary, the coding region of *ZmLEA14tv* was isolated from *Z. mays* cv. Tevang 1 and then cloned into an overexpression cassette. This gene encoded a hydrophobic deduced protein that has a high similarity in structure and a close relationship with other 5C LEA proteins. In addition, the expression of *ZmLEA14tv* in model dicot plants such as tobacco significantly improved the recovery ability, while the enhanced *ZmLEA14tv* transgenic maize showed better germination and growth in drought simulation conditions. These results suggested that *ZmLEA14tv* could act as a potential candidate for genetic engineering to improve drought and other osmotic stress tolerance.

Author Contributions: N.V.H. and H.T.T.H. conceived and designed the experiments; B.M.M., N.T.L., H.H.H., and N.X.T. performed the experiments; B.M.M., L.T.T.H., and H.T.T.H. analyzed the data; B.M.M., N.V.H., and H.T.T.H. prepared the manuscript.

Acknowledgments: The authors thank Doan Thi Bich Thao and Nguyen Thu Hoai for participating in some greenhouse experiments.

References

1. Chandler, P.M.; Robertson, M. Gene expression regulated by abscisic acid and its relation to stress tolerance. *Annu. Rev. Plant Physiol. Plant Mol. Biol.* **1994**, *45*, 113–141. [CrossRef]
2. Hong-Bo, S.; Zong-Suo, L.; Ming-An, S. LEA proteins in higher plants: Structure, function, gene expression and regulation. *Colloids Surf. B Biointerfaces* **2005**, *45*, 131–135. [CrossRef] [PubMed]

3. Kimura, M.; Yamamoto Yoshiharu, Y.; Seki, M.; Sakurai, T.; Sato, M.; Abe, T.; Yoshida, S.; Manabe, K.; Shinozaki, K.; Matsui, M. Identification of arabidopsis genes regulated by high light–stress using cDNA microarray. *Photochem. Photobiol.* **2007**, *77*, 226–233.
4. Tunnacliffe, A.; Wise, M.J. The continuing conundrum of the LEA proteins. *Naturwissenschaften* **2007**, *94*, 791–812. [CrossRef]
5. Baker, J.; Van Dennsteele, C.; Dure, L. Sequence and characterization of 6 LEA proteins and their genes from cotton. *Plant Mol. Biol.* **1988**, *11*, 277–291. [CrossRef] [PubMed]
6. Dure, L.; Galau, G.A. Developmental Biochemistry of Cottonseed Embryogenesis and Germination. *Plant Physiol.* **1981**, *68*, 187–194. [CrossRef] [PubMed]
7. Oliveira, E.; Amara, I.; Bellido, D.; Odena, M.A.; Dominguez, E.; Pages, M.; Goday, A. LC-MSMS identification of *Arabidopsis thaliana* heat-stable seed proteins: Enriching for LEA-type proteins by acid treatment. *J. Mass Spectrom.* **2007**, *42*, 1485–1495. [CrossRef] [PubMed]
8. Hundertmark, M.; Hincha, D.K. LEA (Late Embryogenesis Abundant) proteins and their encoding genes in *Arabidopsis thaliana*. *BMC Genom.* **2008**, *9*, 118–139. [CrossRef] [PubMed]
9. Garay-Arroyo, A.; Colmenero-Flores, J.M.; Garciarrubio, A.; Covarrubias, A.A. Highly hydrophilic proteins in prokaryotes and eukaryotes are common during conditions of water deficit. *J. Biol. Chem.* **2000**, *275*, 5668–5674. [CrossRef] [PubMed]
10. Yale, J.; Bohnert, H.J. Transcript expression in *Saccharomyces cerevisiae* at high salinity. *J. Biol. Chem.* **2001**, *276*, 15996–16007. [CrossRef]
11. Battaglia, M.; Olvera-Carrillo, Y.; Garciarrubio, A.; Campos, F.; Covarrubias, A.A. The enigmatic LEA proteins and other hydrophilins. *Plant Physiol.* **2008**, *148*, 6–24. [CrossRef] [PubMed]
12. Wolkers, W.F.; McCready, S.; Brandt, W.F.; Lindsey, G.G.; Hoekstra, F.A. Isolation and characterization of a D-7 LEA protein from pollen that stabilizes glasses in vitro. *Biochim. Biophys. Acta Protein Struct. Mol. Enzymol.* **2001**, *1544*, 196–206. [CrossRef]
13. Wang, M.; Li, P.; Li, C.; Pan, Y.; Jiang, X.; Zhu, D.; Zhao, Q.; Yu, J. SiLEA14, a novel atypical LEA protein, confers abiotic stress resistance in foxtail millet. *BMC Plant Biol.* **2014**, *14*, 290–305. [CrossRef] [PubMed]
14. Li, X.; Wu, L.; Wang, J.; Sun, J.; Xia, X.; Geng, X.; Wang, X.; Xu, Z.; Xu, Q. Genome sequencing of rice subspecies and genetic analysis of recombinant lines reveals regional yield- and quality-associated loci. *BMC Biol.* **2018**, *16*, 102. [CrossRef] [PubMed]
15. Hirsch, C.N.; Hirsch, C.D.; Brohammer, A.B.; Bowman, M.J.; Soifer, I.; Barad, O.; Shem-Tov, D.; Baruch, K.; Lu, F.; Hernandez, A.G.; et al. Draft assembly of elite inbred line ph207 provides insights into genomic and transcriptome diversity in maize. *Plant Cell* **2016**, *28*, 2700–2714. [CrossRef] [PubMed]
16. Li, F.; Fan, G.; Lu, C.; Xiao, G.; Zou, C.; Kohel, R.J.; Ma, Z.; Shang, H.; Ma, X.; Wu, J.; et al. Genome sequence of cultivated Upland cotton (*Gossypium hirsutum* TM-1) provides insights into genome evolution. *Nat. Biotechnol.* **2015**, *33*, 524–530. [CrossRef] [PubMed]
17. Wang, X.S.; Zhu, H.B.; Jin, G.L.; Liu, H.L.; Wu, W.R.; Zhu, J. Genome-scale identification and analysis of LEA genes in rice (*Oryza sativa* L.). *Plant Sci.* **2007**, *172*, 414–420. [CrossRef]
18. Magwanga, R.O.; Lu, P.; Kirungu, J.N.; Lu, H.; Wang, X.; Cai, X.; Zhou, Z.; Zhang, Z.; Salih, H.; Wang, K.; et al. Characterization of the late embryogenesis abundant (LEA) proteins family and their role in drought stress tolerance in upland cotton. *BMC Genet.* **2018**, *19*, 6–37. [CrossRef] [PubMed]
19. Li, X.; Cao, J. *Late Embryogenesis Abundant* (*LEA*) gene family in maize: Identification, evolution, and expression profiles. *Plant Mol. Biol. Rep.* **2016**, *34*, 15–28. [CrossRef]
20. Galau, G.A.; Wang, H.Y.; Hughes, D.W. Cotton *Lea5* and *Lea14* encode atypical late embryogenesis-abundant proteins. *Plant Physiol.* **1993**, *101*, 695–696. [CrossRef]
21. Maitra, N.; Cushman, J.C. Isolation and characterization of a drought-induced soybean cDNA encoding a D95 family late-embryogenesis-abundant protein. *Plant Physiol.* **1994**, *106*, 805–806. [CrossRef] [PubMed]
22. Zegzouti, H.; Jones, B.; Marty, C.; Lelievre, J.M.; Latche, A.; Pech, J.C.; Bouzayen, M. ER5, a tomato cDNA encoding an ethylene-responsive LEA-like protein: Characterization and expression in response to drought, ABA and wounding. *Plant Mol. Biol.* **1997**, *35*, 847–854. [CrossRef] [PubMed]
23. Kim, H.S.; Lee, J.H.; Kim, J.J.; Kim, C.H.; Jun, S.S.; Hong, Y.N. Molecular and functional characterization of *CaLEA6*, the gene for a hydrophobic LEA protein from *Capsicum annuum*. *Gene* **2005**, *344*, 115–123. [CrossRef] [PubMed]

24. Park, S.C.; Kim, Y.H.; Jeong, J.C.; Kim, C.Y.; Lee, H.S.; Bang, J.W.; Kwak, S.S. Sweetpotato late embryogenesis abundant 14 (*IbLEA14*) gene influences lignification and increases osmotic- and salt stress-tolerance of transgenic calli. *Planta* **2011**, *233*, 621–634. [CrossRef] [PubMed]
25. He, S.; Tan, L.; Hu, Z.; Chen, G.; Wang, G.; Hu, T. Molecular characterization and functional analysis by heterologous expression in *E. coli* under diverse abiotic stresses for *OsLEA5*, the atypical hydrophobic LEA protein from *Oryza sativa* L. *Mol. Genet. Genom.* **2012**, *287*, 39–54. [CrossRef]
26. Sharma, A.; Kumar, D.; Kumar, S.; Rampuria, S.; Reddy, A.R.; Kirti, P.B. Ectopic expression of an atypical hydrophobic group 5 LEA protein from wild peanut, *Arachis diogoi* confers abiotic stress tolerance in tobacco. *PLoS ONE* **2016**, *11*, e0150609. [CrossRef]
27. FAOSTAT. 2017. Available online: http://www.fao.org/faostat/en/#data/QC (accessed on 6 December 2018).
28. Daryanto, S.; Wang, L.; Jacinthe, P.A. Global synthesis of drought effects on maize and wheat production. *PLoS ONE* **2016**, *11*, e0156362. [CrossRef] [PubMed]
29. Liu, Y.; Wang, L.; Xing, X.; Sun, L.; Pan, J.; Kong, X.; Zhang, M.; Li, D. ZmLEA3, a multifunctional group 3 LEA protein from maize (*Zea mays L.*), is involved in biotic and abiotic stresses. *Plant Cell Physiol.* **2013**, *54*, 944–959. [CrossRef]
30. Topping, J.F. Tobacco transformation. In *Plant Virology Protocols: From Virus Isolation to Transgenic Resistance*; Foster, G.D., Taylor, S.C., Eds.; Humana Press: Totowa, NJ, USA, 1998; pp. 365–372.
31. Frame, B.; Main, M.; Schick, R.; Wang, K. Genetic transformation using maize immature zygotic embryos. In *Plant Embryo Culture: Methods and Protocols*; Thorpe, T.A., Yeung, E.C., Eds.; Humana Press: Totowa, NJ, USA, 2011; pp. 327–341.
32. Tamura, K.; Filipski, A.; Peterson, D.; Stecher, G.; Kumar, S. MEGA6: Molecular Evolutionary Genetics Analysis Version 6.0. *Mol. Biol. Evol.* **2013**, *30*, 2725–2729. [CrossRef]
33. Goyal, K.; Walton, L.J.; Tunnacliffe, A. LEA proteins prevent protein aggregation due to water stress. *Biochem. J.* **2005**, *388*, 151–157. [CrossRef]
34. Kumar, M.; Choi, J.; An, G.; Kim, S.R. Ectopic expression of *OsSta2* enhances salt stress tolerance in rice. *Front. Plant Sci.* **2017**, *8*, 316. [CrossRef] [PubMed]
35. Kumar, M.; Lee, S.C.; Kim, J.Y.; Kim, S.J.; Aye, S.S.; Kim, S.R. Over-expression of dehydrin gene, *OsDhn1*, improves drought and salt stress tolerance through scavenging of reactive oxygen species in rice (*Oryza sativa* L.). *J. Plant Biol.* **2014**, *57*, 383–393. [CrossRef]

14

Differential Aquaporin Response to Distinct Effects of Two Zn Concentrations after Foliar Application in Pak Choi (*Brassica rapa* L.) Plants

Hamideh Fatemi [1], Chokri Zaghdoud [2], Pedro A. Nortes [3], Micaela Carvajal [4] and Maria del Carmen Martínez-Ballesta [4,5,*]

1 Department of Horticulture, University of Mohaghegh Ardabili, Ardabil 56199-13131, Iran; ha.fatemi@yahoo.com

2 Dry Land Farming and Oasis Cropping Laboratory, Institute of Arid Regions of Medenine, University of Gabes, Medenine 4119, Tunisia; chokri_zaghdoud@yahoo.fr

3 Departament of Irrigation, Centro de Edafología y Biología Aplicada del Segura (CEBAS-CSIC), P.O. Box 164, 30100 Espinardo (Murcia), Spain; panortes@cebas.csic.es

4 Department of Plant Nutrition, Centro de Edafología y Biología Aplicada del Segura (CEBAS-CSIC), P.O. Box 164, 30100 Espinardo (Murcia), Spain; mcarvaja@cebas.csic.es

5 Department of Agronomy Engineering, Universidad Politécnica de Cartagena, Paseo Alfonso XIII, 48, 30203 Cartagena (Murcia), Spain

* Correspondence: mcarmen.ballesta@upct.es

Abstract: Zinc (Zn) is considered an essential element with beneficial effects on plant cells; however, as a heavy metal, it may induce adverse effects on plants if its concentration exceeds a threshold. In this work, the effects of short-term and prolonged application of low (25 μM) and high (500 μM) Zn concentrations on pak choi (*Brassica rapa* L.) plants were evaluated. For this, two experiments were conducted. In the first, the effects of short-term (15 h) and partial foliar application were evaluated, and in the second a long-term (15 day) foliar application was applied. The results indicate that at short-term, Zn may induce a rapid hydraulic signal from the sprayed leaves to the roots, leading to changes in root hydraulic conductance but without effects on the whole-leaf gas exchange parameters. Root accumulation of Zn may prevent leaf damage. The role of different root and leaf aquaporin isoforms in the mediation of this signal is discussed, since significant variations in *PIP1* and *PIP2* gene expression were observed. In the second experiment, low Zn concentration had a beneficial effect on plant growth and specific aquaporin isoforms were differentially regulated at the transcriptional level in the roots. By contrast, the high Zn concentration had a detrimental effect on growth, with reductions in the root hydraulic conductance, leaf photosynthesis rate and Ca^{2+} uptake in the roots. The abundance of the PIP1 isoforms was significantly increased during this response. Therefore, a 25 μM Zn dose resulted in a positive effect in pak choi growth through an increased root hydraulic conductance.

Keywords: aquaporin; *Brassica rapa*; gas exchange parameters; growth; root hydraulic conductance; zinc

---

1. Introduction

Zinc (Zn) is an essential element for plants and animals [1]. In plants, as a co-factor of at least 300 metalloenzymes [2], Zn plays an important role in protein structure [3]. Another important role of Zn is the scavenging, for oxidation, of reactive oxygen species (ROS) in cells under normal and stress conditions [4]. However, when the Zn concentration exceeds a threshold in the cells, it can induce stress in the plants. Thus, at high concentrations, an important decrease in some physiological functions in plants—such as photosynthesis, respiration and reproductive performance—has been observed [5–7].

Zinc fertilization has beneficial effects for crops, since this microelement is involved in the synthesis of tryptophan, a precursor of indole acetic acid (IAA), responsible for growth stimulation [8]. However, shoot growth was significantly reduced at Zn concentrations above 25 mg L^{-1} in nutrient solution or above 170 mg kg^{-1} in the soil. The sensitivity to Zn toxicity differed among other crops, being sensitivity higher in celery > Chinese cabbage > pak choi. But pak choi can accumulate high levels of Zn in their edible parts with negative impact for human health [9]. The threshold between optimal Zn dose for plant development and the amount in which Zn leads to plant toxicity symptoms or human damage needs to be study in this new crop. However, the application of Zn to the soil results less efficient than foliar supply, due to soil and roots limitations and the poor Zn mobility in the phloem [10]. For this reason, repeated foliar sprays of Zn are frequent during vegetables cultivation.

The aquaporins (AQPs) are a family of small (24–30 kDa), pore-forming, integral membrane proteins, and are involved in the transport of water and small solutes such as urea, CO_2, ammonia, silicic acid and boric acid [11–13]. Plant AQPs are subdivided into plasma membrane intrinsic proteins (PIPs), tonoplast intrinsic proteins (TIPs), NOD26-like intrinsic proteins (NIPs), small basic intrinsic proteins (SIPs) and uncharacterised intrinsic proteins (XIPs) [14–16]. AQPs are responsible for 75%–95% of water passage in the plant [17–19] and are the target of many heavy metals and ions within the cells, including Zn. It has been shown that exposure to heavy metals (Zn, Cd or Cu) decreases root and leaf AQPs expression, in order to avoid water loss and maintain the plant water status [20,21]. Furthermore, in *Mesembryanthemum crystallinum*, exposure to 500 μM $ZnSO_4$ caused differential changes in water status, depending on the duration of the treatment [22]. Thus, reductions in the root to stem water flow, leaf water content, relative water content in leaf tissues, transpiration rate, leaf osmotic potential and expression of AQP genes in the root and leaf tissues were observed after 24 and 72 h of Zn treatment. However, the expression of the *PIP1;4* isoform was unaffected, being continuous and stable, while the expression of other AQP genes—such as those of the plasma membrane (*McPIP1;1; McPIP2;1; McPIP2;3*) or tonoplast AQPs (*McTIP1;2 and McTIP2;2*)—was decreased, showing the roles of different AQPs isoforms in the heavy metal response as key elements with specific involvement. Additionally, Gitto and Fricke [21] reported the effects of 0.1 and 1 mM Zn on water relations in barley (*Hordeum vulgare* L.) plants in relation to AQPs expression. They found that the decline in expression of three AQPs (*HvPIP1;2, HvPIP2;4, HvPIP2;5*) was stronger (46%–77%) with the application of 0.1 mM Zn than with 1 mM Zn (20%–50%); the simultaneous reductions in plant transpiration rate and root hydraulic conductivity (Lpr), of 24% (0.1 mM Zn) and 58% (1 mM Zn), respectively, would have limited the transport of Zn to the shoot to avoid major toxicity.

Zinc affects not only AQPs expression, but also their functionality, through the binding of the metal to the thiol groups of the protein, inducing a conformational change in the structure and the closure of the pore [23]. Additionally, heavy metals have been found to diminish water transport in *Actinidia deliciosa* protoplasts [24]. Yukutake et al. [25] showed that the water permeability mediated by AQPs was rapidly and reversibly regulated by dynamic changes in the intracellular Zn^{2+} concentration linked to a disturbed cellular redox state. In addition, there is evidence that Zn may be an integral part of biomembranes, and thus required for the stability and control of the lateral mobility of membrane molecules [26].

It has been postulated that a hydraulic signal sent from the root to the shoot could be responsible for the reduction in leaf turgor after root abiotic stress perception [27]. Similarly, the disruption of the water status during heavy metal stress could be a consequence of a decreased number of stomata and/or their closure for water preservation. Plants must adjust their whole water status to the constant demand of the aerial parts and AQPs may play an important role in the hydraulic balance [21]. The number of stomata was reduced by heavy metals such as copper, cadmium and Zn in *Phaseolus vulgaris* L. [28] and *Beta vulgaris* L. [29]. In Zn-treated plants, the stomata were round in shape and smaller than in control plants [29]. Generally, heavy metals disrupt water flow not only by reductions in stomatal conductance

(Gs), but also by reducing the flow of solutes; this can be a major cause of heavy metals toxicity and the decline in plant biomass at higher concentrations. Interestingly, decreased Gs values have also been reported under Zn deficiency [30], in chickpea plants; their Gs increased when treated with Zn (2.5 $\mu g\ g^{-1}$ soil) and they were able to maintain membrane integrity [30]. The values of gas exchange parameters were also increased after Zn application in cotton (*Gossypium hirsutum* L.), but intercellular CO_2 decreased in Zn-treated plants, compared to controls [31].

Pak choi (*Brassica rapa* L. ssp. *chinensis*), also known as Chinese cabbage, is a popular leafy vegetable, grown and consumed worldwide. It is an annual crop that has optimal growth at temperatures ranging from 15 to 20 °C [32]. Although this vegetable is cultivated mainly in Asia, it could be grown in areas of Central Europe from spring to winter, due to its short vegetative growth period and lack of thermal requirements. Pak choi is not frequently found on supermarket shelfs, opening new perspectives for small producers [33].

Pak choi is well known to be tolerant of heavy metals such as Cd [34] and Zn [35], and it accumulates metal ions mainly in the leaves. In addition, the involvement of calcium (Ca) ions in its Zn tolerance has been reported [36]. Related to this, coordinated increases in Zn and Ca accumulation have been described in *Silene maritima* L., the tolerance of the Zn-tolerant population being associated with the tissue Ca levels and its presence in the leaf tissues [37].

Although the short- and long-term effects of heavy metals on plant water transport have been reported in different works [38,39], as well as the effects of low and high Zn concentrations [21], there are no reports describing the short- and long-term, concentration-dependent effects of Zn application in pak choi plants. In spite that pak choi plants may accumulate higher amounts of metals in their leaves, the adverse effects of these elements at molecular level are not well documented and often poorly understood, in particular those related to the changes in root water transport properties.

Thus, because of the importance of the Zn concentration and the treatment duration regarding the effects on plant water relations and their regulation by distinct AQPs isoforms, the specific aim of this work was to study: (1) The effect of the long-term application of Zn at two concentrations on the water status of pak choi plants, in order to identify the target AQPs isoforms and evaluate their differential responses in the beneficial or toxic effects that were induced in an ion-concentration-dependent way; and (2) the effect of the short-term application of Zn at two concentrations, in order to compare the water balance response with that under long-term Zn application and discern the hydraulic signals moving from shoot to root, with the involvement of AQP isoforms in the signalling process.

2. Materials and Methods

2.1. Plant Material and Growth Conditions

Seeds of pak choi (*Brassica rapa* L. ssp. *chinensis*) were pre-hydrated with de-ionised water and aerated continuously for 24 h. After this, the seeds were germinated in vermiculite, in the dark at 28 °C, for 2 days. They were then transferred to a controlled-environment chamber with a 16 h light and 8 h dark cycle, and temperatures of 20 and 15 °C and relative humidity of 60% and 80%, respectively. Photosynthetically active radiation (PAR) of 400 $\mu mol\ m^{-2}\ s^{-1}$ was provided by a combination of fluorescent tubes (Philips TLD 36 W/83, Jena, Germany and Sylvania F36 W/GRO, Manchester, NH, USA) and metal halide lamps (OsramHQI, T 400 W, Berlin, Germany). After 5 days, the seedlings were placed in 15 L containers with continuously-aerated Hoagland nutrient solution [40].

2.2. Experimental Design

Two types of experiment were conducted. In the first, after one week of growth, a fixed amount of solution (18 mL per 6 plants container) of low (25 μM) and high (500 μM) Zn concentration (as $ZnSO_4$) was applied by foliar spraying during 15 days (long-term application), three times in total (1, 7 and 14 days before harvesting). Zn was applied using a 1 $mL\ L^{-1}$ Tween-20 spraying solution allowing leaf spray retention and an alumni paper section under each spray leaf was used in order to avoid the sprayed Zn to reach the root nutrient solution. Plants were 21 days-old when they were harvested. Distilled water spraying was used as the control.

In the second experiment, after 20 days of plant growth, low (25 μM) and high (500 μM) Zn concentrations (as $ZnSO_4$) were applied by foliar spraying to half of the leaves (+F), leaving the other half without Zn spraying (−F). After 15 h (short-term application) the 21-day-old plants were harvested for measurements. Leaves and roots were separated for determinations in both experimental procedures. The measurements were made in the middle of the photoperiod in order to obtain the highest values for gas exchanges parameters and all samples were collected at this time for the rest of determinations.

2.3. *Root Hydraulic Conductance*

The root hydraulic conductance (L_0) of the plants was measured by pressurising the roots in a Scholander pressure chamber (UGT: Umwelt-Geräte-Technik GmbH, Freising-Weihenstepha, Germany), as described in Javot et al. [41]. The aerial parts of the plant were removed and the freshly-excised roots were inserted into the pressure chamber, in a plastic tube with the same nutrient solution used for their growth. A gradual increase in pressure (from 0.1 to 0.4 MPa) was applied to the detached roots. The sap that accumulated in this pressure range during a certain time, according to the treatment, was collected in a graduated glass micropipette. The sap flow (Jv) was expressed in mg g (root fresh weight)$^{-1}$ h^{-1} and plotted against pressure (MPa), the slope being the L_0 value in mg g (root fresh weight)$^{-1}$ h^{-1} MPa^{-1}.

2.4. *Gas Exchange Measurements*

A LI-6400XT photosynthesis system (Li-Cor, Inc., Lincoln, NE, USA), equipped with a LI-6400-40 Leaf Chamber Fluorometer (Li-Cor, Inc., Lincoln, NE, USA) and a LICOR 6400-01 CO_2 injector (Licor Bioscience, Lincoln, NE, U.S.A.), was used to measure the net photosynthetic rate (A), stomatal conductance (Gs) and leaf transpiration (T). Leaf gas exchange was measured in a 2 cm^2 leaf cuvette. During these measurements, the air CO_2 concentration was controlled using the injection system and compressed CO_2-cylinders with a CO_2 concentration of 400 μmol mol^{-1} CO_2. Measurements were made at a PAR of 500 μmol m^{-2} s^{-1}, and at ambient air temperature and relative humidity. The air flow was set to 400 μmol s^{-1}. The third fully-expanded leaf was chosen for the analysis, after 15 h or 15 d of treatment. The measurements were made in the middle of the photoperiod in order to obtain the highest values.

2.5. *RNA Extraction and Reverse Transcription*

Total RNA was extracted from fully-expanded leaves of both, short and long-term treated plants and they were frozen and ground to a fine powder in liquid nitrogen using a prechilled mortar and pestle. The ground tissues were stored at −80 °C until use. Total RNA was extracted using the RNeasy Plant Mini Kit (Qiagen, Hilden, Germany) according to the manufacturer's protocol. Contaminated DNA from samples was removed with DNase I, using the DNA-free Kit (Ambion, Applied Biosystems, Austin, TX, USA), and the RNA concentration was quantified with a Nanodrop 1000 Spectrophotometer (Thermo Scientific, Waltham, MA USA). The extracted RNA was stored at −80 °C until use.

cDNA was synthesised from 2 mg of total RNA, using M-MLV reverse transcriptase from the RETROscript Kit (Ambion, Applied Biosystems, Austin, TX, USA). Reverse transcription was carried out with heat denaturation of the RNA, according to the manufacturer's instructions.

2.6. *Quantitative Real-Time PCR (QRT-PCR) Analyses*

To compare the expression of *PIP1* and *PIP2* under different treatments, QRT-PCR analyses were performed as described previously [42], in an Applied Biosystems 7500 Real-Time PCR system (ThermoFisher Scientific S.L, Madrid, Spain).

Gene-specific primers of *Brassica rapa* L. ssp. *pekinensis* genes [43] were used for different PIP isoforms. The primers and the lengths of the amplicons are described in Supplementary Table S1. *B. rapa* var. *italica*, 18s ribosomal RNA [43] was used as the reference gene for standardisation of each sample. After denaturation at 95 °C for 10 min, amplification occurred in a two-step procedure: 15 s of denaturation at 95 °C and 1 min of annealing and extension at 60 °C for 40 cycles, followed by a dissociation stage. Data collection was carried out at the end of each round in step 2. These conditions were used for both target and reference genes and the absence of primer–dimers was checked in controls lacking templates. The amplifications were performed on three independent samples for each treatment (biological replicates) and triplicate reactions were carried out for each sample (technical replicates) in 96-well plates. The transcript levels were calculated using the $2^{-\Delta\Delta Ct}$ method [44]. Standard curves (log of the cDNA dilution vs. Ct) using serial 10-fold dilutions of cDNA were built for each pair of selected primers, obtaining 95% PCR efficiency, corresponding to a slope of −3.44.

2.7. Zinc and Calcium Tissue Analyses

Zinc and Ca were analysed in the oven-dried root and leaf tissues (ca. 100 mg DW). The samples were digested, after HNO_3-H_2O_2 (2:1) addition [45], in a microwave oven (CEM Mars Xpress, North Carolina, USA) and analysed by ICP spectrometry (Iris Intrepid II, Thermo Electron Corporation, Franklin, TN, USA).

2.8. Statistical Analysis

The data were analysed statistically, using the SPSS 13.0 software package (IBM, Armonk, NY, USA), by analysis of variance (ANOVA) and by Tukey's test. Significant differences were determined at $p < 0.05$.

3. Results

3.1. Plant Biomass

A preliminary experiment was carried out in order to determine the effect of different foliar-supplied Zn concentrations on biomass, which allowed the selection of the appropriate Zn concentrations for the subsequent experiments. Plant growth was significantly influenced by long-term (15 days) foliar Zn fertilisation, depending on the concentration (Table 1). Thus, at 25 μM Zn, a marked increase in the total plant fresh (FW) and dry (DW) weights (28.48% and 11.85%, respectively) was observed, compared to the controls. By contrast, significant reductions in these parameters were found at 500 μM Zn (30.62% and 25.69% in plant FW and DW, respectively). At 50 and 100 μM Zn there were no significant changes in the fresh and dry weights with regard to the control and at 1 mM Zn an extreme toxic effect in the plants was observed. Therefore, 25 and 500 μM Zn were chosen as low and high concentrations, respectively, for foliar applications in the short and long term (Figure 1).

Table 1. Effect of long-term exposure to different zinc concentrations on the biomass of pak choi (*Brassica compestris*). Values are means ± SE ($n = 5$).

Zinc Treatments	Shoot Fresh Weight	Root Fresh Weight	Shoot Dry Weight	Root Dry Weight
Control	77.87 ± 5.23b	14.71 ± 1.25b	5.51 ± 0.51b	0.99 ± 0.15a
25 μM	98.83 ± 4.74a	20.12 ± 1.46a	6.02 ± 0.57a	1.25 ± 0.10a
50 μM	66.06 ± 5.09cb	11.08 ± 0.77cb	5.10 ± 0.54cb	0.72 ± 0.08b
100 μM	69.00 ± 5.70cb	11.21 ± 1.38cb	5.28 ± 0.26b	0.72 ± 0.11b
500 μM	54.32 ± 3.08c	9.91 ± 0.47c	4.19 ± 0.20c	0.64 ± 0.08b
1 mM	41.89 ± 4.82c	8.83 ± 1.13c	3.74 ± 0.44d	0.54 ± 0.10c

The different letters indicate significant differences among treatments according to Duncan's test ($p < 0.05$).

Figure 1. Pak choi plants treated with Zn (0, 25 and 500 μM).

3.2. *Effect of Zn on Root Hydraulic Conductance (L_0)*

After short-term (15 h) partial foliar application of Zn, the behaviour of L_0 at the low (25 μM) concentration was the opposite of that at the high (500 μM) concentration (Figure 2A), the values being significantly increased (35.16%) and decreased (25.13%), respectively, compared to control plants. In the long-term foliar Zn application, a significant effect on L_0 was only recorded at the high concentration, with the values being lower than those of the short-term application (the reduction reached 57.51%, relative to the controls) (Figure 2B).

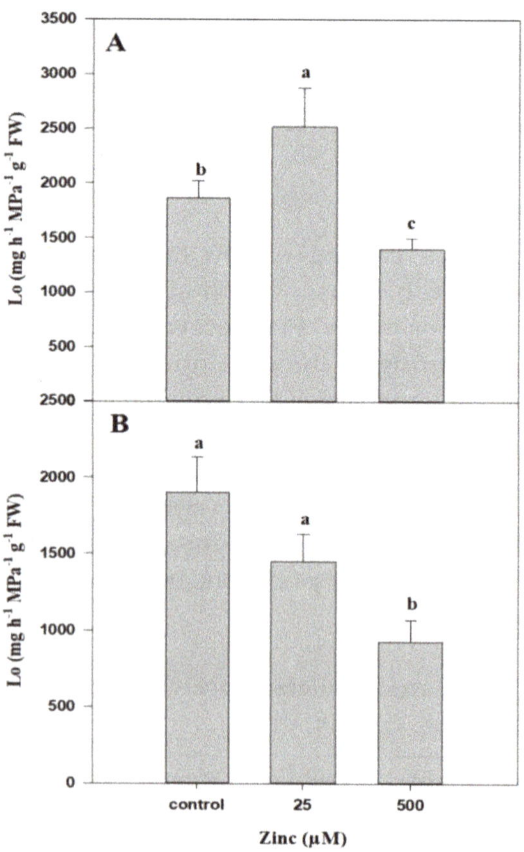

Figure 2. Effect of zinc concentration (0, 25 and 500 μM) on root hydraulic conductance (L_0) of pak choi plants, after foliar application. (**A**) Short-term experiment (15 h) in which half of the leaves were sprayed. (**B**) Long-term experiment (15 days). Values are means ± SE ($n = 5$). Bars with different letters represent significant ($p < 0.05$) differences after ANOVA and an LSD (Least significant difference) test.

3.3. *Effect of Zn on Gas Exchange Parameters*

At both Zn concentrations, the stomatal conductance (Gs), transpiration rate (E) and net photosynthetic rate per unit leaf area (A) remained without significant changes after the short-term treatment, in both types of leaf (with and without foliar Zn spraying) (Figure 3A–C). After the long-term treatment, Gs was decreased (42.46%) at 500 μM $ZnSO_4$, whereas A was increased and decreased at 25 and 500 μM, respectively, but not E, relative to untreated plants (Figure 3D–F).

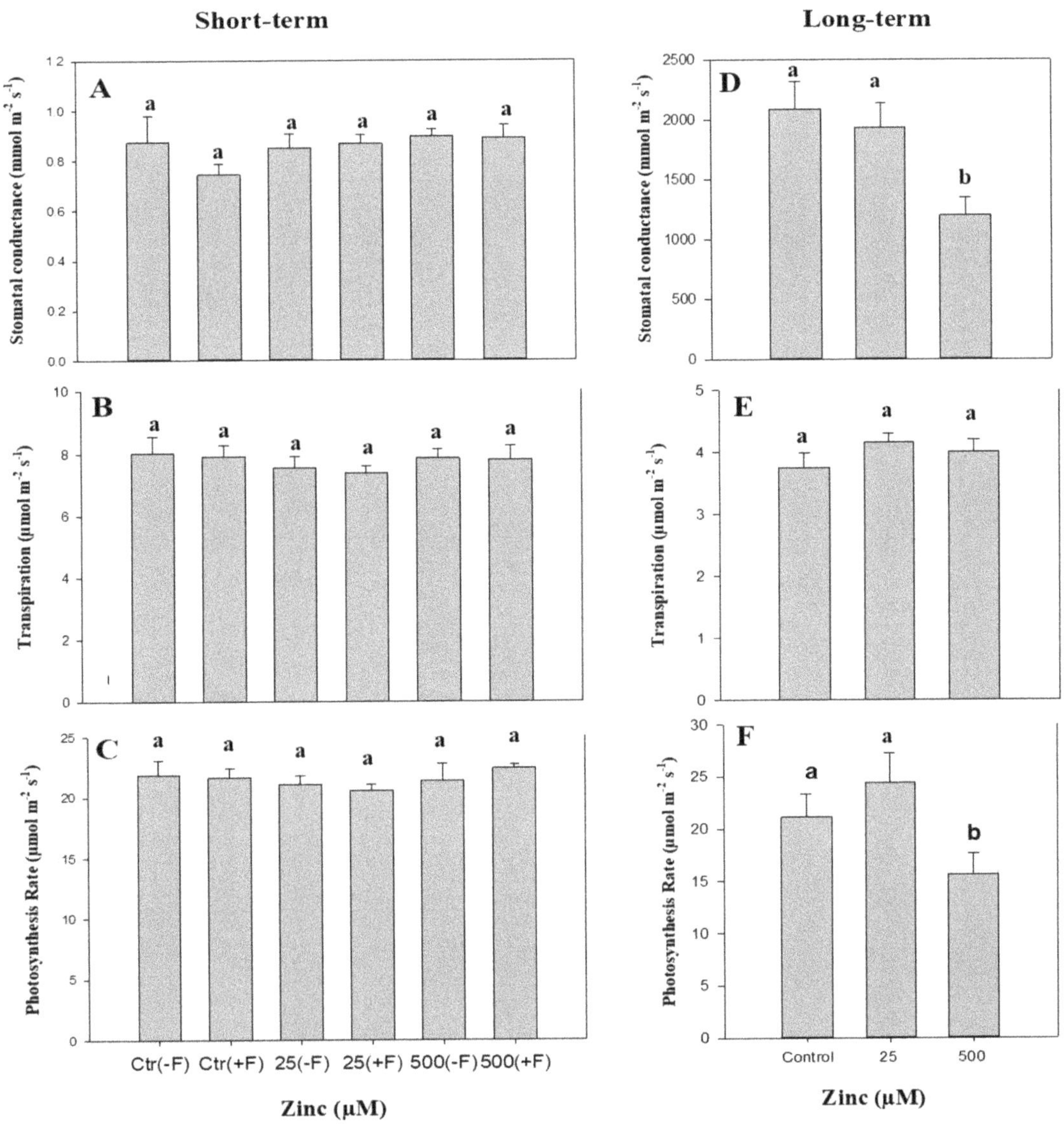

Figure 3. Effect of zinc concentration (0, 25 and 500 μM) on gas exchange parameters of pak choi plants. After short-term application: (**A**) Stomatal conductance, (**B**) transpiration, (**C**) photosynthesis rate, where –F represents the measurements in the leaves that were not sprayed and +F represents the measurements in the leaves that were sprayed. After long-term application: (**D**) Stomatal conductance, (**E**) transpiration, (**F**) carbon assimilation. Values are means ± SE (n = 5). Bars with different letters represent significant ($p < 0.05$) differences after ANOVA and an LSD (Least significant difference) test.

3.4. *Tissue Contents of Zn and Ca*

Macro and micro nutrients were determined in both experiments. However, in addition to Zn, only Ca showed significant differences between treatments.

After 15 h of the partial foliar Zn treatment, the Zn^{2+} content in leaves showed a similar, significant increase at both Zn concentrations, compared to control plants, the values being significantly lower in non sprayed leaves than in sprayed ones (about 2.2- and 2.90-fold at 25 and 500 μM Zn, respectively) (Figure 4A). However, in roots the Zn^{2+} content significantly increased with the spray concentration of Zn; thus, at 500 μM $ZnSO_4$ this increase reached 2.25-fold relative to the control value (Figure 4B). A pattern of variation similar to that of Zn was observed for the leaf Ca content following the short-term fertilisation (Figure 4C), without differences between the control plant leaves and the non-sprayed leaves of the Zn-treated plants. In roots, the Ca content was significantly increased (by 15.88%) and decreased (by 20.97%) at 25 and 500 μM Zn, respectively, after 15 h of treatment (Figure 4D).

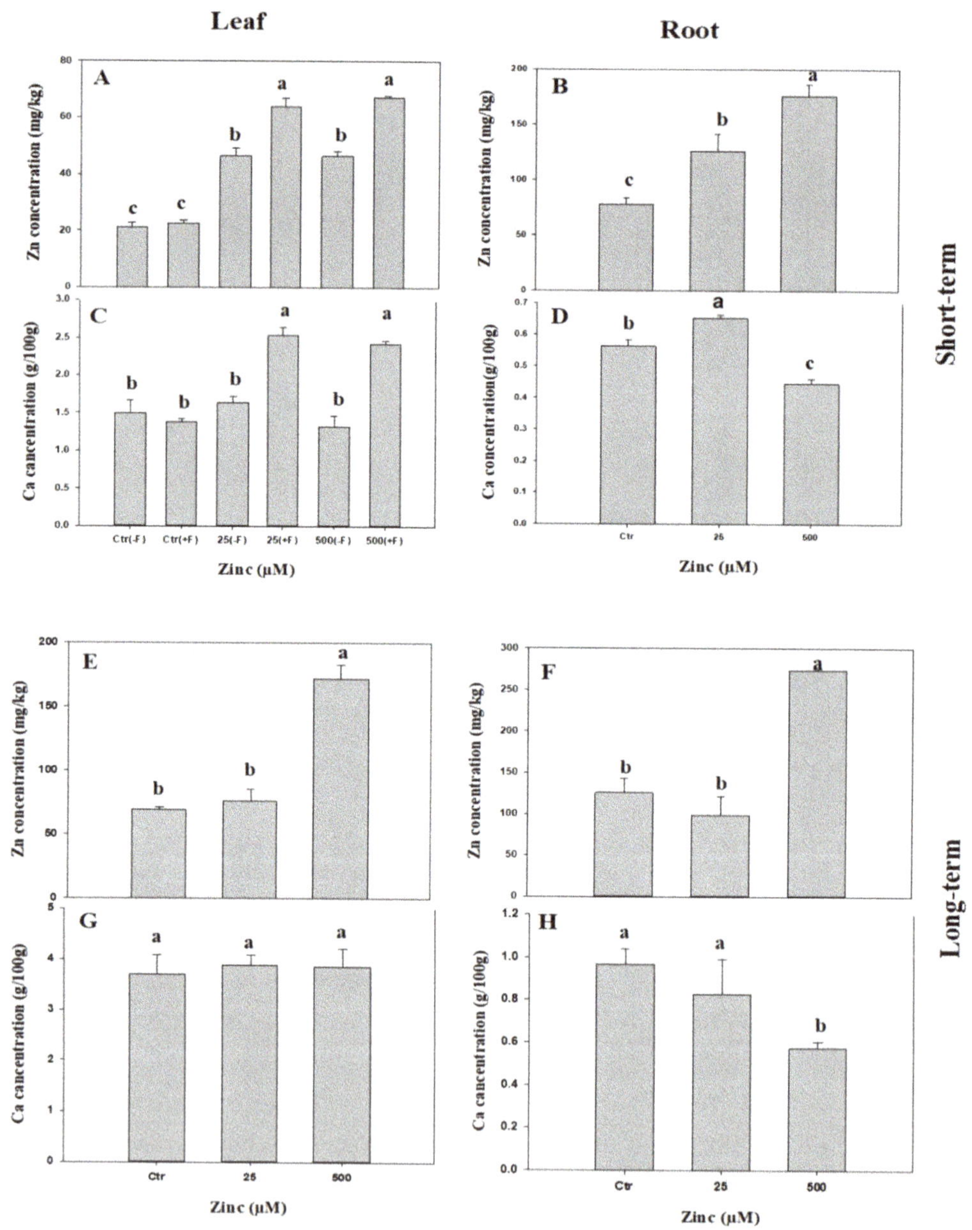

Figure 4. Effect of zinc concentration (0, 25 and 500 μM) on Zn and Ca concentrations of pak choi plants. After short-term application: (**A**) Zn concentration in leaf, (**B**) Zn concentration in root, (**C**) Ca concentration in leaf, (**D**) Ca concentration in root, where –F represents the measurements in the leaves that were not sprayed and +F represents the measurements in the leaves that were sprayed. After long-term application: (**E**) Zn concentration in leaf, (**F**) Zn concentration in root, (**G**) Ca concentration in leaf, (**H**) Ca concentration in root. Values are means ± SE ($n = 5$). Bars with different letters represent significant ($p < 0.05$) differences after ANOVA and an LSD (Least significant difference) test.

In the long-term study the Zn content in the leaves and roots was significantly increased only by the highest Zn concentration, increasing 2.47- and 2.18-fold in leaves and roots, respectively, relative to the controls (Figure 4E,F). Under these conditions, the Ca content was maintained in the leaves, but decreased in the roots (by 40.77%), at 500 μM Zn (Figure 4G,H).

3.5. PIP1 and PIP2 Isoforms Expression in Root and Leaf Tissues

Differences between plant organs in the PIP1 and PIP2 gene expression patterns due to Zn exposure were observed, depending on the experiment duration.

After short-term partial foliar Zn spraying only the highest Zn concentration (500 μM) significantly increased the transcript levels of the aquaporins PIP1;1 and PIP1;2 in leaves (3.71- and 2.42-fold, respectively) (Figure 5A), and of PIP1;4 in roots (3.45-fold), above the control levels (Figure 5B). The changes in these isoforms were specific to the toxic Zn concentration. In addition, while at both Zn concentrations, 25 and 500 μM, the expression levels of PIP2;1 and PIP2;5 in leaves were increased and that of PIP2;2 in roots was decreased, different behaviour of the expression pattern regarding the low and high Zn concentrations was observed for PIP2;2 in leaves and PIP2;1, PIP2;3 and PIP2;7 in roots (Figure 5C,D).

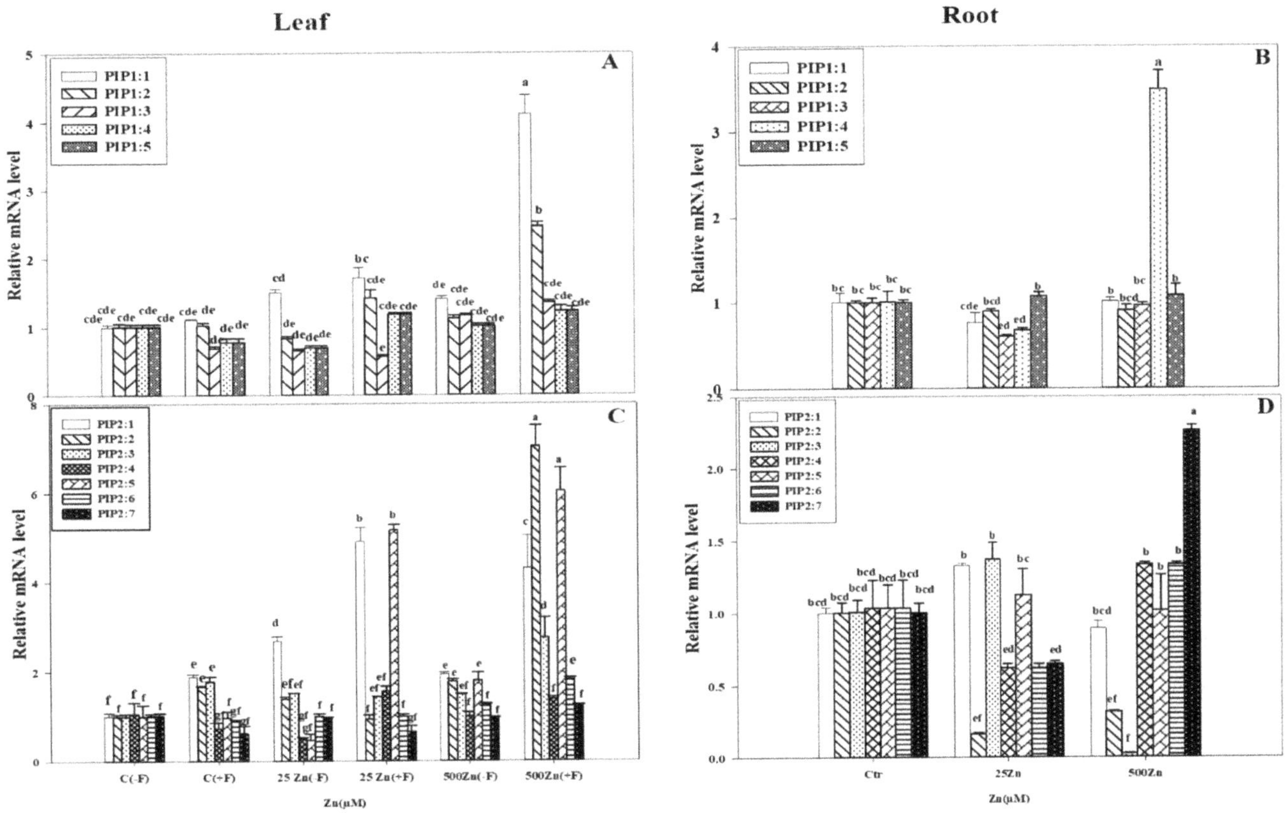

Figure 5. Relative expression level of aquaporin isoforms after short-term application of Zn: (**A**) PIP1 in root, (**B**) PIP1 in leaf, (**C**) PIP2 in root, (**D**) PIP2 in leaf, determined by Q-RT-PCR (quantitative real time-polymerase chain reaction) in control plants and plants treated with 25 or 500 μM Zn. In leaf tissues, gene expression was determined in both non-sprayed (−F) and sprayed (+F) leaves. Mean values and standard errors are shown (n = 3). Bars with different letters represent significant ($p < 0.05$) differences after ANOVA and an LSD test.

The plant hydraulic regulation after long-term foliar Zn spraying showed the implication of other aquaporins isoforms. Thus, in roots, while the PIP1;1 isoform remained up-regulated only at 500 μM Zn (1.84-fold), down-regulation of PIP1;3 and PIP2;4 (1.64-fold) was observed only at 25 μM Zn, relative to the levels of control plants (Figure 6A,B). Interestingly, the expression levels of the

PIP2;6 and PIP2;7 isoforms were the ones most significantly affected by long-term Zn exposure, in both plant organs, without differences between the low and high Zn doses (Figure 6C,D). At 500 μM Zn, PIP2;6 and PIP2;7 were up-regulated, by 4.31- and 2.57-fold, respectively, in leaves; whereas, in roots, they were decreased and increased, respectively, 50- and 5.60-fold, compared to the controls.

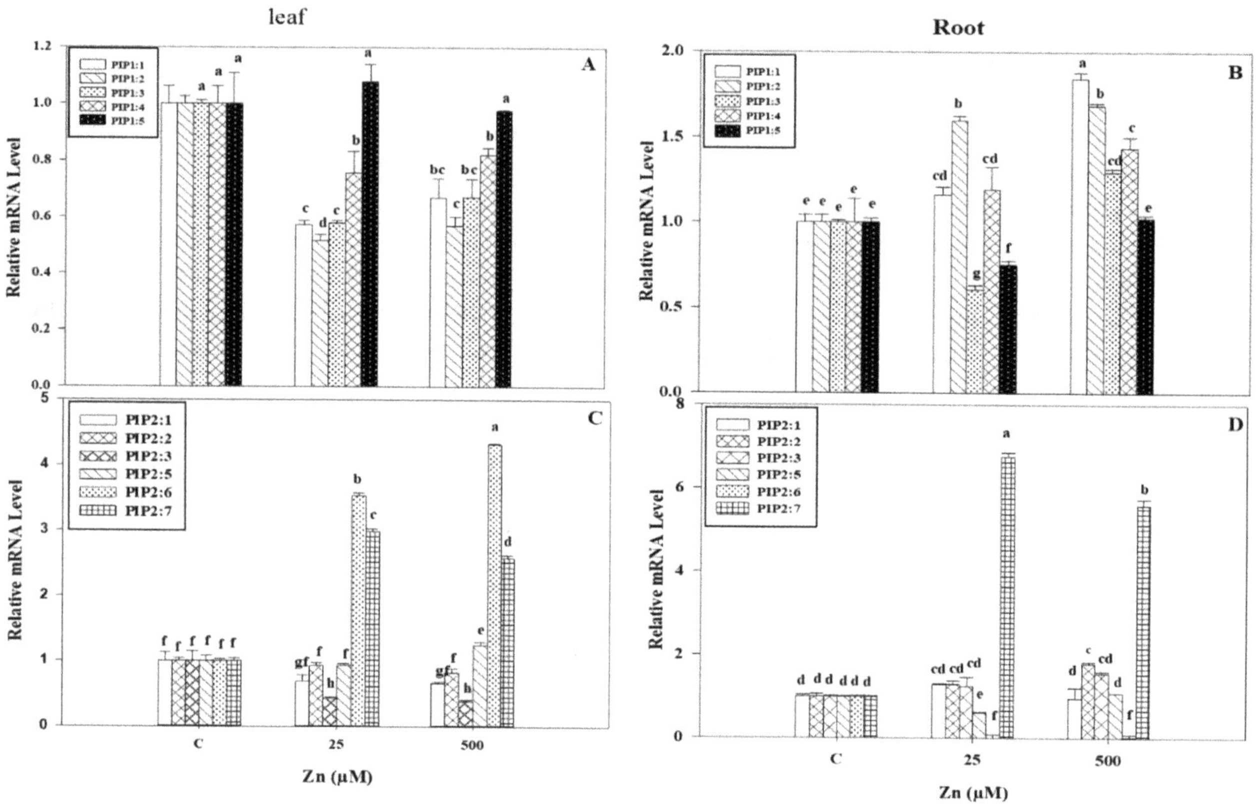

Figure 6. Relative expression level of aquaporin isoforms after long-term application of Zn: (**A**) leaf PIP1, (**B**) leaf PIP2 (**C**)root PIP1 (**D**)root PIP2 genes, determined by Q-RT-PCR in control plants and plants treated with 25 or 500 μM Zn. Mean values and standard errors are shown ($n = 3$). Bars with different letters represent significant ($p < 0.05$) differences after ANOVA and an LSD (Least significant difference) test.

4. Discussion

4.1. *Growth, Root Hydraulic Conductance and Gas Exchange Parameters*

It has been reported that Zn may promote growth at optimal concentration but at higher or low levels induced a growth reduction by interfering with plant metabolic activities [46]. Similarly, in pak choi plants 25 μM Zn had a beneficial effect on plant growth whereas a 500 μM Zn concentration decreased plant biomass at both, shoot and root level. In addition, water relations and photosynthetic rate may result affected by Zn concentration and duration of Zn exposition.

After short-term Zn spraying, no consistent relationship between L_0 and Gs was observed and whereas L_0 was increased and decreased, respectively, at the low and high Zn concentrations, Gs was maintained under all treatments. Thus, in our plants, the Zn applied by foliar spraying was rapidly (after 15 h) and substantially transported to the roots. This supports previous results obtained in wheat, for which foliar-applied Zn was translocated to leaves, both above and below the treated leaf, as well as to the root tips via the phloem [47,48]. Moreover, the higher content of Zn in root tissues relative to leaves was a mechanism to protect photosynthetic tissues, as no changes in the gas exchange parameters were observed in plants after the short-term foliar treatment, independently of the Zn concentration and zone of spraying.

It has been proposed that the movement of water from root to shoot through the xylem was decreased after metal treatment as consequence of structural responses in the cells [49]. However, in the leaves of *A. rubrum*, the relative water content (RWC) was maintained after metal exposition. The authors indicated that plants were able to compensate the effect of the stress and maintain leaf water status [49]. Similar results were found in pak choi leaf tissues after short-term Zn exposition and root hydraulic conductance did not modulate transpiration. This fact indicated that L_0 and Gs were uncoupled processes under these experimental conditions. In *Beta vulgaris*, an ability to rapidly lose (4 h) and gain (24 h) turgor in plants under abiotic stress (200 mM NaCl) was observed, with reductions in Lp_r consistent with Gs decreases [50], but different stresses and genotypes were described. In any case, it is plausible that L_0 in pak choi plants responded to a rapid shoot to root sensing action of Zn that in addition triggers differential regulation of aquaporin gene expression.

In the long-term Zn experiment, no changes were observed in L_0, Gs or E in plants treated with the low Zn concentration, but growth was correlated to an enhanced photosynthetic rate (A) at the optimal Zn (25 μM) dose. It has been previously reported that photosynthesis, carbonic anhydrase activity and chlorophyll concentrations were correlated with Zn nutrition at low levels [51,52]. However, at high Zn concentration a reduced stomatal conductance was in consonance with L_0 reductions and a photosynthetic rate decrease in pak choi. Similarly, it has been reported that high levels of Zn (200–500 μM) inhibited CO_2 assimilation due to structural and functional disturbances in the photosynthetic process [6,53,54]. Thus, photosynthetic response in our plants to long-term Zn could be attributed to an alteration of the photosystem and pigments rather than to the transpiration change. In pea plants, 1 mM cadmium (Cd) had no statistically significant effect on the transpiration rate but it decreased this parameter a 23.1% regarding control in barley plants [55], pointing out the genotype-dependence of the effect of metals on transpiration.

4.2. *Calcium and Zinc Concentrations in Pak Choi Plants after Treatments*

The short-term Zn accumulation in our plants was accompanied by significant increases in the Ca^{2+} contents only in sprayed leaves—to a similar extent at both Zn concentrations—and in roots at the low dose. Calcium is a crucial intracellular messenger and its homeostasis can be modified rapidly by hormonal and environmental stimuli, including metal [56]. The correlation of Zn and Ca in Zn-tolerant and non-tolerant populations of *Silene maritima* was demonstrated by Baker [37]. The important role of Ca in the alleviation of heavy metals toxicity in plants occurs through the prevention of a decrease in the negative charge on the plasma membrane, decreasing the activity of heavy metal ions on the plasma membrane surface, and the maintenance of Ca-related signalling pathways in the presence of the toxicant(s) [57–59]. Davis and Parker [36] also reported that Zn toxicity was highly correlated with the Ca:Zn ratio and reduced stem biomass. According to this, in our pak choi plants the root Ca increment at the low Zn concentration reflects the lack of toxicity of this heavy metal at low doses after a short-term foliar application. By contrast, the decrease in the root Ca content in response to the short-term application of a high Zn concentration is indicative of Zn toxicity and is related to the antagonistic relationship between Zn and Ca in plants [60].

In the long-term experiment the Zn and Ca contents in both organs of plants sprayed with low Zn returned towards control values, compared to the short-term treatment. These results suggest the involvement of an efficient Zn-detoxification mechanism, maintained over time after the low Zn-dose application. Other brassicas, such as *Brassica campestris* L. and *Raphanus sativus* L., have been found to act as hyper-accumulator plants, with higher Zn concentrations in their leaves—relative to the levels of other heavy metals—after irrigation with sewage water [61].

4.3. *PIP Aquaporin Expression Remove*

With the short-term Zn application, no aquaporin expression changes were found in non-spraying leaves and only in the sprayed ones there were significant increases in *PIP1* and *PIP2* expression after 15 h and depending on the Zn dose. This was concomitant with a similar Ca accumulation at both Zn doses. The maintenance of critical Ca levels in the apoplast may be required to keep membrane stability and the water channels open through phosphorylation [62–64], reinforcing previous view that AQP phosphorylation may be among the initial targets of Zn toxicity [65]. In fact, Przedpelska-Wasowicz and Wierzbicka [66] showed that, in *Allium cepa* L., the fast (10 min) toxic effects of heavy metals at the cellular level involved an AQP gating mechanism and this response cannot be ruled out in pak choi plants, being gene expression regulation a compensatory mechanism of the water channel gating.

Our results showed a short-term modulation of leaf *PIP2;1* and *PIP2;5*, which were up-regulated at both Zn concentrations. Exposure of plants to heavy metals (Cd, Ni and Zn) is known to induce water deficit in plant organs [67]. Thus, the leaf *PIP2;1* and *PIP2;5* isoforms may play a key role in water dynamics as part of a rapid Zn-induced sensing response and osmotic stress prevention. The over-expression of these two AQPs isoforms was previously observed to favour water transport into the inner leaf tissues, preventing a fall in leaf water potentials and reducing xylem tensions [68,69]. By contrast, in this study, an up-regulation of *PIP1;1*, *PIP1;2* and *PIP2;2* in pak choi leaves was only observed after short-term application of 500 μM Zn. This fact suggests the implication of these isoforms in the regulation of water homeostasis under a rapid and toxic Zn effect, as reported in several studies for different abiotic stresses [70–73]. Thus, co-expression of *PIP1;2* and *PIP2;5* resulted in increased water-channel activity in *Zea mays* plants [74]. In addition, transgenic rice plants over-expressing *PIP1;1* or *PIP2;2* developed enhanced tolerance to 200 mM mannitol and mild salt stress [70]. In our plants, upregulation of a few individual PIPs in the leaves may redirect water flow into specific cells—which is crucial for plant survival, maintaining gas exchange mechanisms, as it was observed from our data and according to those observed by Alexandersson et al. [75] under water deficit.

Regarding root AQPs expression, in our short-term experiment, among the root *PIP1* genes, only *PIP1;4* showed overexpression at the highest Zn dose. In *Arabidopsis thaliana*, the transcription level of *PIP1;4* isoforms increased more than five-fold over the first 48 h of drought stress (250 mM mannitol), in leaves and roots, as well as in response to salt (150 mM NaCl) and cold stresses [14]. However, expression of this gene was unaffected during *M. crystallinum* adaptation to 1 d of Cu, Zn [22] or salinity [76] stress. After rapid application of a high Zn dose to pak choi plants, *PIP1;4* may function as promotor of water transport from roots to the aerial parts, to maintain water relations and gas exchange in the plants, even at low L_0. The fact that only AQPs located in the plasma membrane of sprayed leaves and roots had modified gene expression suggests the involvement of these proteins in the rapid shoot-to-root hydraulic communication, previous to the Zn effect on leaf transpiration [77].

Additionally, in short-term experiment, root *PIP2;2* and *PIP2;3* showed significant sensitivity to Zn application, especially at high Zn concentration. However, the functional redundancy within AQPs [78] represents a challenge to the determination of the overall role of any given type of AQP in the response to Zn stress. In fact, the decline of both *PIP2;2* and *PIP2;3*—which had high homology, showing 96.8% amino acid identity—could have been compensated by the enhancement of *PIP2;7*. In other reports, a transient up-regulation of *PIP2;3* gene expression upon short-term (2 to 96 h) salt stress has been also reported, implying a possible influence of *PIP2;3* on the short-term response to ionic or purely-osmotic stress [14,79,80]. In addition, rapid (1 to 24 h) up-regulation of *PIP2;7* was observed to be involved in the osmoregulation in plants only under high-toxicity stress [81,82], and the efflux of metalloids from roots to avoid toxicity under excessive levels [83,84], which could explain its overexpression in our pak choi roots only at 500 μM Zn, when root accumulation of Zn^{2+} ions was high enough. In any case, similar to that which occurs with other stresses and genotypes, *PIP2;2*, *PIP2;3* and *PIP2;7* showed a coordinated expression in pak choi plants exposed to Zn.

In the long-term application, by contrast to the short-term experiment, *PIP2;7* was up-regulated in both plant organs, independently of the Zn dose. Apart from its role in plant growth, facilitating water transport into the rapidly-elongating root cells [85], *PIP2;7* facilitates H_2O_2 diffusion across the plasma membrane [86,87], which is an important signal in the regulation of multiple genes associated with abiotic stress tolerance [88,89]. Thus, in pak choi plants *PIP2;7* acts as a sensitive target element for Zn in prolonged heavy metal exposure and its role as part of this signalling response to Zn must be elucidated.

In addition, reductions in root *PIP1;3, PIP1;5, PIP2;5* and *PIP2;6* expression in plants treated with 25 μM Zn were not correlated with the maintenance of L_0, but other significant increments in the AQP mRNA levels were observed such as for *PIP1;1; PIP1;2, PIP1;4* and *PIP2;7* which may contribute to L_0. After the long-term 500 μM -Zn treatment, when Zn^{2+} ions were largely accumulated in leaves and to a greater extent in roots, the majority of *PIP1* isoforms were upregulated in roots. However, L_0 was significantly decreased in these conditions, concomitant with a reduction in the root Ca^{2+} content. This fact pointing out the role of PIP2 subfamily to water transport contribution and the aquaporin regulation through other mechanisms alternative to a transcriptional control as Ca^{2+} dependent-protein phosphorylation. There were also concomitant reductions in A and Gs after long-term exposure to high Zn, indicating severe injury to the photosynthetic apparatus, which finally induced a reduction in plant FW and DW. These results were in consonance with the similar effects of excess of Zn in *Jatropha* plants [90].

5. Conclusions

Pak choi plants have a differential response to Zn depending on the treatment intensity and duration. A rapid shoot-to-root hydraulic signal was involved in the response to short-term partial foliar Zn application, affecting root hydraulic conductance, that it was increased or decreased depending on the Zn dose applied. However, in these plants, L_0 was a no-coordinated process with leaf transpiration and the closure of the stoma under these conditions. Similar root patterns of variation for L_0 and Ca ions with distinct Zn concentrations pointed out the effect of Zn on Ca availability and the importance that this element could have in processes such as aquaporin gating. Leaf *PIP2;1* and *PIP2;5* aquaporin isoforms play a key role sensing the rapid Zn-induced response at leaf level and may act in controlling water dynamics as part of this rapid Zn-induced sensing response preventing osmotic stress. By contrast, *PIP1;1* and *PIP2;2* could be involved in the regulation of leaf water homeostasis, specifically under a toxic Zn effect.

Long-term exposition to low-Zn dose had a beneficial effect on plant growth through an increased leaf photosynthesis rate and maintaining plant water balance. However, high Zn concentrations induced a stomatal closure that together with a decrease in the photosynthesis rate and water transport uptake led to a reduced plant biomass. Differential pattern of aquaporin isoforms reveals that almost all *PIP1* isoforms and *PIP2;6* and *PIP2;7* were involved in sensing the long-term response to Zn treatments reflecting a major Zn dose dependence the PIP1 subfamily. Based on the transcriptional response, other mechanisms of aquaporin regulation at protein level need to be elucidated for the isoforms involved in the response to low and high Zn concentrations and particular function of each isoform in the Zn response have to be deeper studied.

Author Contributions: Experimental design, M.C. and M.d.C.M.-B.; funding acquisition, M.C.; experimental measurements and investigation, H.F., P.A.N. and M.d.C.M.-B.; statistical analysis and investigation, C.Z.; interpretation and validation of the results, H.F., C.Z., P.A.N., M.C. and M.d.C.M.-B.; writing—review and editing, H.F., C.Z., P.A.N., M.C. and M.d.C.M.-B.; All authors have read and agreed to the published version of the manuscript.

Acknowledgments: The authors thank David Walker for correction of the English language and style.

References

1. Prasad, A.S. Zinc: The biology and therapeutics of an ion. *Ann. Intern. Med.* **1996**, *125*, 142–144. [CrossRef] [PubMed]
2. Coleman, J.E. Zinc enzymes. *Curr. Opin. Chem. Biol.* **1998**, *2*, 222–234. [CrossRef]
3. Andreini, C.; Banci, L.; Bertini, I.; Rosato, A. Zinc through the three domains of life. *J. Proteome Res.* **2006**, *5*, 3173–3178. [CrossRef]
4. Broadley, M.R.; White, P.J.; Hammond, J.P.; Zelko, I.; Lux, A. Zinc in plants. *New Phytol.* **2007**, *173*, 677–702. [CrossRef]
5. Ali, G.; Srivastava, P.S.; Iqbal, M. Influence of cadmium and zinc on growth and photosynthesis of *Bacopa monniera* cultivated in vitro. *Biol. Plant.* **2000**, *43*, 599–601. [CrossRef]
6. Khudsar, T.; Uzzafar, M.; Iqbal, M.; Sairam, R.K. Zinc-induced changes in morpho-physiological and biochemical parameters in *Artemisia annua*. *Biol. Plant.* **2004**, *48*, 255–260. [CrossRef]
7. Kholodova, V.P.; Volkov, K.S.; Kuznetsov, V.V. Adaptation of the common ice plant to high copper and zinc concentrations and their potential using for phytoremediation. *Russ. J. Plant Physiol.* **2005**, *52*, 748–757. [CrossRef]
8. Mallick, M.F.R.; Muthukrishnan, C.R. Effect of micronutrients on tomato (*Lycopersicon esculentum* Mill). II. Effect on flowering, fruit-set and yield. *South Indian Hort.* **1980**, *28*, 14–20.
9. Long, X.X.; Yang, X.E.; Ni, W.Z.; Ye, Z.Q.; He, Z.L.; Calvert, D.V.; Stoffella, J.P. Assessing zinc thresholds for phytotoxicity and potential dietary toxicity in selected vegetable crops. *Commn. Soil Sci. Plant. Anal.* **2003**, *34*, 1421–1434. [CrossRef]
10. Swietlik, D. Zinc nutrition of fruit trees by foliar sprays. *Acta Hort.* **2002**, *93*, 123–129. [CrossRef]
11. Deshmukh, R.; Bélanger, R.R. Molecular evolution of aquaporins and silicon influx in plants. *Funct. Ecol.* **2016**, *30*, 1277–1285. [CrossRef]
12. Kitchen, P.; Day, R.E.; Salman, M.M.; Conner, M.T.; Bill, R.M.; Conner, A.C. Beyond water homeostasis: Diverse functional roles of mammalian aquaporins. *Biochim. Biophys. Acta* **2015**, *1850*, 2410–2421. [CrossRef] [PubMed]
13. Kreida, S.; Törnroth-Horsefield, S. Structural insights into aquaporin selectivity and regulation. *Curr. Opin. Struct. Biol.* **2015**, *33*, 126–134. [CrossRef] [PubMed]
14. Jang, J.Y.; Kim, Y.D.; Kim, J.S.; Kang, H. An expression analysis of a gene family encoding plasma membrane aquaporins in response to abiotic stresses in *Arabidopsis thaliana*. *Plant Mol. Biol.* **2004**, *54*, 713–725. [CrossRef] [PubMed]
15. Danielson, J.Å.H.; Johanson, U. Unexpected complexity of the aquaporin gene family in the moss *Physcomitrella patens*. *BMC Plant Biol.* **2008**, *8*, 45. [CrossRef]
16. Ishibashi, K.; Hara, S.; Kondo, S. Aquaporin water channels in mammals. *Clin. Exp. Nephrol.* **2009**, *13*, 107–117. [CrossRef]
17. Henzler, T.; Ye, Q.; Steudle, E. Oxidative gating of water channels (aquaporins) in Chara by hydroxyl radicals. *Plant Cell Environ.* **2004**, *27*, 1184–1195. [CrossRef]
18. Maurel, C. Aquaporins and water permeability of plant membranes. *Annu. Rev. Plant Physiol. Plant Mol. Biol.* **1997**, *48*, 399–429. [CrossRef]
19. Tyerman, S.D.; Bohnert, H.J.; Maurel, C.; Steudle, E.; Smith, J.A.C. Plant aquaporins: Their molecular biology, biophysics and significance for plant water relations. *J. Exp. Bot.* **1999**, *50*, 1055–1071. [CrossRef]
20. Yamaguchi, H.; Fukuoka, H.; Arao, T.; Ohyama, A.; Nunome, T.; Miyatake, K.; Negoro, S. Gene expression analysis in cadmium-stressed roots of a low cadmium-accumulating solanaceous plant, *Solanum torvum*. *J. Exp. Bot.* **2010**, *61*, 423–437. [CrossRef] [PubMed]
21. Gitto, A.; Fricke, W. Zinc treatment of hydroponically grown barley plants causes a reduction in root and cell hydraulic conductivity and isoform-dependent decrease in aquaporin gene expression. *Physiol. Plant.* **2018**, *164*, 176–190. [CrossRef]
22. Kholodova, V.; Volkov, K.; Abdeyeva, A.; Kuznetsov, V. Water status in *Mesembryanthemum crystallinum* under heavy metal stress. *Environ. Exp. Bot.* **2011**, *71*, 382–389. [CrossRef]
23. Tazawa, M.; Asai, K.; Iwasaki, N. Characteristics of Hg- and Zn-sensitive water channels in the plasma membrane of Chara cells. *Bot. Acta* **1996**, *109*, 388–396. [CrossRef]

24. Qiu, Q.S.; Wang, Z.Z.; Cai, Q.G.; Jiang, R.X. Characterization of aquaporins at the plasma membrane of leaf callus protoplasts from *Actinidia deliciosa* cv. Hayward. *Acta Bot. Sin.* **2000**, *42*, 143–147.
25. Yukutake, Y.; Hirano, Y.; Suematsu, M.; Yasui, M. Rapid and reversible inhibition of aquaporin-4 by zinc. *Biochemistry* **2009**, *48*, 12059–12061. [CrossRef]
26. Rygol, J.; Arnold, W.M.; Zimmermann, U. Zinc and salinity effects on membrane transport in *Characonnivens*. *Plant Cell Environ.* **1992**, *15*, 11–23. [CrossRef]
27. Muries, B.; Carvajal, M.; Martínez-Ballesta, M.C. Response of three broccoli cultivars to salt stress, in relation to water status and expression of two leaf aquaporins. *Planta* **2013**, *237*, 1297–1310. [CrossRef]
28. Kasim, W.A. Changes induced by copper and cadmium stress in the anatomy and grain yield of *Sorghum bicolor* (L.) Moench. *Int. J. Agri. Biol.* **2006**, *15*, 123–128.
29. Sagardoy, R.; Vázquez, S.; Florez-Sarasa, I.D.; Albacete, A.; Ribas-Carbó, M.; Flexas, J.; Abadía, J.; Morales, F. Stomatal and mesophyll conductances to CO_2 are the main limitations to photosynthesis in sugar beet (*Beta vulgaris*) plants grown with excess zinc. *New Phytol.* **2010**, *187*, 145–158. [CrossRef]
30. Khan, H.R.; McDonald, G.K.; Rengel, Z. Zinc fertilization and water stress affects plant water relations, stomatal conductance and osmotic adjustment in chickpea (*Cicer arientinum* L.). *Plant Soil* **2004**, *267*, 271–284. [CrossRef]
31. Ahmed, N.; Ahmad, F.; Abid, M.; Aman Ullah, M. Impact of zinc fertilisation on gas exchange characteristics and water use efficiency of cotton crop under arid environment. *Pak. J. Bot.* **2009**, *41*, 2189–2197.
32. Acikgoz, F.E. Seasonal variations on quality parameters of Pak Choi (*Brassica rapa* L. subsp. chinensis L.). *Adv. Crop Sci. Technol.* **2016**, *4*, 4. [CrossRef]
33. Dixon, G.R. Vegetable brassicas and related crucifers. In *Crop Production Science in Horticulture*, 14th ed.; CABI: Walllingford, Oxfordshire, 2007; p. 327.
34. Fang, X.Z.; Zhu, Z.J.; Sun, G.W. Effects of different concentrations of cadmium on growth and antioxidant system in *Brassica campestris* ssp. *Chinensis*. *J. Agro-Environ. Sci.* **2004**, *23*, 877–880.
35. Chen, X.L.; Xu, Y.R.; Cui, X.M.; Wu, X.B. Studies on zinc tolerance and accumulation characteristic of pakchoi (*Brassica campestris* L. ssp. *chinensis* (L.) Makino var. communis Tsen et Lee). *China Vegetables* **2010**, *14*, 19–25.
36. Davis, J.G.; Parker, M.B. Zinc toxicity symptom development and partioning of biomass and zinc in peanut plants. *J. Plant Nutr.* **1993**, *16*, 2353–2369. [CrossRef]
37. Baker, A.J.M. Ecophysiological aspects of zinc tolerance in *Silene maritime*. *New Phytol.* **1978**, *80*, 635–642. [CrossRef]
38. Salah, S.A.; Barrington, S.F. Effect of soil fertility and transpiration rate on young wheat plants (*Triticum aestivum*) Cd/Zn uptake and yield. *Agr. Water Manag.* **2006**, *82*, 177–192. [CrossRef]
39. Llamas, A.; Ullrich, C.I.; Sanz, A. Ni^{2+} toxicity in rice: Effect on membrane functionality and plant water content. *Plant Physiol. Biochem.* **2008**, *46*, 905–910. [CrossRef]
40. Hoagland, D.T.; Arnon, D.I. The water culture method for growing plants without soil. *Calif. Agr. Expt. Sta. Circ.* **1938**, *347*, 1–39.
41. Javot, H.; Lauvergeat, V.; Santoni, V.; Martin-Laurent, F.; Güclü, J.; Vinh, J.; Heyes, J.; Franck, K.I.; Schäffner, A.R.; Bouchez, D.; et al. Role of a single aquaporin isoform in root water uptake. *Plant Cell* **2003**, *15*, 509–522. [CrossRef]
42. Muries, B.; Faize, M.; Carvajal, M.; Martínez-Ballesta, M.C. Identification and differential induction of the expression of aquaporins by salinity in broccoli plants. *Mol. BioSyst.* **2011**, *7*, 1322–1335. [CrossRef] [PubMed]
43. Kayum, M.A.; Park, J.I.; Nath, U.K.; Biswas, M.K.; Kim, H.-T.; Nou, I.-S. Genome-wide expression profiling of aquaporin genes confer responses to abiotic and biotic stresses in *Brassica rapa*. *BMC Plant Biol.* **2017**, *17*, 23. [CrossRef]
44. Livak, K.J.; Schmittgen, T.D. Analysis of relative gene expression data using real-time quantitative PCR and the 2(-Delta Delta C(T)) Method. *Methods* **2001**, *25*, 402–408. [CrossRef]
45. Hansen, T.H.; Laursen, K.H.; Persson, D.P.; Pedas, P.; Husted, S.; Schjoerring, J.K. Micro-scaled high-throughput digestion of plant tissue samples for multi-elemental analysis. *Plant Method.* **2009**, *5*, 1–11. [CrossRef] [PubMed]
46. Mukhopadhyay, M.; Das, A.; Subba, P.; Bantawa, P.; Sarkar, B.; Ghosh, P.D.; Mondal, T.K. Structural, physiological and biochemical profiling of tea plants (*Camellia sinensis* (L.) O. Kuntze) under zinc stress. *Biol. Plant.* **2013**, *57*, 474–480. [CrossRef]

47. Haslett, B.S.; Reid, R.J.; Rengel, Z. Zinc mobility in wheat: Uptake and distribution of zinc applied to leaves or roots. *Ann. Bot.* **2001**, *87*, 379–386. [CrossRef]
48. Rengel, Z. Genotypic difference in micronutrient use efficiency in crops. *Commn. Soil Sci. Plant. Anal.* **2001**, *323*, 1163–1186. [CrossRef]
49. De Silva, N.D.G.; Cholewa, E.; Ryser, P. Effects of combined drought and heavy metal stresses on xylem structure and hydraulic conductivity in red maple (*Acer rubrum* L.). *J. Exp. Bot.* **2012**, *63*, 5957–5966. [CrossRef]
50. Vitali, V.; Bellati, J.; Soto, G.; Ayub, N.D.; Amodeo, G. Root hydraulic conductivity and adjustments in stomatal conductance: Hydraulic strategy in response to salt stress in a halotolerant species. *AoB Plants* **2015**, *7*, plv136. [CrossRef]
51. Ohki, K. Effect of Zinc nutrition on photosynthesis and carbonic anhydrase activity in cotton. *Physiol. Plant.* **1976**, *38*, 300–304. [CrossRef]
52. Arif, N.; Yadav, V.; Singh, S.; Singh, S.; Ahmad, P.; Mishra, R.K.; Sharma, S.; Tripathi, D.K.; Dubey, N.K.; Chauhan, D.K. Influence of high and low levels of plant-beneficial heavy metal ions on plant growth and development. *Front. Environ. Sci.* **2016**, *4*, 69. [CrossRef]
53. Sagardoy, R.; Morales, F.; Lopez-Millan, A.F.; Abadıa, A.; Abadıa, J. Effects of zinc toxicity on sugar beet (*Beta vulgaris* L.) plants grown in hydroponics. *Plant Biol.* **2009**, *11*, 339–350. [CrossRef]
54. Vassilev, A.; Nikolova, A.; Koleva, L.; Lidon, F. Effects of excess Zn on growth and photosynthetic performance of young bean plants. *J. Phytol.* **2011**, *3*, 58–62.
55. Januškaitienė, I. Impact of low concentration of cadmium on photosynthesis and growth of pea and barley. *Environ. Res. Eng. Manag.* **2010**, *3*, 24–29.
56. White, P.; Broadley, M.R. Calcium in plants. *Ann. Bot.* **2003**, *92*, 487–511. [CrossRef]
57. Kinraide, T.B.; Pedler, J.F.; Parker, D.R. Relative effectiveness of calcium and magnesium in the alleviation of rhizotoxicity in wheat induced by copper, zinc, aluminum, sodium, and low pH. *Plant Soil* **2004**, *259*, 201–208. [CrossRef]
58. Kobayashi, Y.; Kobayashi, Y.; Watanabe, T.; Shaff, J.F.; Ohta, H.; Kochian, L.V.; Wagatsuma, T.; Kinraide, T.B.; Koyama, H. Molecular and physiological analysis of Al^{3+} and H^+ rhizotoxicities at moderately acidic conditions. *Plant Physiol.* **2013**, *163*, 180–192. [CrossRef]
59. Pandey, P.; Srivastava, R.K.; Dubey, R.S. Salicylic acid alleviates aluminum toxicity in rice seedlings better than magnesium and calcium by reducing aluminum uptake, suppressing oxidative damage and increasing antioxidative defense. *Ecotoxicology* **2013**, *22*, 656–670. [CrossRef]
60. Prasad, R.; Shivay, Y.S.; Kumar, D. Interactions of zinc with other nutrients in soils and plants—A review. *Indian J. Fertil.* **2016**, *12*, 16–26.
61. Khan, I.; Ghani, A.; Rehman, A.U.; Awan, S.A.; Noreen, A.; Khalid, I. Comparative analysis of heavy metal profile of *Brassica campestris* (L.) and *Raphanus sativus* (L.) irrigated with municipal waste water of Sargodha city. *J. Clin. Toxicol.* **2016**, *6*, 307. [CrossRef]
62. Azad, A.K.; Sawa, Y.; Ishikawa, T.; Shibata, H. Phosphorylation of plasma membrane aquaporin regulates temperature-dependent opening of tulip petals. *Plant Cell Physiol.* **2004**, *45*, 608–617. [CrossRef]
63. Jia, X.Y.; Xu, C.Y.; Jing, R.L.; Li, R.Z.; Mao, X.G.; Wang, J.P.; Chang, X.P. Molecular cloning and characterization of wheat calreticulin (CRT) gene involved in drought stressed responses. *J. Exp. Bot.* **2008**, *59*, 739–751. [CrossRef]
64. Martínez-Ballesta, M.C.; Cabañero, F.; Olmos, E.; Periago, P.M.; Maurel, C.; Carvajal, M. Two different effects of calcium on aquaporins in salinity-stressed pepper plants. *Planta* **2008**, *228*, 15–25. [CrossRef]
65. Ariani, A.; Barozzi, F.; Sebastiani, L.; Di Toppi, L.S.; Di Sansebastiano, G.P.; Andreucci, A. AQUA1 is a mercury sensitive poplar aquaporin regulated at transcriptional and post-translational levels by Zn stress. *Plant Physiol. Biochem.* **2019**, *135*, 588–600. [CrossRef]
66. Przedpelska-Wasowicz, E.M.; Wierzbicka, M. Gating of aquaporins by heavy metals in *Allium cepa* L. epidermal cells. *Protoplasma* **2011**, *248*, 663–671. [CrossRef]
67. Rucińska-Sobkowiak, R. Water relations in plants subjected to heavy metal. *Acta Physiol. Plant.* **2016**, *38*, 257. [CrossRef]
68. Aroca, R.; Ferrante, A.; Vernieri, P.; Chrispeels, M.J. Drought, abscisic acid and transpiration rate effects on the regulation of PIP aquaporin gene expression and abundance in *Phaseolus vulgaris* plants. *Ann. Bot.* **2006**, *98*, 1301–1310. [CrossRef]

69. Maurel, C.; Verdoucq, L.; Luu, D.T.; Santoni, V. Plant aquaporins: Membrane channels with multiple integrated functions. *Annu. Rev. Plant Biol.* **2008**, *59*, 595–624. [CrossRef]
70. Guo, L.; Wang, Z.Y.; Lin, H.; Cui, W.E.; Chen, J.; Liu, M.; Chen, Z.L.; Qu, L.J.; Gu, H. Expression and functional analysis of the rice plasma-membrane intrinsic protein gene family. *Cell Res.* **2006**, *16*, 277–286. [CrossRef]
71. Sreedharan, S.; Shekhawat, U.K.S.; Ganapathi, T.R. Transgenic banana plants overexpressing a native plasma membrane aquaporin MusaPIP1;2 display high tolerance levels to different abiotic stresses. *Plant Biotechnol. J.* **2013**, *11*, 942–952. [CrossRef]
72. Xu, Y.; Hu, W.; Liu, J.; Zhang, J.; Jia, C.; Miao, H.; Xu, B.; Jin, Z. A banana aquaporin gene, MaPIP1;1, is involved in tolerance to drought and salt stresses. *BMC Plant Biol.* **2014**, *14*, 59. [CrossRef]
73. Wang, L.; Liu, Y.; Feng, S.; Yang, J.; Li, D.; Zhang, J. Roles of plasmalemma aquaporin gene *StPIP1* in enhancing drought tolerance in potato. *Front. Plant Sci.* **2017**, *8*, 616. [CrossRef]
74. Fetter, K.; Van, V.W.; Moshelion, M.; Chaumont, F. Interactions between plasma membrane aquaporins modulate their water channel activity. *Plant Cell* **2004**, *16*, 215–228. [CrossRef]
75. Alexandersson, E.; Fraysse, L.; Sjövall-Larsen, S.; Gustavsson, S.; Fellert, M.; Karlsson, M.; Johanson, U.; Kjellbom, P. Whole gene family expression and drought stress regulation of aquaporins. *Plant Mol. Biol.* **2005**, *59*, 469–484. [CrossRef]
76. Abdeeva, A.R.; Kholodova, V.P.; Kuznetsov, V.V. Expression of aquaporin genes in the common ice plant during induction of the water-saving mechanism of CAM photosynthesis under salt stress conditions. *Dokl. Biol. Sci.* **2008**, *418*, 30–33. [CrossRef]
77. Vandeleur, R.K.; Sullivan, W.; Athman, A.; Jordans, C.; Gilliham, M.; Kaiser, B.N.; Tyerman, S.D. Rapid shoot-to-root signalling regulates root hydraulic conductance via aquaporins. *Plant Cell Environ.* **2014**, *37*, 520–538. [CrossRef]
78. Da Ines, O. Functional Analysis of PIP2 Aquaporins in *Arabidopsis thaliana*. Ph.D. Thesis, Ludwig Maximilians Universität, Munich, Germany, 2008.
79. Boursiac, Y.; Chen, S.; Luu, D.T.; Sorieul, M.; van den Dries, N.; Maurel, C. Early effects of salinity on water transport in Arabidopsis roots. Molecular and cellular features of aquaporin expression. *Plant Physiol.* **2005**, *139*, 790–805. [CrossRef]
80. Maathuis, F.J.; Filatov, V.; Herzyk, P.; Krijger, G.C.; Axelsen, K.B.; Chen, S.; Green, B.J.; Li, Y.; Madagan, K.L.; Sánchez-Fernández, R.; et al. Transcriptome analysis of root transporters reveals participation of multiple gene families in the response to cation stress. *Plant J.* **2003**, *35*, 675–692. [CrossRef]
81. Weig, A.; Deswarte, C.; Chrispeels, M.J. The major intrinsic protein family of *Arabidopsis* has 23 members that form three distinct groups with functional aquaporins in each group. *Plant Physiol.* **1997**, *114*, 1347–1357. [CrossRef]
82. Pih, K.T.; Kabilan, V.; Lim, J.H.; Kang, S.G.; Piao, H.L.; Jin, J.B.; Hwang, I. Characterization of two new channel protein genes in *Arabidopsis*. *Mol. Cells* **1999**, *9*, 84–90.
83. Mosa, K.A.; Kumar, K.; Chhikara, S.; Mcdermott, J.; Liu, Z.; Musante, C.; White, J.C.; Dhankher, O.P. Members of rice plasma membrane intrinsic proteins subfamily are involved in arsenite permeability and tolerance in plants. *Transgenic Res.* **2012**, *21*, 1265–1277. [CrossRef] [PubMed]
84. Kumar, K.; Mosa, K.; Chhikara, S.; Musante, C.; White, J.C.; Dhankher, O.P. Two rice plasma membrane intrinsic proteins, OsPIP2;4 and OsPIP2;7, are involved in transport and providing tolerance to boron toxicity. *Planta* **2014**, *239*, 187–198. [CrossRef] [PubMed]
85. Markakis, M.N.; De Cnodder, T.; Lewandowski, M.; Simon, D.; Boron, A.; Balcerowicz, D.; Doubbo, T.; Taconnat, L.; Renou, J.P.; Höfte, H.; et al. Identification of genes involved in the ACC-mediated control of root cell elongation in *Arabidopsis thaliana*. *BMC Plant Biol.* **2012**, *12*, 208. [CrossRef] [PubMed]
86. Dynowski, M.; Schaaf, G.; Loque, D.; Moran, O.; Ludewig, U. Plant plasma membrane water channels conduct the signalling molecule H_2O_2. *Biochem. J.* **2008**, *414*, 53–61. [CrossRef]
87. Bienert, G.P.; Møller, A.L.B.; Kristiansen, K.A.; Schulz, A.; Møller, I.M.; Schjoerring, J.K.; Jahn, T.P. Specific aquaporins facilitate the diffusion of hydrogen peroxide across membranes. *J. Biol. Chem.* **2007**, *282*, 1183–1192. [CrossRef]
88. Neill, S.J.; Desikan, R.; Hancock, J.T. Hydrogen peroxide signalling. *Curr. Opin. Plant Biol.* **2002**, *5*, 388–395. [CrossRef]

89. Ben Rejeb, K.; Benzarti, M.; Debez, A.; Bailly, C.; Savouré, A.; Abdelly, C. NADPH oxidase-dependent H_2O_2 production is required for salt-induced antioxidant defense in *Arabidopsis thaliana*. *J. Plant Physiol.* **2015**, *174*, 5–15. [CrossRef]
90. Luo, Z.B.; He, X.J.; Chen, L.; Tang, L.; Gao, S.; Chen, F. Effects of zinc on growth and antioxidant responses in *Jatropha curcas* seedlings. *Int. J. Agric. Biol.* **2010**, *12*, 119–124.

PERMISSIONS

All chapters in this book were first published in MDPI; hereby published with permission under the Creative Commons Attribution License or equivalent. Every chapter published in this book has been scrutinized by our experts. Their significance has been extensively debated. The topics covered herein carry significant findings which will fuel the growth of the discipline. They may even be implemented as practical applications or may be referred to as a beginning point for another development.

The contributors of this book come from diverse backgrounds, making this book a truly international effort. This book will bring forth new frontiers with its revolutionizing research information and detailed analysis of the nascent developments around the world.

We would like to thank all the contributing authors for lending their expertise to make the book truly unique. They have played a crucial role in the development of this book. Without their invaluable contributions this book wouldn't have been possible. They have made vital efforts to compile up to date information on the varied aspects of this subject to make this book a valuable addition to the collection of many professionals and students.

This book was conceptualized with the vision of imparting up-to-date information and advanced data in this field. To ensure the same, a matchless editorial board was set up. Every individual on the board went through rigorous rounds of assessment to prove their worth. After which they invested a large part of their time researching and compiling the most relevant data for our readers.

The editorial board has been involved in producing this book since its inception. They have spent rigorous hours researching and exploring the diverse topics which have resulted in the successful publishing of this book. They have passed on their knowledge of decades through this book. To expedite this challenging task, the publisher supported the team at every step. A small team of assistant editors was also appointed to further simplify the editing procedure and attain best results for the readers.

Apart from the editorial board, the designing team has also invested a significant amount of their time in understanding the subject and creating the most relevant covers. They scrutinized every image to scout for the most suitable representation of the subject and create an appropriate cover for the book.

The publishing team has been an ardent support to the editorial, designing and production team. Their endless efforts to recruit the best for this project, has resulted in the accomplishment of this book. They are a veteran in the field of academics and their pool of knowledge is as vast as their experience in printing. Their expertise and guidance has proved useful at every step. Their uncompromising quality standards have made this book an exceptional effort. Their encouragement from time to time has been an inspiration for everyone.

The publisher and the editorial board hope that this book will prove to be a valuable piece of knowledge for researchers, students, practitioners and scholars across the globe.

LIST OF CONTRIBUTORS

Hongmin Hou, Lingling Lv and Yugang Zhang
College of Horticulture, Qingdao Agricultural University, Qingdao 266109, China
Qingdao Key Laboratory of Genetic Development and Breeding in Horticultural Plants, Qingdao Agricultural University, Qingdao 266109, China

Heqiang Huo
Mid-Florida Research and Education Center, University of Florida, Apopka, FL 32703, USA

Hongyi Dai
College of Horticulture, Qingdao Agricultural University, Qingdao 266109, China

Jiahui Liang, Jing Zheng and Hongqing Wang
Department of Fruit Science, College of Horticulture, China Agricultural University, Beijing 100193, China

Ze Wu
Key Laboratory of Landscaping Agriculture, Ministry of Agriculture and Rural Affairs, College of Horticulture, Nanjing Agricultural University, Nanjing 210095, China

Charles Nelimor
West African Science Service Centre on Climate Change and Adapted Land-use (WASCAL), Graduate Research Program on Climate Change and Biodiversity, Université Felix Houphouët Boigny, Abidjan 31 B.P. 165, Cote d'Ivoire
International Institute of Tropical Agriculture, Ibadan 200001, Nigeria
Department of Bioscience, Université Felix Houphouët Boigny, Abidjan 22 B.P. 461, Cote d'Ivoire

Baffour Badu-Apraku and Ana Luísa Garcia-Oliveira
International Institute of Tropical Agriculture, Ibadan 200001, Nigeria

Antonia Yarney Tetteh
Department of Biochemistry and Biotechnology, Kwame Nkrumah University of Science and Technology, University Kumasi 00233, Ghana

Assanvo Simon-Pierre N'guetta
Department of Bioscience, Université Felix Houphouët Boigny, Abidjan 22 B.P. 461, Cote d'Ivoire

Zehao Hou, Junliang Yin, Yifei Lu, Jinghan Song, Shuping Wang, Yingxin Zhang and Zhengwu Fang
College of Agriculture, Yangtze University, Jingzhou 434000, Hubei, China

Shudong Wei
College of Life Science, Yangtze University, Jingzhou 434000, Hubei, China

Zhixiong Liu
College of Horticulture and Gardening, Yangtze University, Jingzhou 434000, Hubei, China

Kazi Khayrul Bashar, Md. Ruhul Amin, Ummay Honi, Md. Abu Sadat and Quazi Md. Mosaddeque Hossen
Bangladesh Jute Research Institute, Manik Mia Avenue, Dhaka 1207, Bangladesh

Md. Zablul Tareq
Bangladesh Jute Research Institute, Jagir, Manikganj 1800, Bangladesh

Md. Tahjib-Ul-Arif
Department of Biochemistry and Molecular Biology, Bangladesh Agricultural University, Mymensingh 2202, Bangladesh

Olga Mayoral
Department of Science Education, Universitat de València (UV), Avda. Tarongers, 4, 46022 Valencia, Spain
Botanical Garden UV, Universitat de València, Calle Quart, 80, 46008 Valencia, Spain

Jordi Solbes, José Cantó and Tatiana Pina
Department of Science Education, Universitat de València (UV), Avda. Tarongers, 4, 46022 Valencia, Spain

Sugenith Arteaga
Institute for Plant Molecular and Cell Biology (IBMCP, UPV-CSIC), Universitat Politècnica de València, Camino de Vera s/n, 46022 Valencia, Spain

Lourdes Yabor
Institute for Plant Molecular and Cell Biology (IBMCP, UPV-CSIC), Universitat Politècnica de València, Camino de Vera s/n, 46022 Valencia, Spain
Permanent Address: Laboratory for Plant Breeding and Conservation of Genetic Resources, Bioplant Center, University of Ciego de Avila, Ciego de Ávila 69450, Cuba

María José Díez, Jaime Prohens and Oscar Vicente
Institute for the Conservation and Improvement of Valencian Agrodiversity (COMAV, UPV), Universitat Politècnica de València, Camino de Vera s/n, 46022 Valencia, Spain

Monica Boscaiu
Mediterranean Agroforestry Institute (IAM, UPV), Universitat Politècnica de València, Camino de Vera s/n, 46022 Valencia, Spain

Cuiyu Liu, Yujie Zhao, Xueqing Zhao, Jinping Wang and Zhaohe Yuan
Co-Innovation Center for Sustainable Forestry in Southern China, Nanjing Forestry University, Nanjing 210037, China
College of Forestry, Nanjing Forestry University, Nanjing 210037, China

Mengmeng Gu
Department of Horticultural Sciences, Texas A&M AgriLife Extension Service, College Station, TX 77843-2134, USA

Eszter Nemeskéri and Lajos Helyes
Institute of Horticulture, Szent István University, H-2100 Gödöllő, Hungary

Aleck Kondwakwenda, Julia Sibiya, Rebecca Zengeni, Cousin Musvosvi and Samson Tesfay
School of Agriculture, Earth and Environmental Sciences, University of KwaZulu-Natal, Private Bag X01, Scottsville 3209, Pietermaritzburg, South Africa

María Isabel Martínez-Nieto, Elena Estrelles, Josefa Prieto-Mossi and Pilar Soriano
Jardí Botànic-ICBiBE, Universitat de València, Quart 80, 46008 Valencia, Spain

Josep Roselló
Estación Experimental Agraria de Carcaixent, Carretera CV-5950 (Camino Del Barranquet), Carcaixent, 46740 Valencia, Spain

Bui Manh Minh, Nguyen Thuy Linh and Ha Hong Hanh
Institute of Genome Research, Vietnam Academy of Science and Technology, 18 Hoang Quoc Viet Street, Cau Giay District, Hanoi 100000, Vietnam

Le Thi Thu Hien, Nong Van Hai and Huynh Thi Thu Hue
Institute of Genome Research, Vietnam Academy of Science and Technology, 18 Hoang Quoc Viet Street, Cau Giay District, Hanoi 100000, Vietnam
Graduate University of Science and Technology, Vietnam Academy of Science and Technology, 18 Hoang Quoc Viet Street, Cau Giay District, Hanoi 100000, Vietnam

Nguyen Xuan Thang
Maize Research Institute of Vietnam, Vietnam Academy of Agricultural Sciences, 229 Nguyen Thai Hoc Street, Dan Phuong District, Hanoi 100000, Vietnam

Hamideh Fatemi
Department of Horticulture, University of Mohaghegh Ardabili, Ardabil 56199-13131, Iran

Chokri Zaghdoud
Dry Land Farming and Oasis Cropping Laboratory, Institute of Arid Regions of Medenine, University of Gabes, Medenine 4119, Tunisia

Pedro A. Nortes
Departament of Irrigation, Centro de Edafología y Biología Aplicada del Segura (CEBAS-CSIC), 30100 Espinardo (Murcia), Spain

Micaela Carvajal
Department of Plant Nutrition, Centro de Edafología y Biología Aplicada del Segura (CEBAS-CSIC), 30100 Espinardo (Murcia), Spain

Maria del Carmen Martínez-Ballesta
Department of Plant Nutrition, Centro de Edafología y Biología Aplicada del Segura (CEBAS-CSIC), 30100 Espinardo (Murcia), Spain
Department of Agronomy Engineering, Universidad Politécnica de Cartagena, Paseo Alfonso XIII, 48, 30203 Cartagena (Murcia), Spain

Index

Printed in the USA
CPSIA information can be obtained
at www.ICGtesting.com
LVHW080051211023
761596LV00010BA/2008